ENCYCLOPAEDIA OF
ELECTRICAL ENGINEERING

ENCYCLOPAEDIA OF
ELECTRICAL ENGINEERING

Manoj Jaiswal

ANMOL PUBLICATIONS PVT. LTD.
NEW DELHI - 110 002 (INDIA)

ANMOL PUBLICATIONS PVT. LTD.
H.O.: 4374/4B, Ansari Road, Darya Ganj,
New Delhi-110 002 (India)
Ph.: 23278000, 23261597

B.O.: No. 1015, Ist Main Road, BSK IIIrd Stage
IIIrd Phase, IIIrd Block
Bangalore - 560 085 (India)
Visit us at: www.anmolpublications.com

Encyclopaedia of Electrical Engineering

First Published, 2006

ISBN 81-261-3040-7

PRINTED IN INDIA

Printed at Mehra Offset Press, Delhi.

Preface

The Encyclopaedia of Electrical Engineering covers all aspects of Electrical Engineering and its role in society. This Encyclopaedia covers a range of various Electrical engineering disciplines such as Electromechanics, Electronic and Microprocessor Power Apparatuses, Electrical Apparatuses, Protection, Energy Distribution, Electromechanical Apparatuses for Automation, Technology of Electronic and Electrical Apparatuses, Design of Electrical Apparatuses, Running of Electrical Apparatuses, Household Devices and Engineering Design, Cable Engineering, Insulation Engineering, Engineering Design in Computers, Plasma Installations and Systems, Electrical Engineering Installations and Systems, Management in Electrical Engineering, Electrical Equipment and Automation of Autonomous Electric Transport, Electrical Equipment and Automation of Trunk-Railway Electric Transport , Urban Electric Transport and High-Speed Over-Land Transport.

In recent years there have been dramatic changes in Electrical engineering on both national scale and international scale. New techniques and products have found global recognition. These changes have had a major impact not only in industry but also on all our lives. Not only Technology is becoming more complex, but even terminology is becoming complex day by day.

This new encyclopaedia integrates all of the diverse approaches that have gone into the making of the modern field of Electrical engineering. It is unique among available reference works in that it covers the most recent topics of research. Encyclopaedia of Electrical engineering is intended for all those interested in this field. It will be of immense value to students of all kinds, especially students of different fields of Engineering.

It will also be a major source of reference for professionals and students, who require an authoritative and up-to-date guide to assist them in their work. In preparation of this book, the author has freely consulted large number of books and journals so no authenticity is claimed. Author is especially thankful to Anmol Publications Pvt. Ltd., New Delhi for shaping this book in its final form. Suggestions for further improvement of this book are not only welcome but also greatly appreciated.

Author

A Abbreviation for "ampere" a unit of electrical current.

A C potentiometer An apparatus for the comparison of A.C. Voltages. Balance requires both the magnitude and the phase angle of the unknown voltage to be balanced with the known voltage. This may be done either in Cartesian form or in polar form.

A scope A radar display on which slant range is shown as the distance along a horizontal trace.

A-B Test A test by which an observor subjectively compares the performance of two components of the same type; for example, a test between two different speakiers. For the test to be scientifically valid, the inputs, levels, and listening conditions should be matched.

Abrasive A hard and wear-resistant material (such as a ceramic) used to wear, grind, or cut away other material.

Absolute pressure Gage pressure plus atmospheric pressure.

Absolute pressure transducer A transducer which measures pressure in relation to zero pressure (a vacuum on one side of the diaphragm).

Absolute Refers to a feedback device such as a resolver or an absolute encoder that provides unique position information for each discrete shaft location; unlike an incremental feedback device which requires a known reference point, an absolute feedback device retains position information when power to the system is momentarily lost.

Absolute value Value of an expression without regard to sign or phase angle.

Absolute ventilation efficiency A quantity which expresses the ability of a ventilation system to reduce a pollution concentration relative to the feasible theoretical maximum performance.

Absolute zero The temperature at which substances possess minimal energy. Absolute zero is 0° Kelvin or 0° Rankine and is estimated to be -459.67° F (-273.15° C).

Absorbing media The media in which the process of absorption and reflection of energy incident is completed.

Absorption The process of one substance entering into the inner structure of another. Likewise, the process of turning the energy incident into inner energy of the absorbing media.

Absorption capacity / absorptivity Absorptivity of a surface is the fraction of radiant energy incident which is absorbed. It is numerically equal to the emmisivity for any given surface.

Absorption wavemeter An instrument used to measure audio frequencies.

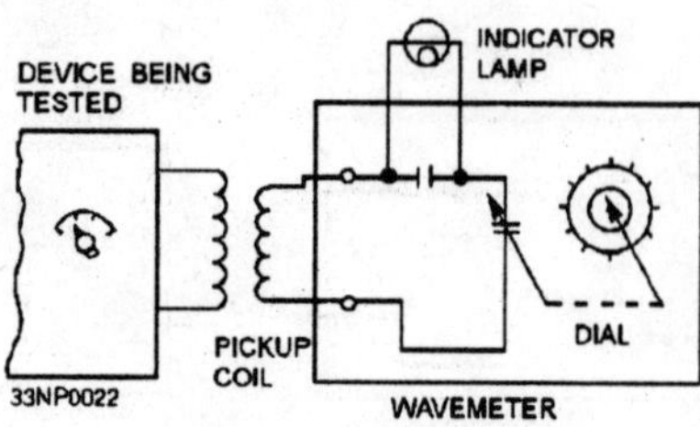

Fig. Absorption wavemeter

Absorption, law of In Boolean algebra, the law which states that the odd term will be absorbed when a term is combined by logical multiplication with the logical sum of that term and another term, or when a term is combined by logical addition with the logical product of one term and another term (for example, A(A + B) = A + AB = A).

ABX Comparator A device that randomly selects between two components being tested. The listener doesn't know which device is selected.

AC Alternating current A periodically varying current whose average value over a period is zero. Unless distinctly specified otherwise, the term refers to a current which reverses at regular recurring intervals of time and which has alternately positive and negative values.

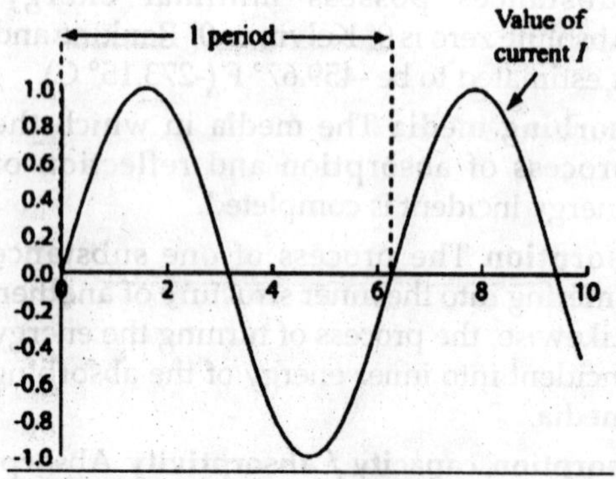

Fig. Alternating current

AC coupling Circuit that passes an AC signal while blocking a DC voltage.

AC generator Device used to transform mechanical energy into AC electrical power.

AC induction motor Class of motors that derives its name from the fact that current is induced into the rotor windings without any physical connection with the stator windings (which are directly connected to an AC power supply); adaptable to many different environments and capable of providing considerable power as well as variable speed control.

AC line frequency The frequency of the alternating current power line measured in Hertz (Hz), usually 50 or 60Hz.

AC load line A graph representing all possible combinations of AC output voltage and current for an amplifier.

AC motors A Motor operating on AC current that flows in either direction (AC current). There are two general types: Induction, and Synchronous.

AC power supply Power supply that delivers an AC voltage.

AC Pressurization technique This technique allows building airtightness to be examined at small (<4 Pa) pressure differentials with minimal interference from climatic forces. The air flow through the building envelope can be evaluated by using a piston assembly to vary the effective volume of the structure. Measuring the amplitude of the pressure response inside the building and phase relationship between this pressure and the velocity of the piston, enables the air flow through the building to be determined.

AC synchronous motor Class of motors that has its phases in correct relationship with each other at every rotor position so that the phases of the stator are synchronous with the rotor poles; adaptable to many different environments and capable of providing considerable power as well as variable speed control.

AC voltage A voltage in which the polarity alternates.

AC/DC Equipment that will operate on either an AC or DC power source.

AC-3 (Audio Coding 3) Dolby's digital audio data compression algorithm adopted for HDTV transmission and used in DVDs, laserdiscs and CDs for 5.1 multichannel home theater use and automotive surround application.

Accelerating anode An electrode charged several thousand volts positive and used to accelerate electrons toward the front of a cathode-ray tube.

Accelerometer A device which converts the effects of mechanical motion into an electrical signal that is proportional to the acceleration value of the motion. A sensor. A transducer.

Acceptor atoms Trivalent atoms that accept free electrons from pentavalent atoms.

Acceptor impurity An impurity which, when added to a semiconductor, accepts one electron from a neighboring atom and creates a hole in the lattice structure of the crystal. Also called trivalent impurity.

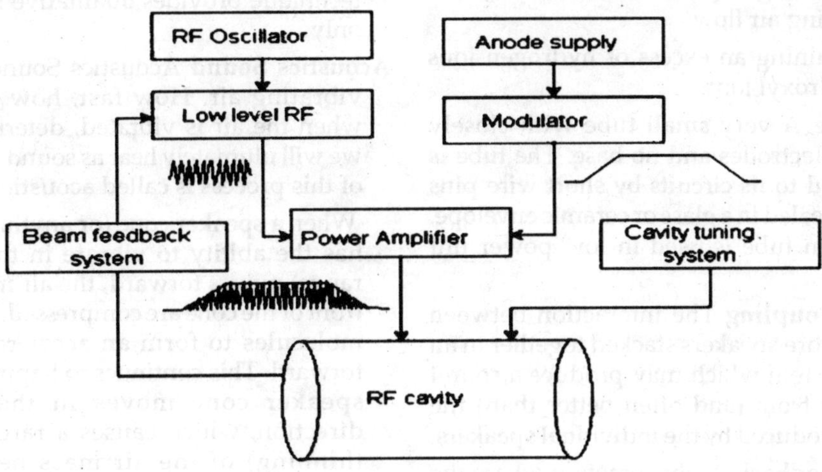

Fig. Accelerating anode

Acceleration servosystem A servosystem that controls the acceleration (rate of change in velocity) of a load.

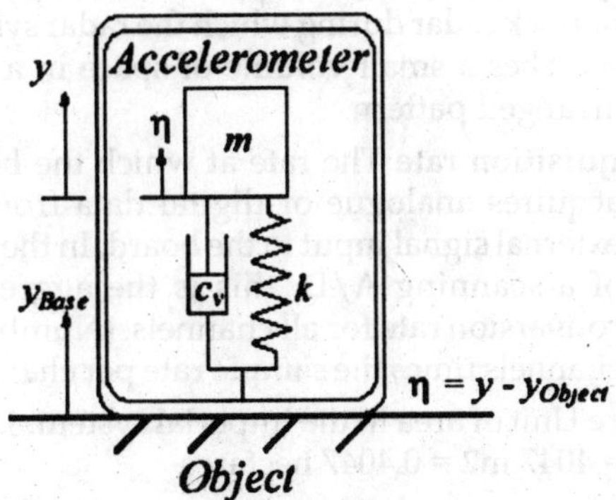

Fig. Accelerometer

Access time The time interval between the request for imformation and the instant that this information is available. This is often the time taken for one complete access of a peripheral by the central processing unit.

Accessory A device, other than current using equipment, associated with such equipment or with the wiring of an installation.

Accumulated temperature difference The concept of accumulated temperature difference or degree days allows the need for heating in a building to be assessed.

Accumulator Storage battery or secondary cell for storing electricity.

Accuracy The degree of correctness with which the measuring system yields the "true value" of a measured quantity, where the"true value" refers to an accepted standard, such as a standard meter or volt. Typically described in terms of a maximum percentage of deviation expected based on a full-scale reading.

Accuracy of an input The maximum expected deviation of the indicated value from the true value.

Accuracy of an output The maximum expected deviation of the actual value from the desired value.

Ach (Air changes per hour) Unit used for quantifying air flow.

Acid Containing an excess of hydrogen ions over hydroxyl ions.

Acorn tube A very small tube with closely spaced electrodes and no base. The tube is connected to its circuits by short wire pins that are sealed in a glass or ceramic envelope. The acorn tube is used in low power uhf circuits.

Acoustic Coupling The interaction between two or more speakers stacked together in an audio system which may produce a sound different from (and often better than) the sound produced by the individual speakers.

Acoustic feedback A phenomenon where the sound from a loudspeaker is picked up by the microphone or other transducer, like a phono cartridge feeding it, and re-amplifys it through the same loudspeaker only to return to the same microphone to be re-amplified again, etc. Each time the signal becomes larger until the system runs away and rings, or feeds back on itself producing the characteristic scream or squeal found in sound (mostly, PA) systems. These buildups often occur at particular frequencies called feedback frequencies.

Acoustic phase The time alignment of sound waves. Phase is important at crossover points, where phase shift (time shift) occurs between the drivers. A difficult technical challenge with crossover design is precisely achieving the right amount of phase shift in each driver at the crossover point, so that the frequencies sum together properly.

Acoustic Suspension (enclosure) A Sealed box system that uses the resistance of the internal air of a sealed enclosure to control the motion of the cone. Sometimes known as an infinite baffle type.

Acoustic technique A method of detecting cracks in a building where leakage may occur. A steady source of high pitched sound is placed within the building and a microphone is used outside the building as a detector. Leaks correspond to an increase in volume of the sound transmitted. This technique provides qualitative information only.

Acoustics Sound Acoustics Sound is simply vibrating air. How fast, how much, and when the air is vibrated, determines what we will ultimately hear as sound. The science of this process is called acoustics.

When a speaker cone (or anything else that has the ability to vibrate in the auditory range) moves forward, the air molecules in front of the cone are compressed, causing the molecules to form an accelerating wave forward. This continues to happen until the speaker cone moves in the opposite direction, which causes a rarefaction (or thinning) of the air mass between the speaker and the listener. This is the basic concept of how sound waves are produced.

Acoustics The degree of sound. The nature, cause, and phenomena of the vibrations of elastic bodies; which vibrations create compressional waves or wave fronts which are transmitted through various media, such as air, water, wood, steel, etc.

Acquisition Operational phase of a fire-control or track radar during which the radar system searches a small volume of space in a pre-arranged pattern.

Acquisition rate The rate at which the board acquires analogue or digital data from an external signal input to the board. In the case of a scanning A/D, this is the aggregate conversion rate for all channels. (Number of channels times the sample rate per channel.)

Acre Unit of area in the Imperial system. 1 acre = 4047 m2 = 0.4047 hectares

Active A word prefacing certain circuits in which the processing is performed by use of transistor or tube juctions, rather than passive componenets such as resistors, capacitors, and coils. Such items as crossovers and equalizers may be constructed either way. Active processing usually affords more more options, and greater precision, albeit at greater cost.

Active attack An attack which results in an unauthorized state change, such as the manipulation of files, or the adding of unauthorized files.

Active component A component that provides gain or amplification such as transistor, integrated circuit, valve - such as a triode value.

Active Display A special feature for front panel receiver displays that generates animated patterns for both segmented and dot matrix LCDs that proceed the sequential display of information such as clock, CD titles, and radio station call letters and frequencies.

Active element An element capable of generating electrical energy.

Active filter A filter that uses an amplifier in addition to reactive components to pass or reject selected frequencies.

Active iron The amount of steel (iron) in the stator and rotor of a motor. Usually the amount of active iron is increased or decreased by lengthening or shortening the rotor and stator (they are generally the same length).

Active region The region of BJT operation between saturation and cutoff used for linear amplification.

Active satellite A satellite that amplifies the received signal and retransmits it back to earth.

Active video The portion of a video signal that contains the visible picture information.

Activity (ai) A thermodynamic term for the apparent or active concentration of a free ion in solution. It is related to concentration by the activity coefficient.

Activity Coefficient (fi) A ratio of the activity of species i(ai) to its molality (C). It is a correction factor which makes the thermodynamic calculations correct. This factor is dependent on ionic strength, temperature, and other parameters. Individual ionic activity coefficients, f+ for cation and f- for an anion, cannot be derived thermodynamically. They can be calculated only by using the Debye-Huckel law for low concentration solutions in which the interionic forces depend primarily on charge, radius, and distribution of the ions and on the dielectric constant of the medium rather than on the chemical properties of the ions. Mean ionic activity coefficient (f±) or the activity of a salt, on the other hand, can be measured by a variety of techniques such as freezing point depression and vapor pressure as well as paired sensing electrodes. It is the geometric mean of the individual ionic activity coefficients $f_\pm = (f_+^{n+} f_-^{n-})^{1/n}$

Actual The present value of the controlled variable.

Actuation force The amount of force required to complete the contact switching action. When this term is applied to thermostats, it refers to the amount of force generated by the bimetal disc, which opens or closes the contacts.

Actuation time The time between application of the nominal relay coil voltage and the final closure of the relay contacts after the contact bounce interval. Also called operate time.

Actuator A means provided to execute the output from the controller, (i.e., the control

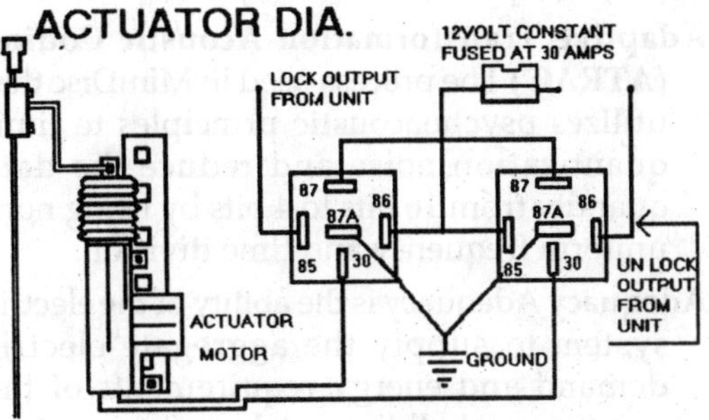

Fig. Actuator

action), through operating a final control element. This control action is determined such as to reduce the error between the measured value and desired value of the controlled variable.

Acute angle An angle of less than 90o.

Adaptation changes in an organism's structure or behavior that help it adjust to its surroundings. An increase or decrease in sensitivity to a given stimulus that results from exposure to that stimulus.

Adapter A mechanism or device for attaching non-mating parts.

Adaptive control The primary objective of adaptive, or 'self-learning', or 'heuristic' control is to reduce uncertainties concerning knowledge of the environment and system dynamics in an on-line or real time fashion and to alter controller parameters (gain,integral such as to cause system operation to continuously seek better performance. Thus more accurate control is achieved over a wide range of external conditions since control parameters are automatically adjusted as conditions vary. The need for manual re-tuning to adjust for seasonal variations is eliminated.

Adaptive Reception A circuit that provides adjustable delay such that the time at which a sound wave is produced from various signal channels can be changed. This allows synchronization for each channel of a multi-channel output device, so that sound waves from multiple speakers will converge at approximately the same time at a single point in space (the listener).

Adaptive Transformation Acoustic Coding (ATRAC) The process used in MiniDisc that utilizes psychoacoustic principles to limit quantization noise and reduce the data quantity from 16 bits to 4 bits by using non-uniform frequency and time division.

Adequacy Adequacy is the ability of the electric system to supply the aggregate electric demand and energy requirements of the consumers at all times, taking into account scheduled and unscheduled outages of system facilities. Security is the ability of the electric system to withstand sudden disturbances such as electric short circuits or unanticipated loss of system facilities.

Adiabatic process taking place without heat entering or leaving the system.

Administrative Security The management constraints and supplemental controls established to provide an acceptable level of protection for data.

Admittance method The CIBSE method of determining the peak inside environmental temperature.

Admittance The reciprocal of the impedance. The admittance is the complex ratio of the current flowing through divided by the voltage across a device, circuit element, or network.

Adsorption The adhesion of a thin film of liquid or gases to the surface of a solid substance.

Adventitious opening (Also known as unintentional opening & fortuitous leakage) An opening within the building envelope which, in terms of ventilation, is unintentional, for example, cracks around doors and windows.

Aerial A length of wire designed to transmit or receive radio waves.

Aerodynamics Study of air moving around solid objects, or air flowing around a stationary structure.

AF (Audio frequency) Generally in the range 20Hz to 20KHz.

Affinity A thermodynamic measurement of the strength of binding between molecules, say between an antibody and antigen. Each antibody/antigen pair has an association constant, Ka, expressed in L/mol.

AGC Abbreviation for "automatic gain control".

Agent A software program that crawls around the Web to find, catalogue and report information about where it has been.

Air change A quantity of fresh air equal to the volume of the room or building being ventilated.

Air change efficiency A measure of how quickly the air in the room is replaced. It represents, the ratio between the nominal time constant, and the air change time, for the room.

Air change rate The volumetric rate at which air enters (or leaves) a building or zone expressed in units of building or zone volume.

Air conditioning (Also known as environmental control) Treatment or control of the air in buildings so as to render it more comfortable or healthful for human beings or more suitable for manufacturing process. Alternatively, air conditioning refers to the control of temperature, moisture content, cleanliness, air quality, and air circulation as required by occupants, a process or a product in the space; Air conditioning is always associated with the cooling and dehumidification process of air and is always therefore identified with refrigeration equipment.; Control over relative humidity by the addition of moisture constitutes full air conditioning. However, the more often used partial or comfort air conditioning (cooling and drying) which uses refrigeration equipment only is still referred to as air conditioning; In the context of HVAC systems, ventilation should not be confused with air conditioning as refrigeration equipment is not necessarily provided with ventilation equipment.

Air control panel Panel that monitors the dry-air input at each user equipment.

Air core transformer A transformer composed of two or more coils that are wound around a non-metallic core.

Aliasing If the sample rate of a function (fs) is less than two times the highest frequency value of the function, the frequency is ambiguously presented. The frequencies above (fs/2) will be folded back into the lower frequencies producing erroneous data.

Alkaline Containing an excess of hydroxyl ions over hydrogen ions.

Alkaline cell A primary cell that delivers more current than a carbon-zinc cell. Also known as an "alkaline manganese cell".

Allergen a substance that induces allergic reaction.

Alligator clip Spring clip on the end of a test lead used to make a temporary connection.

Allowance parts list (APL) Repair parts required for units having the equipment/component listed.

Alloy A composition of two or more metals.

Alternate sweep A vertical mode of operation for a dual-trace oscilloscope. The signal from the second channel is displayed after the signal from the first channel. Each trace has a complete trace, and the display continues to alternate.

Alternator Whine A siren-like whining that appears as the rotational speed of an engine increases. The noise is usually the result of a voltage differential created by more than one ground path or a poor ground path to the affected equipment.

Aluminum creep 1. The movement of aluminum wire from a point where pressure is applied.

2. The "retreat" of heated aluminum wire as it cools.to that of sunlight at the earth's surface when the sun is at zenith.

AM Abbreviation for "amplitude modulation".

Ambient air the air surrounding an object.

Ambient compensation The design of an instrument such that changes in ambient temperature do not affect the readings of the instrument.

Ambient conditions The conditions around the transducer (pressure, temperature, etc.).

Ambient pressure Pressure of the air surrounding a transducer.

Ambiguous returns Echoes that exceed the prt of a radar and appear at incorrect ranges.

American wire gauge (AWG) The standards adopted in the United States for the measurement of wire sizes. AWG table

Ammeter An instrument used for measuring the electrical current flow in a portion of a circuit.

Amp (Ampere) The unit of electrical current. Also milliamp (one thousandth of an amp) and microamp (one millionth of an amp). One amp corresponds to the flow of about 6 x 1018 electrons per second.

Amp hour (AH or Amp-hr) A unit of measurement which quantifies the amount of current flow for an amount of time. For example, a current of 1 amp drawn from a battery for 10 hours would consume 10 Amp-hrs of charge from the battery. Fundamentally, a measurement of electrical charge.

Ampacity The current-carrying capacity of conductors or equipment, expressed in ampere.

Ampere hour capacity The quantity of electricity measured in ampere-hour which may be delivered by a cell or battery under specified conditions.

Ampere turn (AT) Formerly used as the unit of magnetomotive force (mmf). It is the product of the number of turns in a coil and the current in amperes which flows through it. (Note: Since turns is not a unit, the SI Unit of mmf is the ampere)

Amperes (A) Ampere is a unit measurement of current of electrical energy equal to one coulomb of charge per second. Most DC applications refer to positive current - current which flows from a positive potential to a more negative potential, with respect to a reference point which is designated as zero or neutral potential (usually ground).

The electrons in a circuit flow in the opposite direction as the current itself. Ampere is commonly abbreviated as "amp", not to be confused with amplifiers, of course, which are also commonly abbreviated "amp". In computation, the abbreviation for amperes is commonly, "I".

Amperite (Ballast) tube A current-controlling resistance device designed to maintain substantially constant current over a specified range of variation in applied voltage or resistance of a series circuit.

Amperometric sensor Amperometric sensors involve a heterogeneous electron transfer as a result of an oxidation/reduction of an electro-active species at a sensing electrode surface. A current is measured at a certain imposed voltage of the sensing electrode with respect to the reference electrode. Analytical information is obtained from the current-concentration relationship at that given applied potential.

Amplidyne A special dc generator in which a small dc voltage applied to field windings controls a large output voltage from the generator. In effect, an amplidyne is a rotary amplifier that often times produces gain of approximately 10,000.

Amplification 1. The process of enlarging a signal in amplitude (as of voltage or current). 2. The ratio of output magnitude to input magnitude in a device that is intended to produce an output that is an enlarged reproduction of its input.

Amplification factor The voltage gain of an amplifier with no load on the output.

Amplitude A measurement of the distance from the highest to the lowest excursion of motion, as in the case of mechanical body in oscillation or the peak-to-peak swing of an electrical waveform.

Amplitude distortion Distortion that is present in an amplifier when the amplitude of the output signal fails to follow exactly any increase or decrease in the amplitude of the input signal.

Amplitude Modulation (AM) In radio broadcasting, amethod of modulation in

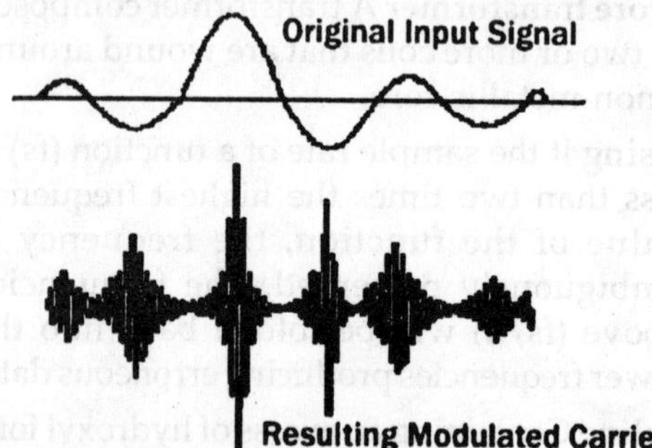

Fig. Amplitude Modulation

which the amplitude of the carrier voltage is varied in proportion to the changing frequency value of an applied (audio) voltage.

Amplitude span The Y-axis range of a graphic display of data in either the time or frequency domain. Usually a log display (db) but can also be linear.

Amplitude stability Amplitude stability refers to the ability of the oscillator to maintain a constant amplitude in the output waveform.

Analogue A way to represent data by means of continuously variable quantities. Analogue Output A continuously variable signal that is used to represent a value, such as the process value or set point value. Typical hardware configurations are 0-20mA, 4-20mA or 0-5VDC.

Analogue output A voltage or current signal that is a continuous function of the measured parameter.

Analogue ramp BA voltage output of constant slope, dV/dt (Volts/second).

Analogue signal A continuously changing variable.

Analogue Switch A hardware-oriented switch that only passes signals that are faithful analogues of transducer parameters.

Analogue to digital converter, A/D converter, A to D converter A device or circuit used to convert an analogue signal to a digital signal across a pair of terminals.

Analogue trigger An event that occurs at a user-selected point on an analogue input signal. The polarity, sensitivity, and hysteresis of the analogue trigger can often be programmed.

Analogueue A system in which data is represented as a continuously varying voltage.

Analogueue energy model The analoguey that exists between electrical flow and heat flow may be used to construct electrical analogueue devices for the study of complex heat flow phenomena. The technique, although having had its use as a research tool, has little application in a design context.

Analogueue meter Show a particular (continuous variable) deflection for a given input quantity.

Analoguey A lik- The angle between the normal and the path of a wave through the second medium .

Angled (box or enclosure) A type of speaker enclosure wherein the woofers, or Drivers, are situated in a box that is angled on one side. This is usually done to facilitate a more direct coupling of the upper part of the speaker's range to the listening environment.

Angstrom [Å°] Measure of length: 1 Å°= 1x10-10 m.

Angular frequency The motion of a body or a point moving circularly, referred to as the circular frequency O which is the frequency in cycles per second (cps) multiplied by the term

2. and expressed in radians per second (2pf).

Angular velocity w Rate of rotation about an axis. It is the rate of change of angle with time. It is measured either in revolutions per second, revolutions per minute (r.p.m.) Or radians per second (rad/s).

Anion A negatively charged ion (Cl-, NO3-, S2- etc.)

Anisotropic The property of a radiator that allows it to emit strong radiation in one direction.

Ankle biter A person who aspires to be a hacker/cracker but has very limited knowledge or skills related to AIS's. Usually associated with young teens who collect and use simple malicious programs obtained from the Internet.

Anneal Heat process used to remove stress, crystallize or render deposited material more uniform.

Anode 1. A positive electrode of an electrochemical device (such as a primary or secondary electric cell) toward which the negative ions are drawn.

2. The semiconductor-diode terminal that is positive with respect to the other terminal when the diode is biased in the forward direction.

Anode foil An aluminum electrolytic capacitor is typically composed of an anode foil, a cathode foil and separator material which are wound together and impregnated with an electrolyte. The anode foil has an aluminum oxide layer acting as the dielectric. After the aluminum foil (65 to 100 microns) is electrochemically etched to increase it's surface area, the dielectric is produced by anodic oxidation of its surface.

Increasing the surface area of the foil serves to increase its unit capacitance after it is anodized.

Anodize When immersed in certain electrolyte solutions and connected to a DC power source as an anode with the solution as a cathode, the surface layer of the "valve" metal forms an impervious, amorphous film of oxide having the property of restricting the flow of current in one direction and permitting it to flow in the opposite direction (i.e., a rectifing characteristic).

Common capacitor metals (and dielectrics) are aluminum (Al2O3) and tantalum (Ta2O5). Other valve metals are zinc and magnesium.

Anomaly detection model A model where intrusions are detected by looking for activity that is different from the user's or system's normal behavior.

Antenna A conductor or set of conductors used to radiate RF energy into space or to collect RF energy from space or to do both.

Antenna beam width Width of a radar beam measured between half-power points.

Antenna coupler A device used for impedance matching between an antenna and a transmitter or receiver.

Antenna receiving A device that converts a radiated electromagnetic wave into an electrical wave.

Antenna system Routes RF energy from the transmitter, radiates the energy into space, receives echoes, and routes the echoes to the receiver.

Antenna transmitting A device that converts an electrical wave into an electromagnetic wave that radiates away from the antenna.

Anti friction bearing An anti-friction bearing is a bearing utilizing rolling elements between the stationary and rotating assemblies.

Anti reset windup This is a feature in a three-mode PID controller which prevents the integral (auto reset) circuit from functioning when the temperature is outside the proportional band.

Antiseize compound A silicon-based, high-temperature lubricant applied to threaded components to aid in their removal after they have been subjected to rapid heating and cooling.

Antitransmit receive tube (ATR) A tube that isolates the transmitter from the antenna and receiver. Used in conjunction with a tr tube.

Apartment A separate room or suite of rooms in a building occupied by one party. Many of these apartments form apartment blocks, or blocks of flats.

Aperiodic damping A system of damping so large that after having being subject to a single constant or instantaneous disturbance, the system tends to a state of equilibrium, without oscillating around it.

Aperture Opening. In optical instruments, it is the size of the opening admitting light.

Apogee The point in the orbit of a satellite the greatest distance from the earth.

Apparent drift The effect of the earth's rotation on a gyro that causes the spinning axis to appear to make one complete rotation in one day.

Apparent power That power apparently available for use in an ac circuit containing a reactive element. It is the product of effective voltage times effective current expressed in volt-amperes. It must be multiplied by the power factor to obtain true power available.

Appliance An Item of current using equipment other than a luminaire or an independent motor. These are generally not industrial and normally built in standardized sizes or types, which are installed or connected as a unit to perform one or more functions.

Application A program that performs a specific task such as word processing, database management or web browsing.

Application level gateway (Firewall) A firewall system in which service is provided by processes that maintain complete TCP connection state and sequencing. Application level firewalls often re-address traffic so that outgoing traffic appears to have originated from the firewall, rather than the internal host.

Application program A computer program used to perform a particular kind of work, such as data acquisition.

Applications software In a BEMS, programs that provide functions such as DDC algorithms, energy management, and lighting control; cf. Operating Software.

Arc Highly luminous discharge at a very high temperature (around 3000oc). An electric arc is produced when an electric current flows between two electrodes.

Archie A program that locates files that are freely available on anonymous ftp sites across the Internet. To use Archie, telnet to one of these sites and login as archie. Type help to obtain full instructions. archie.internic.net archie.ans.net archie.rutgers.edu archie.sura.net archie.unl.edu

Architecture The product, created in the process of applying art and science in the designing buildings.

Archiving The storage of data in a retrievable data set.

Are A metric unit of area, especially with land. Equal to 100 square meter.

Argand diagram A plot of the cartesian co-ordinates (real and imaginary components) or the polar co-ordinates of a complex number on the x-y plane.

Arithmetic Logic Unit (ALU) The part of a CPU where binary data is acted upon with mathematical operations.

Arithmetic progression A series of quantities in which each term differs from the preceding term by a constant difference.

Arithmetic summation of uncertainties The sum of the moduli of uncertainties. The most pessimistic method of combining uncertainties. If three uncertainties are estimated at ±3% each then the arithmetic sum is ±9%.

Arm's reach A zone of accessibility to touch, extending from any point on a surface where persons usually stand or move about to the limits which a person can reach with a hand in any direction without assistance.

Armature 1. In a relay, the movable portion of the relay.

2. The windings in which the output voltage is generated in a generator or in which input current creates a magnetic field that interacts with the main field in a motor.

Armature current (AMPS) Rated full load armature circuit current.

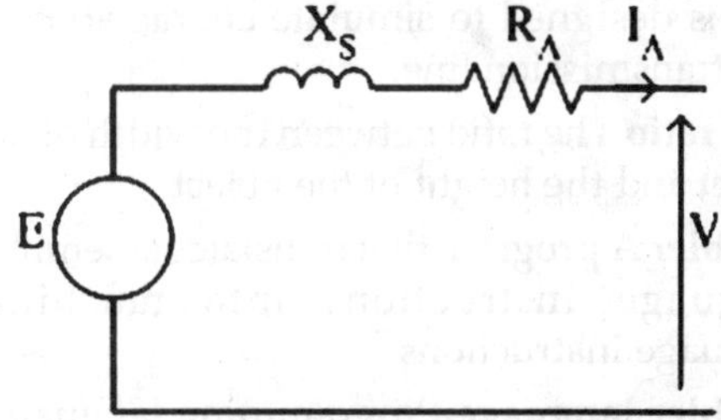

Fig. Armature current

Armature inductance (MH) Armature inductance in milli-henries (saturated).

Armature losses Copper losses, eddy current losses, and hysteresis losses that act to decrease the efficiency of armatures.

Armature reaction The current that flows in the armature winding of a DC motor tends to produce magnetic flux in addition to that produced by the field current. This effect, which reduces the torque capacity, is called armature reaction and can effect the commutation and the magnitude of the motor's generated voltage.

Armature resistance (OHMS) Armature resistance is measured in ohms at 25*f* C. (cold) .

Armoured cable Cable with a metal protective covering.

Armstrong oscillator An oscillator that uses an isolation transformer to achieve positive feedback from output to input.

Arrester A device placed from phase to ground, or phase to phase, whose nonlinear impedance characteristics provide a path for high-amplitude transients.

Arrhenius equation The equation representing the rate constant as k = AeEa/RT where A represents the product of the collision frequency and a steric factor, and eEa/RT is fraction of collisions with sufficient energy to produce a reaction.

Artificial intelligence The process of enabling computers to mimic human learning and decision-making.

Artificial transmission line An LC network that is designed to simulate characteristics of a transmission line.

Aspect ratio The ratio between the width of an object and the height of the object.

Assembler A program that translates assembly language instructions into machine language instructions.

Assembler language Programming language which allows the programmer to define labels and fixed values to then use these labels with a mnemomic instruction to produce a machine code computer program.

Assembly A number of parts or subassemblies, or any combination thereof, joined together to perform a specific function.

Assembly language A machine oriented language in which mnemonics are used to represent each machine language instruction. Each CPU has its own specific assembly language.

Assessment Surveys and Inspections; an analysis of the vulnerabilities of an AIS. Information acquisition and review process designed to assist a customer to determine how best to use resources to protect information in systems.

Assurance A measure of confidence that the security features and architecture of an AIS accurately mediate and enforce the security policy.

Astable A circuit that has no stable state and thus oscillates at a frequency dependent on component values.

Astable multivibrator A multivibrator that has no stable state. Also called free-running because it alternates between two different output voltage levels during the time it is on. The frequency is determined by the RC time constant of the coupling circuit.

Astigmatism An aberration of a lens with spherical surfaces such that the image of a point not lying on the optical axis is a pair of short lines normal to each other and at slightly different distances from the lens. Radial and tangential lines are in focus in different image planes.

Asymmetric Not possessing symmetry. Unequal distribution about one or more axes.

Asymmetrical multivibrator A multivibrator that generates rectangular waves.

Asymmetry potential The potential developed across the glass membrane with identical solutions on both sides. Also a term used when comparing glass electrode potential in ph 7 buffer.

Asynchronous A communication method where data is sent when it is ready without being referenced to a timing clock, rather than waiting until the receiver signals that it is ready to receive.

Asynchronous communication Data communication in which the timing of information transmission is not related to any sytem frequency, e.g. 50 Hz.

Asynchronous orbit One where the satellite does not rotate or move at the same speed as the earth.

ATC Automatic temperature compensation.

Atmosphere Unit of pressure corresponding to standard atmospheric pressure. It is taken as the pressure that will support a column of mercury 760 mm high.

Atmospheric pressure Standard Pressure exerted by the earth's atmosphere on bodies located within it. Standard atmospheric pressure is 14.7 psia (1.013 bar abs.) measured at sea level and 60¼F (15¼C).

Atom The smallest particle that an element can be broken down into and still maintain its unique identity.

Atomic mass unit (AMU) A unit of mass used to express relative atomic masses. It is equal to 1/12 of the mass of an atom of the isotope carbon-12 and is equal to 1.66033x10-27 kg

Atomic number (also proton number Z) The number of protons within the atomic nucleus of a chemical element.

Atomic orbital The region in space around the nucleus of an atom in which an electron with a given set of quantum numbers is most likely to be found.

Atomic weight The weighted average mass of the atoms in a naturally occurring element.

Atrium An open space in the middle or at the edge of a building, usually enclosed, but still allowing the penetration of light.

Attack An attempt to bypass security controls on a computer. The attack may alter, release, or deny data. Whether an attack will succeed depends on the vulnerability of the computer system and the effectiveness of existing countermeasures.

Attenuate The act of reducing the Amplitude or intensity of a signal. In speaker systems, high frequency drivers are commonly more efficient than low frequency drivers. This creates a need to adjust the driver levels to create a uniform overall frequency response. L-pads are commonly used for many passive systems

Attenuation The ability of a filter circuit to reduce the amplitude of unwanted frequencies to a level below that of the desired output frequency.

Attenuator A passive device used to reduce signal strength.

Attic A low storey or structure above the main part of a dwelling. Alternatively known as a loft or roofspace.

Atto (a) Decimal sub-multiple prefix corresponding to 10-18

Attraction The force that tends to make two objects approach each other. Attraction exists between two unlike magnetic poles (north and south) or between two unlike static charges.

Audio 1. Of, or relating to, humanly audible sound, i.e., audio is all the sounds that humans hear.

2. a. Relating to the broadcasting or reception of sound.

Audio amplifier An amplifier designed to amplify frequencies between 15 hertz (15 Hz) and 20 kilohertz (20 kHz).

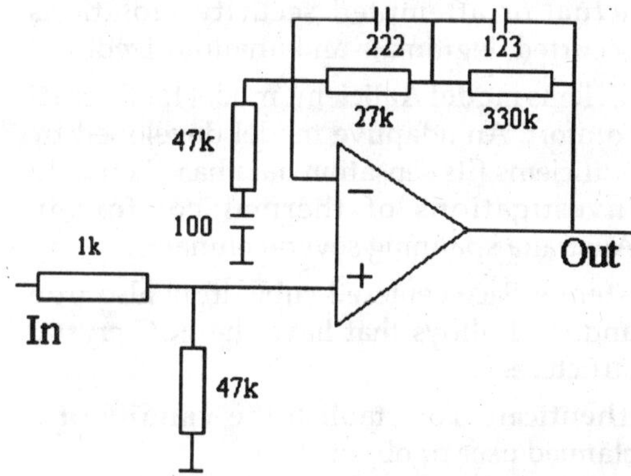

Fig. Audio amplifier

Audio follow video (AFV) A control mode in a routing switcher (switching array) in which the audio inputs associated with a video input are automatically selected when the video source is selected. That is, audio and video are always switched together. Audio may be either single channel or multi-channel (stereo).

Audio frequency A frequency corresponding to audible sound waves, and thus corresponds to a frequency between 20 Hz and 20 khz.

Audio Frequency Spectrum The band of frequencies extending roughly from 20 Hz to 20 kHz and encompassing the full range of normal human hearing.

Audio frequency tone shift A system that uses amplitude modulation to change dc mark and space impulses into audio impulses.

Audio taper (aka A-taper) Usually 15% resistance at the 50% rotation point .

Audiophile A species of dedicated audio nut who actually reads definitions like this.

Audit The independent examination of records and activities to ensure compliance with established controls, policy, and operational procedures, and to recommend any indicated changes in controls, policy, or procedures.

Audit trail In computer security systems, a chronological record of system resource usage. This includes user login, file access, other various activities, and whether any actual or attempted security violations occurred, legitimate and unauthorized.

Auliciems model Auliciems model for thermal comfort. An adaptive model developed by Auliciems fits sensation data based on field investigations of thermal comfort in Australia spanning several climates.

Austenite Face-centered cubic iron; also iron and steel alloys that have the FCC crystal structure.

Authenticate To establish the validity of a claimed user or object.

Authentication header (AH) A field that immediately follows the IP header in an IP datagram and provides authentication and integrity checking for the datagram.

Authentication To positively verify the identity of a user, device, or other entity in a computer system, often as a prerequisite to allowing access to resources in a system.

Auto Memory A tuner feature that automatically finds the strongest stations in the local area, and places them in preset memories

Auto ranging A circuit to switch between 85 - 132VAC and 170 - 264VAC input automatically.

Auto zero An automatic internal correction for offsets and/or drift at zero voltage input.

Automated information system (AIS) Any equipment of an interconnected system or subsystems of equipment that is used in the automatic acquisition, storage, manipulation, control, display, transmission, or reception of data and includes software, firmware, and hardware.

Automated security incident measurement (ASIM) Monitors network traffic and collects information on targeted unit networks by detecting unauthorized network activity.

Automated security monitoring All security features needed to provide an acceptable level of protection for hardware, software, and classified, sensitive, unclassified or critical data, material, or processes in the system.

Automatic control system (also automatic regulating system) A system that reacts to a change in the variable it controls by adjusting other variables to restore the system to the desired balance.

Automatic frequency control (AFC) Similar to Automatic Fine Tune (AFT). A circuit that keeps a receiver in tune with the wanted transmission.

Automatic gain control A circuit used to vary radar receiver gain for best reception of signals that have widely varying amplitudes.

Automatic Music Search (AMS) A feature that allows a CD or cassette mechanism to skip forward or backwards to another track.

Automatic reset 1. A feature on a limit controller that automatically resets the controller when the controlled temperature returns to within the limit bandwidth set.

2. The integral function on a PID controller which adjusts the proportional bandwidth with respect to the set point to compensate for droop in the circuit, i.e., adjusts the controlled temperature to a set point after the system stabilizes.

Automatic tracking Tracking done by equipment that compares the direction of the antenna axis and the direction of the received signal and uses the difference (error) signal to reposition the antenna.

Automatic transfer switch A switch that automatically transfers electrical loads to alternate or emergency-standby power sources.

Automatic Tuner Activation(ATA) A feature that allows the tuner to be accessed while a tape deck is rewinding or fast forwarding.

Automatic volume/gain control A circuit used to limit variations in the output signal strength of a receiver.

Automation The application of mechanical or electronic techniques to minimise the use of manpower in any process.

Autoranging The ability of an instrument to switch among ranges automatically. The ranges are usually in decade steps.

Autoranging time For instruments with autoranging capability, the time interval between application of a step input signal and its display, including the time for determining and changing to the correct range.

Autoreclose A feature of certain circuit breakers where they close automatically after a predetermined time after an automatic opening due to a transient fault.

Autotransformer A transformer in which both the primary and the secondary windings share common turns. It provides no isolation.

Autotransformer starter A starter that includes an auto-transformer to furnish reduced voltage for starting an alternating current motor.

Auxiliary contacts The contacts of a switching device, in addition to the main current contacts, that operate with the movement of the latter. They can be normally open (NO) or normally closed (NC) and change state when operated.

Auxiliary output An output that controls external activities that are not directly related to the primary, control output. For example, door latches, gas purges, lights and buzzers.

Availability The proportion of the switched-on time during which the equipment is available for work; cf. maintainability and reliability.

Avalanche A build up of particles caused by the collision of a high energy particle with any other form of matter. (Note: The term is derived from the avalanches occurring in a mountain)

Avalanche effect A reverse breakdown effect in diodes that occurs at reverse voltages beyond 5 volts. The released electrons are accelerated by the electric field, which results in a release of more electrons in a chain or "avalanche" effect.

AVC Abbreviation for "automatic volume control".

Average control If the value of the control variable depends on the location of the sensor, it may be necessary to apply corrector action in proportion to the average deviation measured in several locations.

Average or apparent power The result of multiplying the rms value of the voltage by the rms value of the current in an electronic circuit. It is expressed in watts (W) for resistive loads and in voltamperes (VA) for reactive loads. The real power is usually less because of losses when the power factor is accounted.

Average power 1. The peak power value averaged over the pulse-repetition time.

2. Output power of a transmitter as measured from the start of one pulse to the start of the next pulse.

Average value (OF AC) The average of all the instantaneous values of one-half cycle of alternating current.

Avionics Aviation electronics.

Avogadro's number The number of atoms in exactly 12 grams of pure 12C, equal to 6.022 × 1023.

Axial or vaneaxial fan A fan in which the airflow, at all times, from entry to exit is predominately parallel to the axis of rotation.

Axial thrust The force or loads that are applied to the motor shaft in a direction parallel to the axis of the shaft. (Such as from a fan or pump).

Axis Line about which a given body or system is considered to rotate.

Axis of rotation (Spin Axis) The axis of rotation (spin axis) is that straight line about which a body rotates.

Axis of symmetry Line about which a given figure is symmetrical.

Azimuth Angular measurement in the horizontal plane in a clockwise direction.

B

Back door A hole in the security of a computer system deliberately left in place by designers or maintainers. Synonymous with trap door; a hidden software or hardware mechanism used to circumvent security controls.

Back electromotive force (back emf) The emf that opposes the normal flow of current in a circuit.

Back end Back end refers to the package assembly and test stages of production in semiconductor manufacturing. It includes burn-in and environmental test functions.

Back flashover Flash-over occurring from an object usually at earth potential (such as a tower) to a line conductor due to the potential of the earthed object rising due to lightning.

Back Light In receivers, a display may be lit from the rear to create better visibility under a wide range of ambient light conditions.

Back Plate The part of the woofers metal Basket or frame on which the Magnet structure is mounted.

Back resistance The larger resistance value observed when you are checking the resistance of a semiconductor.

Backbone 1. The high-traffic-density connectivity portion of any communications network.

2. In packet-switched networks, a primary forward- direction path traced sequentially through two or more major relay or switching stations. Note: In packet-switched networks, a backbone consists primarily of switches and inter-switch trunks.

Backdraughting The reversed flow of polluted air (or flue gases) in a chimney, flue or other air outlet, back into the room or building.

Background leakage Unidentified openings or gaps in a building envelope through which infiltration can take place.

Background noise The total noise floor from all sources of interference in a measurement system, independent of the presence of a data signal.

Background task An operation that can take place while another program or processing routine is running without apparent interruption to that program or routine. For example, an interrupt or DMA operation.

Backup A system, device, file or facility that can be used as an alternative in case of a malfunction or loss of data.

Baffle A flat panel that divides the front and rear sound waves produced by a woofer. Sometimes baffle is used to mean an enclosure or the front panel on which the speaker is mounted .

Bag sampling method May refer to:

A method of measuring the air change rate using tracer gas and a two channel pump. One channel dispenses tracer gas from a bag of known volume, the other draws room air into a sample bag. The final concentration of tracer in the room air and the amount of tracer gas used, enables the calculation of the average inverse air change rate.

A method of measuring the air change rate by which tracer is discharged into a sample volume and mixed. Sample bags are inflated with room air at intervals and the concentration of tracer measured.

Balanced Referring to wiring: Audio signals require two wires. In an unbalanced line the shield is one of those wires. In a balanced line, there are two wires plus the shield. For the system to be balanced requires balanced electronics and usually employs XLR connectors. Balanced lines are less apt to pick up external noise. This is usually not a factor in home audio, but is a factor in professional audio requiring hundreds or even thousands of feet of cabling. Many higher quality home audio cables terminated with RCA jacks are balanced designs using two conductors and a shield instead of one conductor plus shield.

Balanced bridge Condition that occurs when a bridge circuit is adjusted to produce a zero output.

Balanced code In PCM systems, a code constructed so that the frequency spectrum resulting from the transmission of any code word has no dc component. In PCM, a code that has a finite digital sum variation.

Balanced fan pressurization A measurement technique using two or more blower doors to evaluate the leakage of individual internal partitions and external walls of multizone buildings. The technique involves using the fans to induce a zero pressure difference across certain building components, thus eliminating their leakage from the measurement.

Balanced line A transmission line consisting of two conductors in the presence of ground, capable of being operated in such a way that when the voltages of the two conductors at all transverse planes are equal in magnitude and opposite in polarity with respect to ground, the currents in the two conductors are equal in magnitude and opposite in direction.

Balanced mixer A waveguide arrangement that resembles a T and uses crystals for coupling the output to a balanced transformer.

Balanced output A differential output circuit pair with equal source impedance on each side.

Balanced phase detector A circuit that controls the oscillator frequency.

Balanced supply/extract ventilation system A ventilation system in which fans both supply and extract air from an enclosed space at equal rates.

Balanced three phase A three phase voltage or current is said to be balanced when the magnitude of each phase is the same, and the phase angles of the three phases differ from each other by 120°. A star-connected load or a delta-connected load is said to be balanced when the three arms of the star or the delta have equal impedances in magnitude and phase.

Balanced wiring Audio line signals require two conductors. In an unbalanced line, the shield is one of those. In a balanced line, there are two internal wires plus the shield. For the system to be balanced requires output transformers and usually employs XLR connectors. Balanced lines are less apt to pick up external noise. This is usually not a factor in home or car audio, but is a factor in professional audio requiring hundreds or even thousands of feet of cabling.

Ballast A Ballast is an electrical device which is required for all discharge lamps. It limits the current through the lamp, preventing damage to both the lamp and the electrical supply.

Ballistic galvanometer Instrument for measuring the total quantity of electricity passing through a circuit due to a momentary current. The period of oscillation of the galvanometer must be long compared with the time during which the current flows.

Ballistics The study of the flight path of projectiles.

Balun An acronym for Balanced/Unbalanced. A device used to couple a balanced system {impedance} to an un-balanced system. A wide-band impedance matching transformer providing a 4:1 impedance ratio.

Banana Jacks & Plugs A set of connectors in which 4 spring contacts are wrapped vertically around a central pin like a banana peel. When inserted into the receptacle jack it maintains a strong and consistent contact. This type of connector is highly regarded as an excellent and reliable interconnector for cables between amplifiers and speakers.

Band A collection of orbitals, each delocalized throughout the solid, that are so closely spaced in energy as to be nearly continuous.

Band gap The energy separation between the top of the valence band and the bottom of the conduction band.

Band gap energy (Eg) For semiconductors and insulators, the energies that lie between the valence and conduction bands; for intrinsic materials, electrons are forbidden to have energies within this range.

Band limiting filters A low-pass and a high-pass filter in series, acting together to restrict (limit) the overall bandwidth of a system. Many audio amplifiers and processors, having switches labeled as "Rumble" or "Hiss," are filters of this type.

Band pass filter A tuned circuit designed to pass a band of frequencies between a lower cut-off frequency (f1) and a higher cut-off frequency (f2). Frequencies above and below the pass band are heavily attenuated.

Band reject filter A tuned circuit that does not pass a specified band of frequencies.

Band stop filter A tuned circuit designed to stop frequencies between a lower cut-off frequency (f1) and a higher cut-off frequency (f2) of the amplifier while passing all other frequencies.

Bandpass (box or enclosure) An enclosure that is specifically tuned to give maximum energy to a very limited range of frequencies, usually the lowest. In this arrangement, the woofers are fully enclosed in the box with the sound pressure being vented through one or more ports.

Bandpass filter A filter that allows a narrow band of frequencies to pass through the circuit. Rejects or attenuates frequencies that are either higher or lower than the desired band of frequencies. A filter that ideally passes all frequencies between two non-zero finite limits and bars all frequencies not within the limits. The cutoff frequencies are usually taken to be the 3-dB points.

Bandpass Gain The increase (or decrease) in efficiency of loudspeakers, due to the enclosure size and tuning. This is measured by the midband sensitivity of the speaker as a whole.

Bandstop filter A filter designed to eliminate all frequencies within a band of frequencies.

Bandwidth A term used to describe how much data you can send through a connection to the net. The transmission capacity of a given medium, in terms of how much data the medium can transmit in a given amount of time. The greater the bandwidth, the faster the rate of data transmission. Information carrying capacity of a communication channel.

Bandwidth Abbr. BW The numerical difference between the upper and lower -3 dB points of a band of audio frequencies. Used to figure the Q, or quality factor, for a filter.

Barium Ferrite A speaker magnet material made from an alloy with iron and barium for improved magnetic strength.

Barometer Instrument for measuring atmospheric pressure.

Barretter A type of bolometer characterized by an increase in resistance as the dissipated power rises.

Barrier A part providing a defined degree of protection against contact with live **parts** from any usual direction of access.

Barrier potential The natural difference of potential that exists across a forward biased pn junction.

Base 1. A reference value.

2. A number that is multiplied by itself as many times as indicated by an exponent.

3. Same as radix.

4. The region between the emitter and collector of a transistor that receives minority carriers injected from the emitter. It is the element that corresponds to the control grid of an electron tube.

Base biasing A method of biasing a BJT in which the bias voltage is supplied to the base by means of a resistor.

Base case model (standard) Computer model of a particular building. The base case model can be used to assess the relative performance of a certain (new) feature of the building by changing the model parameters associated with that feature. Comparison of the results for the base case model with those for the changed model will reveal the relative performance of the feature.

Base injection modulator Similar to a control-grid modulator. The gain of a transistor is varied by changing the bias on its base.

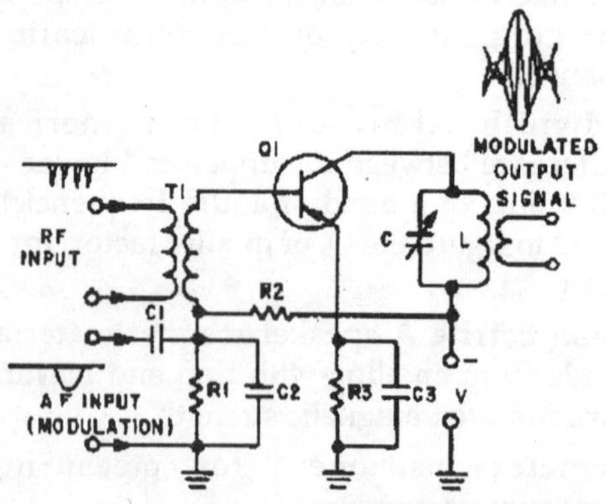

Fig. Base injection modulator

Base load The minimum load experienced by an electric system over a given period of time, which must be supplied at all times.

Base load capacity Capacity of generating equipment operated to serve loads 24-hours per day.

Base load plant A power plant built to operate around-the-clock. Such plants tend to have low operating costs and high capital costs and are best utilized by running continuously. Coal fired and nuclear fuelled plants are typical base load plants.

Base speed, RPM The speed which a DC motor develops at rated armature and field voltage with rated load applied.

Baseband An un-modulated signal or band of signals. The video signal seen on a waveform monitor is a baseband video signal.

Baseline forecast A prediction of future energy needs which does not take into account the likely effects of new conservation programs that have not yet been started.

Basic A high-level programming language designed at Dartmouth College as a learning tool. Acronym for Beginner's All-purpose Symbolic Instruction Code.

Basic insulation Insulation applied to live parts to provide basic protection against electric shock and other hazards, which do not necessarily include insulation used exclusively for functional purposes.

Basic insulation level BIL It defines the insulation level of power system equipment. It is a statement of the impulse (lightning or switching as appropriate) withstand voltage and the short duration power frequency withstand voltage.

Basket The metal frame structure of a standard dynamic loudspeaker. In larger, heavier speakers, this may be made of cast metal for extra strength and rigidity. All the other elements of the speaker are mounted on this structure.

Bass The portion of the audible sound spectrum that contains the lowest frequencies. These frequencies have the longest wavelength and require considerably greater electrical power to render them at their original strength. In a good modern speaker system, the bass portion of the response curve extends from as high as 500 hertz, down to 20 Hz.

Bass Boost/Enhancer Circuit An active low pass amplifier section added to some receivers, equalizers, and amplifiers that allows as much as an 18 decibel boost to be applied to an audio signal in the low frequency 35 to 90 Hertz range.

Bass Reflex (box or enclosure) A speaker box design that makes use of a port or Passive Radiator which allows the energy derived from the motion of the back of speaker cone to be redirected in such a way as to reinforce the front radiation. This smooths and extends the low frequency response, but the effect is sharply Rolled Off on the low end, as the port signal goes back out of phase with the front. The overall effect of this is to tune the bass response to a particular point on the lower end of the spectrum, below which it rolls off sharply.

Battery A number of primary or secondary cells arranged in series or parallel. A device for turning chemical energy into electrical energy.

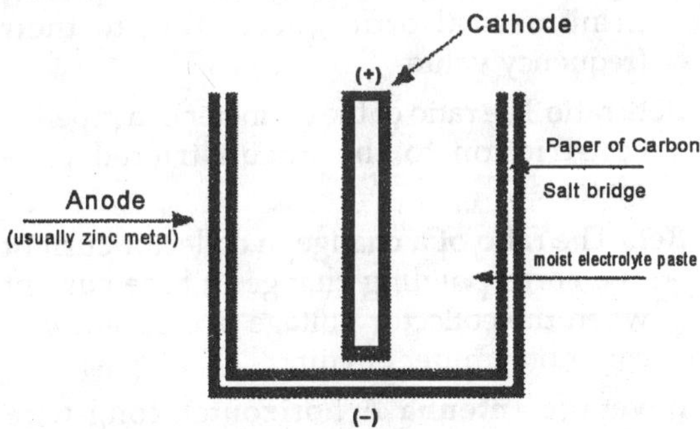

Fig. Battery

Battery backup A battery or a set of batteries in a UPS system. Its purpose is to provide an alternate source of power if the main source is interrupted.

Battery capacity The amount of energy available from a battery. Battery capacity is expressed in ampere-hours.

Battery charger A device or a system which provides the electrical power needed to keep the battery backup fully charged.

Baud A unit of data transmission speed equal to the number of bits (or signal events) per second; 300 baud = 300 bits per second.

Baud rate The signaling or symbol rate of a digital transmission path or device. A symbol can represent more than one bit of information, depending on the encoding or modulation scheme used to create the symbol. Often used interchangeably with bits per second (BPS), although incorrectly.

Beam lead chip Semiconductor chip with electrodes (leads) extended beyond the wafer.

Beam power tube An electron tube in which the grids are aligned with the control grid. Special beam-forming plates are used to concentrate the electron stream into a beam. Because of this action, the beam-power tube has high power-handling capabilities.

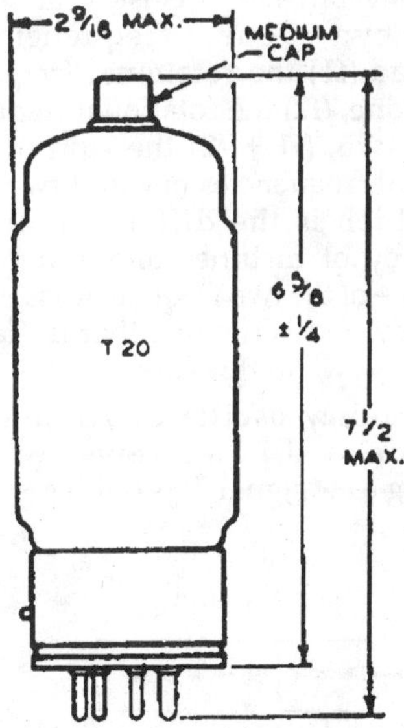

Fig. Beam power tube

Bearing resolution Ability of a radar to distinguish between targets that are close together in bearing.

Bearing RTD A probe used to measure bearing temperature to detect an overheating condition. The RTD's resistance varies with the temperature of the bearings.

Bearings Are used to reduce friction and wear while supporting rotating elements. For a motor it must provide a relatively rigid support for the output shaft. The bearing acts as the connection point between the rotating and stationary elements of a motor. There are various types such as roller, ball, sleeve (journal), and needle. The ball bearing is used in virtually all types and sizes of electric motors. It exhibits low friction loss, is suited for high speed operation and is compatible in a wide range of temperatures. There are various types of ball bearings such as open, single shielded or sealed. Reliance Electric offers a unique PLS bearing system.

Beat frequency Beat frequencies are periodic vibrations that result from the addition and subtraction of two or more sinusoids. For example, in the case of two turbine aircraft engines that are rotating at nearly the same frequency but not precisely at the same frequency; Four frequencies are generated:(f1) the rotational frequency of turbine one, (f2) the rotational frequency of turbine two, (f1 + f2) the sum of turbine rotational frequencies one and two, and (f1 - f2) which is the difference or "beat" frequency of turbines one and two. The difference of the two frequencies is the lower frequency and is the one that is "felt" as a beat or "wow" in this case.

Beat frequency oscillator An additional oscillator used in a receiver when it is receiving a cw signal. It provides an audible tone.

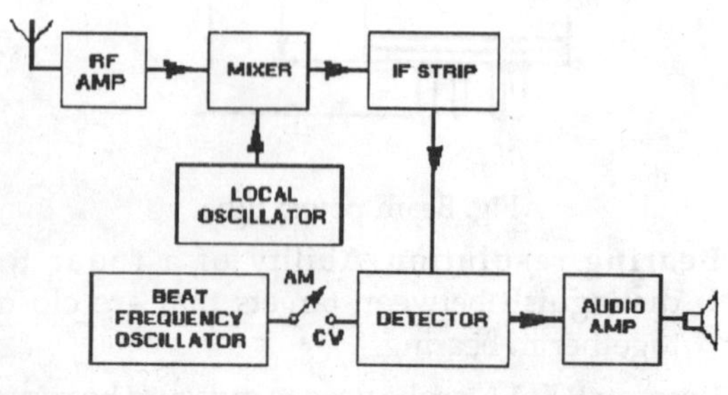

Fig. Beat frequency oscillator

Beckmann thermometer Sensitive thermometer for measuring small differences or changes in temperature.

Bel A unit of measure of ratios of power levels. The unit that expresses the logarithmic ratio between the input and output of any given component, circuit, or system. bel; [B] = log 10(P1/P2) where P1 and P2 are power levels. The dB, equal to 0.1 B, is a more commonly used unit.

Beryllia Beo (Beryllium Oxide) A high-temperature mineral insulation material; toxic when in powder form.

Bessel Alignment A particular crossover configuration which offers superior phase coherence in exchange for slightly lower output level match.Bessel: A design that places emphasis on phase and transient response over reducing ripple.

Best fit straight line (BFSL) A line midway between two parallel straight lines enclosing all output vs. Pressure values.

Best Tuning Memory (BTM) A feature in which the tuner selects radio stations by signal strength, and assigns them to presets in numerical order, according to their frequency value.

Beta ratio The ratio of the diameter of a pipeline constriction to the unconstricted pipe diameter.

Beta The ratio of a change in collector current to a corresponding change in base current when the collector voltage is constant in a common-emitter circuit.

Beverage antenna A horizontal, long-wire antenna designed for reception and transmission of low-frequency, vertically polarized ground waves.

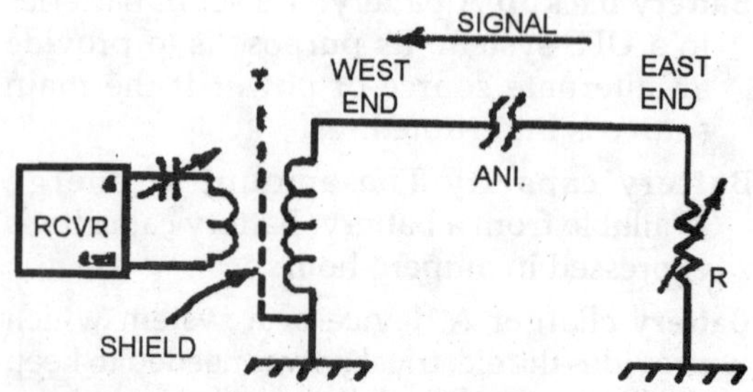

Fig. Beverage antenna

Bewley lattice diagram This is a convenient diagram devised by Bewley, which shows at a glance the position and direction of motion of every incident, reflected, and transmitted surge on the system at every instant of time. The diagram overcomes the difficulty of otherwise keeping track of the multiplicity of successive reflections at the various junctions.

Bias current Current that flows through the base-emitter junction of a transistor and is adjusted to set the operating point of the transistor.

Biasing Applying a voltage, often done to alter the electrical and optical output of a device such as a light emitting diode (LED).

Biba integrity model A formal security model for the integrity of subjects and objects in a system.

Bidirectional array An array that radiates in opposite directions along the line of maximum radiation.

Bifilar winding Indicates two distinct windings in the same physical arrangement; these windings are usually wired together, either in series or in parallel, to form one phase.

Bilateral contract A direct contract between the power producer and user or broker outside of a centralized power pool.

Bilayer lipid membrane (BLM) The structure found in most biological membranes, in which two layers of lipid molecules are so arranged that their hydrophobic parts interpenetrate, whereas their hydrophilic parts form the two surfaces of the bilayer.

Billion A thousand million or 109 (US). Also a million million (British)

Bimetallic strip A strip composed of two different metals welded together in such a way that a rise of temperature will cause it to deform as a result of the unequal expansion. It is used in switches for control of temperature.

Binary 1. A number system that uses a base, or radix, of 2. Two digits (1 and 0) are used in the binary system.

2. Pertaining to a characteristic that involves the selection, choice, or condition in which there are only two possibilities.

3. A bistable multivibrator (flip-flop) is one example of a binary device.

Binary cell An element in a computer which can store information by virtue of its ability to remain stable in one of two possible states.

Binary code A method of representing two possible conditions (on or off, high or low, one or zero, the presence of a signal or absence of a signal). Electronic circuits designed to work in such a way that only two conditions are possible.

Binary number system A number system using two digits, symbols, or characters (usually 1 and 0). A number that is expressed in binary notation and is usually characterized by the arrangement of bits in sequence, with the understanding that successive bits are interpreted as coefficients of successive powers of the base 2.

Binary point The radix point that separates powers of two and fractional powers of two in a binary number.

Binary switching A facility which can be provided by a micorprocessor based step controller. Each physical step controls a load which is twice the size of the one preceeding it, i.e. the loads are spli in the ratios 1:2:3:4 etc. The order in which the physical steps are switched is such that a four step controller can control in 10 stages, a six step in 63 stages, and so on. The most common application is in the control of electric heater batteries.

Biosensor The term" biosensor" is a general designation that denotes either a sensor to detect a biological substance or a sensor which incorporates the use of biological molecules such as antibodies or enzymes. Biosensors are a subcategory of chemical sensors.

Biot number Dimensionless number, equal to Bi =(alo)/la , where a is the heat transfer coefficient from the surface to the environment (or from the environment to the surface), lois specific dimension and la is the thermal conductivity coefficient of the body.

Bipolar chopper drive Drive that uses the switch mode method to control motor current and polarity.

Bipolar junction transistor(BJT) Transistor with n-type and p-type semiconductors having base-emitter and collector-base junctions.

Bipolar power supply A special power supply which responds to the polarity as well as the magnitude of a control instruction and is able to linearly pass through zero to produce outputs of either positive or negative polarity.

Bipolar Transistor A older but still effectively used transistor type that contains two p or n junctions or diodes between two layers of opposite polarity material (emitter and collector) . In handling large power, mostly replaced by MOSFET types.

Bisection Division into two equal parts.

Bistable A device that is capable of assuming either one of two stable states.

Bistable multivibrator A multivibrator with two stable states. An external signal is required to change the output from one state to the other. Also called a latch.

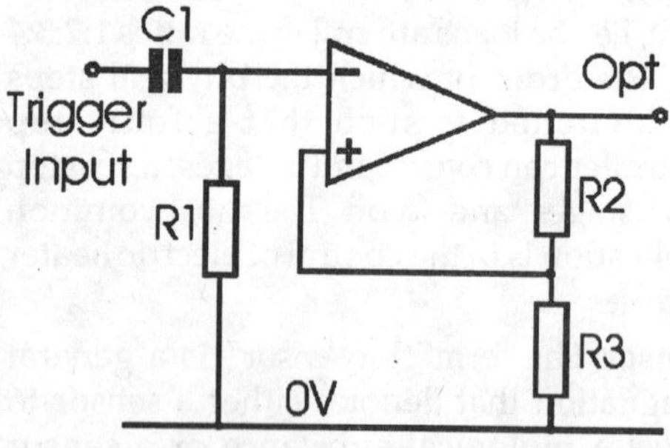

Fig. Bistable multivibrator

Bit The unit of information in information theory. The amount of information required to specify one of two alternatives 0 and 1.

Bit rate In a bit stream, the number of bits occurring per unit time, usually expressed in bits per second.

Bit slip In digital transmission, the loss of a bit or bits, caused by variations in the respective clock rates of the transmitting and receiving devices.

Bitumen A term covering numerous "tarry" mixtures of hydrocarbons.

BL (measured in Tesla meters) The product of a speaker driver's gap flux density and the length of the voice coil conductor in the gap.

Black The reference colour of equipment that passes unclassified information. It normally refers to patch panels.

Blackbody A theoretical object that radiates the maximum amount of energy at a given temperature, and absorbs all the energy incident upon it. A blackbody is not necessarily black. (The name blackbody was chosen because the colour black is defined as the total absorption of light energy.)

Blackout A total loss of the commercial electrical power lasting for more than one cycle. Blackouts can result from any of a number of problems, ranging from Acts of God (high winds, storms, lightning, falling trees, floods, etc.

Blank Skip A cassette feature that automatically detects blank areas of the tape over a set number of seconds in length and activates Fast Forward, until either the end of the tape, or audio information is reached.

Blanking/Blanking interval The period of time when a television monitor is "blanked" while the electron beam retraces from right to left or bottom to top. In a baseband video signal, the intervals between active video lines and between the last active line in a field and the first active line in the next. Ideally, a video switcher would sense when a blanking period occurs and would switch the video signal during this time. This prevents any visually unpleasant video effects on a monitor. This requires the video switcher to actively monitor each of the user's video sources.

Bleaching Removing the colour from coloured materials by chemical transformation.

Bleeder current The current through a bleeder resistor. In a voltage divider, bleeder current is usually determined by the 10 percent rule of thumb.

Bleeder resistor A resistor used to draw a fixed current.

Block diagram A diagram in which the major components of an equipment or a system are represented by squares, rectangles, or other geometric figures, and the normal order of progression of a signal or current flow is represented by lines.

Board mount An AC/DC power supply or a DC/DC converter designed for mounting on a PCB. SIP terminals, DIP terminals, or SMD terminals used for electrical connections and solder points.

Board of trade unit or BOT unit. Unit of electrical energy (British) supplied to the consumer. Equal to 1 kwh. Energy obtained when a power of 1 kw of power is maintained for 1 hour.

Bode plot Semi-log plots of the magnitude (in decibel) and phase angle of a transfer function (or performance) against frequency.

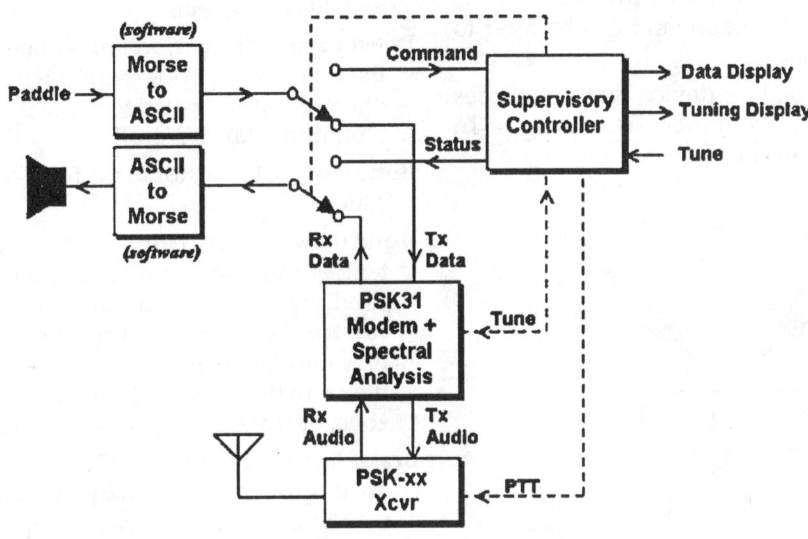

Fig. Block diagram

Block transfer The process, initiated by a single action, of transferring one or more blocks of data.

Blocked grid keying A method of keying in which the bias is varied to turn plate current on and off.

Blower door A device that fits into a doorway of a building, containing a powerful fan, for supplying or extracting a measured rate of air flow. It is normally used for testing air leakage by pressurization or depressurization.

Body centered cubic (BCC) A common crystal structure found in some elemental metals. Within the cubic unit cell, atoms are located at corner and cell center positions.

Boiler compensation An operation which changes the operating temperature of a boiler usually according to the outside air temperature.

Boiler optimisation An energy management function which acts to balance boiler operation to loads and control combustion air.

Boiling The state of a liquid at its boiling point when the maximum vapour pressure of the liquid is equal to the external pressure to which the liquid is subject, and the liquid is freely converted into vapour.

Boiling point The temperature at which a substance in the liquid phase transforms to the gaseous phase; commonly refers to the boiling point of water which is 100°C (212°F) at sea level.

Boiling water reactor or BWR A nuclear reactor in which water is used as coolant and moderator. Steam is thus produced in the reactor under pressure and can be used to drive a turbine.

Bolometer A loading device that undergoes changes in resistance as changes in dissipated power occur.

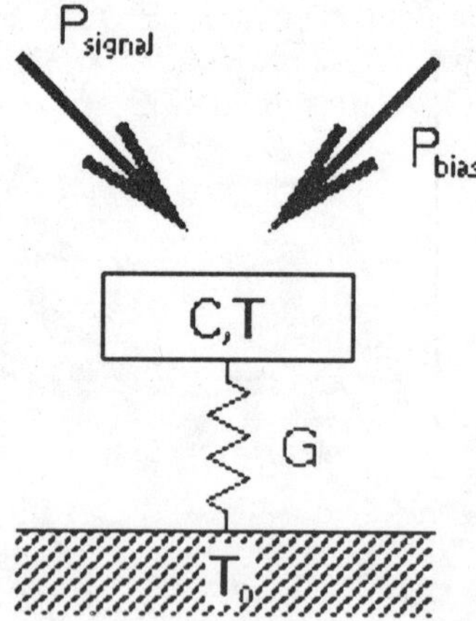

Fig. Bolometer

Bomb A general synonym for crash, normally of software or operating system failures.

Bonding A complete and permanent electrical connection. The permanent joining of metallic parts to form an electrically conductive path that ensures electrical continuity and the capacity to conduct safely any current likely to be imposed.

Bonding conductor A protective conductor providing equipotential bonding.

Bonding wires Fine wires connecting the bonding pads of the chip to the external leads of the package.

Bookmark A way of storing web addresses so you can find them easily (Bookmarks can be found on Netscape - they are called "favorites" on Microsoft Internet Explorer).

Boolean 1. Pertaining to the process used in the algebra formulated by George Boole.

2. Pertaining to the operations of formal logic.

Boolean algebra A system of logic dealing with on-off circuit elements associated by such operators as the AND, OR, NAND, NOR, and NOT functions.

Boost period The period immediately prior to the occupancy period during which plant is operated at its full rate capacity. See also Optimum Start Controller.

Bottoming A transistor in the fully conducting state.

Boundary condition These are the temperature, flux and other environmental conditions that pertain on either side of a surface. According to the particular surface, they may be obtained from the climate data file, from the calculated values in an adjacent zone, or from user-specified values.

Brake horse power bhp, BHP Horse power of an engine measured by the degree of resistance offered by a brake. Represents the useful power that the machine can develop.

Brakes An external device or accessory that brings a running motor to a standstill and/ or holds a load. Can be added to a motor or incorporated.

Braking torque The torque required to bring a motor down to a standstill. The term is also used to describe the torque developed by a motor during dynamic braking conditions.

Branch An individual current path in a parallel circuit.

Branch current The portion of total current flowing in one path of a parallel circuit.

Brazing A metal joining technique that uses a molten filler metal alloy having a melting temperature greater than about 425oC (800oF).

Breach The successful defeat of security controls which could result in a penetration of the system. A violation of controls of a particular information system such that information assets or system components are unduly exposed.

Break In a switch, the number of breaks refers to the number of points at which the switch opens the circuit; for example, single break and double break.

Break before make Disconnecting the present circuit before connecting a new circuit. Also known as Break/Make.

Breakaway A routing control mode wherein an audio source can be selected independently of the video source and vice versa.

Breakdown The phenomenon occurring in a reverse-biased semiconductor diode. The start of the phenomenon is observed as a transition from a high dynamic resistance to one of substantially lower dynamic resistance. This is done to boost the reverse current.

Breakdown torque The maximum torque a motor will develop at rated voltage without a relatively abrupt drop or loss in speed.

Breakdown voltage Voltage at which the breakdown of a dielectric or insulator occurs.

Breakdown voltage rating The dc or ac voltage which can be applied across insulation portions of a transducer without arcing or conduction above a specific current value.

Breaker Short for circuit breaker.

Breakover voltage Minimum voltage required to cause a diac to break down and conduct.

Breeder reactor A nuclear reactor which produces the same kind of fissile material as it burns. For example, a reactor using plutonium as a fuel can produce more plutonium than it uses by conversion of Uranium-238.

Bridge generally a short-circuit on a PC board caused by solder joining two adjacent tracks.

Bridge Mounted (2 & 3-way speakers) In combined 2 and 3-way speakers, which have woofers together with a Tweeter, or a Tweeter and a Midrange Driver, the smaller drivers are attached to the woofer by either one of two standard mountings.

The bridge-mount method has a metal or plastic bridge running from one side of the woofer's outer perimeter to the other. The smaller driver or drivers, together with the Crossover network, is then mounted on the bridge. The advantage of this arrangement is that the woofer is left intact with no exposure of the voice coil or other internal elements. On the negative side, there is more covered surface area that marginally reduces the output at certain frequencies, and makes it a little more difficult to mount the unit.

The post-mount method provides a post that is attached to the center of the inner magnetic pole, to which the other drivers are then mounted. It has a reduced obstruction to woofer dispersion, but it does require a hole in the Dust Cover/Cap that can give environmental access to the woofer's inner workings. This creates the possibility in some less well designed units of operational degradation from airborne particles over time.

Bridge rectifier A full-wave rectifier where the diodes are connected in a bridge circuit (two of them are always conducting at any given time). This allows the current to the load during both the positive and negative alternation of the supply voltage. This is the most common type of rectifier circuit to produce a unidirectional voltage for an alternating input.

Bridged Power Bridging An amplifier, combines the power output of two channels into one channel. Bridging allows the amplifier to drive one speaker with more power than the amp could produce for two speakers. Because of this high power output, bridging is the best way to drive a single subwoofer.

Bridged Transformer Less (BTL) A circuit design wherein two small Integrated Circuit (IC) amplifier channels are bridged together to provide a single, larger output circuit.

These circuits are limited by their current capabilities and the amount of heat they generate.

Brightness Brightness is the quotient of the luminous intensity of a small element of the source and the area of the element projected on to a plane perpendicular to the given direction. (Unit: candela per unit area or Cd/m2)

Brightness control The name given to the potentiometer used to vary the potential applied to the control grid of a CRT.

Broadband communications The result of utilities forming partnerships to offer consumers "one-stop-shopping" for energy-related and high-tech telecommunications services.

Broadside array An array in which the direction of maximum radiation is perpendicular to the plane containing the elements.

Brownout A long duration reduction in the voltage of the ac supply without complete loss of power. Brownouts are usually caused by heavy usage during peak hours and sometimes may even be planned as an energy conservation strategy.

Brush A piece of current conducting material (usually carbon or graphite) which rides directly on the commutator of a commutated motor and conducts current from the power supply to the armature windings.

Brushed DC motor Class of motors that has a permanent magnet stator and a wound iron-core armature, as well as mechanical brushes for commutation; capable of variable speed control, but not readily adaptable to different environments.

Brushes Sliding contacts, usually carbon, that make electrical connection to the rotating part of a motor or generator.

Brushless servomotor Class of servomotors that uses electrical feedback rather than mechanical brushes for commutation; durable and adaptable to many different environments.

Buffer 1. A storage area for data that is used to compensate for a speed difference, when transferring data from one device to another. Usually refers to an area reserved for I/O operations, into which data is read, or from which data is written.

2. Any substance or combination of substances which, when dissolved in water, produces a solution which resists a change in its hydrogen ion concentration on the addition of an acid or alkali.

Buffer amplifier An amplifier that isolates one circuit from another. It decreases the loading effect on an oscillator by reducing the interaction between the load and the oscillator.

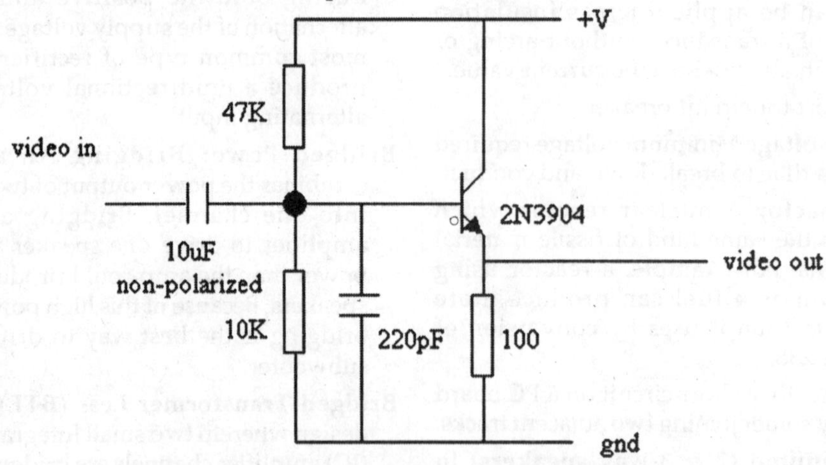

Fig. Buffer amplifier

Buffer capacity (B) A measure of the ability of the solution to resist ph change when a strong acid or base is added.

Buffer memory Temporary storage area for acquired or generated data.

Buffer overflow This happens when more data is put into a buffer or holding area, then the buffer can handle. This is due to a mismatch in processing rates between the producing and consuming processes. This can result in system crashes or the creation of a back door leading to system access.

Buffer registerthe register that holds digital data temporarily.

Bug An unwanted and unintended property of a program or piece of hardware, especially one that causes it to malfunction.

Building envelope The total area of the boundary surfaces of a building through which heat, light, air and moisture are transferred between the internal spaces and the outside environment.

Building related illness Diagnosable illnesses whose symptoms can be identified and whose cause can be directly attributed to airborne building pollutants (e.g., Legionnaire's disease, hypersensitivity, pneumonitis). •

Building service entrythe point where commercial power enters the building.

Built in Crossovers Frequently Used to limit the high-frequencies reaching a subwoofer, a low-pass filter crossover allows only frequencies below the crossover point to be amplified. A high-pass crossover allows only frequencies above the crossover point to be amplified – used to keep destructive low bass away from small speakers, so they can played safely. Crossovers may be variable or selectable. Continuously Variable means the crossover circuit can be adjusted to any frequency between the listed end points. Selectable means that any of several preset crossover points can be chosen to accomodate variuous driver (speaker) designs.

Built in test equipment (BITE) A permanently mounted device that is used expressly for testing an equipment or system.

Bulb (Liquid-in-Glass Thermometer) The area at the tip of a liquid-in-glass thermometer containing the liquid reservoir.

Bulk power market Wholesale purchases and sales of electricity.

Bulk power supply Commonly used interchangeably with wholesale power supply. In broader terms, it refers to the aggregate of electric generating plants, transmission lines, and related equipment.

Bulk resistance The natural resistance of a "P" type or "N" type semiconductor material.

Bullet Horn (tweeter) A type of tweeter in which the radiator has a large passive, bullet-shaped device above its center that extends the nominal dispersion angle of the sound, thus allowing it to cover a greater area with high frequency radiation

Bulletin board An area on a computer where people can read or write messages. Certain bulletin boards are private, requiring an access code.

Bumped A method of woofer construction in which the rear suspension system is anchored a little further back by designing the back plate so that it is press stamped, or cast, outward. This allows greater Excursion of the voice coil, and prevents "bottoming out," which is very destructive to the coil form when large signals move the voice coil beyond its range limits. This technique does not eliminate the problem, but does help to reduce it.

Buncher cavity The input resonant cavity in a conventional klystron oscillator.

Buncher grid In a velocity-modulated tube, the grid that concentrates the electrons in the electron beam into bunches.

Burn in A long term screening test (either vibration, temperature or combined test) that is effective in weeding out infant mortalities because it simulates actual or worst case operation of the device, accelerated through a time, power, and temperature relationship.

Burnishing tool A tool used to clean and polish contacts on a relay.

Burst mode A data acquisition mode in which a group of analogue input channels are scanned at an interval determined by the pacer clock and the signal from each channel within the scan is converted at a higher rate determined by the burst mode conversion clock. This mode minimizes the skew between channels.

Burst pressure The maximum pressure applied to a transducer sensing element or case without causing leakage.

Burst proportioning A fast-cycling output form on a time proportioning controller (typically adjustable from 2 to 4 seconds) used in conjunction with a solid state relay to prolong the life of heaters by minimizing thermal stress.

Bus A cable that is connected to a number of different devices, sensors, controllers, outstations, etc., that acts a means of data exchange. There are two main types, serial and parallel. In building services it is most common to use serial types where data flows on just two cables.

Bus bar A heavy copper strap or bar used to connect several circuits together when a large current carrying capacity is required.

Bushing Bushings are insulators which are used to take high voltage conductors through earthed barriers such as walls, floors, metal, and tanks.

Buss or Bus A signal-carrying conductor or electrical pathway designed to carry multiple signals. A mixing console auxiliary bus may carry signals derived from several channels on that console. The term is sometimes used to refer to a power distribution circuit, or "mains".

Butterworth crossover A type of crossover circuit utilizing low-pass filter design characterized by having a maximally flat magnitude response, i.e., no variation in the amplitude response in the domain of the passband.

Butterworth filter A type of active filter characterized by a constant gain (flat response) across the midband of the circuit and a 20 db per decade roll-off rate for each pole contained in the circuit.

Butyl A type of rubber used for speaker surrounds. Butyl has very good damping characteristics and is resistant to UV contamination from the sun.

BW Abbreviation for bandwidth.

Bypass capacitor A capacitor placed from a dc signal to ground to remove any ac component of the signal by creating an ac short circuit to ground.

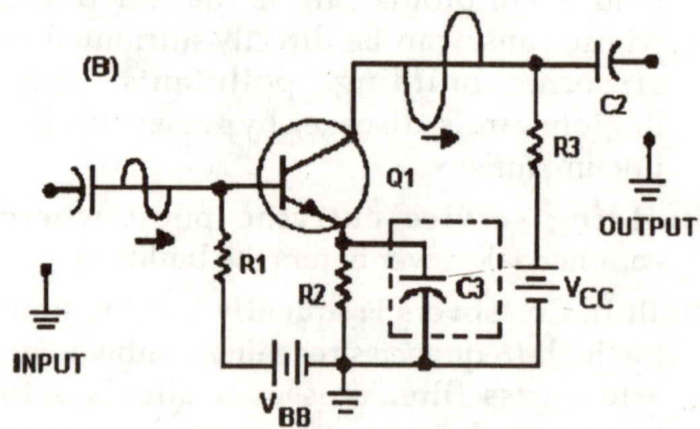

Fig. Bypass capacitor

C

Cabin Gain A low frequency boost normally obtained inside a vehicle interior when woofers are optimally in phase, and with the proper enclosures.

Cable An assembly of one or more insulated conductors, or optical fibers, or a combination of both, within an enveloping jacket. Either a stranded conductor (single-conductor cable) or a combination of conductors insulated from one another (multiple conductor cable). Small sizes are commonly referred to as stranded wire or as cords.

Cable assembly A cable that is ready for installation in specific applications and usually terminated with connectors.

Cable channel An enclosure situated above or in the ground, ventilated or closed, and having dimensions which do not permit the access of persons but allow access to the conduits and/or cables throughout their length during and after installation. A cable channel may or may not form part of the building construction.

Cable coupler A means of enabling the connection or disconnection, at will, of two flexible cables. It consists of a connector and a plug.

Cable ducting An enclosure of metal or insulating material, other than conduit or cable trunking, intended for the protection of cables which are drawn in after erection of the ducting.

Cable harness A group of wires or ribbons of wiring used to interconnect electronic systems and subsystems.

Cable ladder A cable support consisting of a series of transverse supporting elements rigidly fixed to main longitudinal supporting members.

Cable tray A rigid structure use to support cables. A raceway consisting of a continuous base with raised edges and no covering. A cable tray may or may not be perforated.

Cable trunking A closed enclosure normally of rectangular cross section, of which one side is removable or hinged, used for the protection of cables and for the accommodation of other electrical equipment.

Cable tunnel A corridor containing supporting structures for cables and joints and/or other elements of wiring systems and whose dimensions allow persons to pass freely throughout the entire length.

Calibration The graduation or confirmation of the graduation of an instrument to enable measurements in definite units to be made with it. Thus for example the deflection of a meter can be calibrated to read the current causing the deflection.

Calibration offset An adjustment to eliminate the difference between the indicated value and the actual process value.

Calorie (cal) Unit of quantity of heat. The amount of heat required to raise the temperature of 1 gram of water through 1o C. 1 calorie = 4.184 joule.

Calorific value The calorific value of a fuel is the quantity of heat produced by a given weight of the fuel on complete combustion.

Candela (Cd) The candela is the SI unit of luminous intensity. It is defined as the luminous intensity, in a given direction, of a source that emits monochromatic radiation of frequency 540 x 1012 hertz and that has a radiant intensity in that direction of 1/683 watt per steradian.

Candle power The candle power of a light source, in a given direction, is the luminous intensity of the source in that direction expressed in terms of the candela.

Capability Maximum load that a generating unit can carry without exceeding ratings.

Capacitance(C) In a capacitor or system of conductors and dielectrics, the property that permits the storage of electrically separated charges when potential differences exist between the conductors. Capacitance is related to charge and voltage as follows: C = Q/V, where C is the capacitance in farads, Q is the charge in coulombs, and V is the voltage in volts.

Capacitance, electrostatic Ability of conductors, separated by dielectrics, to store electrostatic charges. The charge Q is directly related to the product of capacitance C and voltage V as

Q = CV.

Capacitance is directly proportional to the area of either conductor A (called plates in the parallel-plate capacitor) and indirectly proportional to the distance between the plates d :

C = epsilonA / 4pid

where epsilon is the dielectric constant of the medium between the plates.

Capacitive reactance The opposition, expressed in ohms, offered to the flow of an alternating current by capacitance. The symbol for capacitive reactance is XC.

Capacitor (Power audio) Power stabilizing capacitors store the necessary power amplifiers need to punch larger bass notes while limiting clipping. They store energy during intervals when it is not required, which is most of the time, and release it when demand exceeds what is available from the car's power system. The amount of capacitance to be used is half (.5) farad per 500 watts of available RMS power. Capacitors are not used with amplifiers that supply less than 300 watts RMS in total.

Capacitor A device which, when connected in an alternating-current circuit, causes the current to lead the voltage in time phase. The peak of the current wave is reached ahead of the peak of the voltage wave. This is the result of the successive storage and discharge of electric energy used in 1 phase motors to start or in 3 phase for power factor correction.

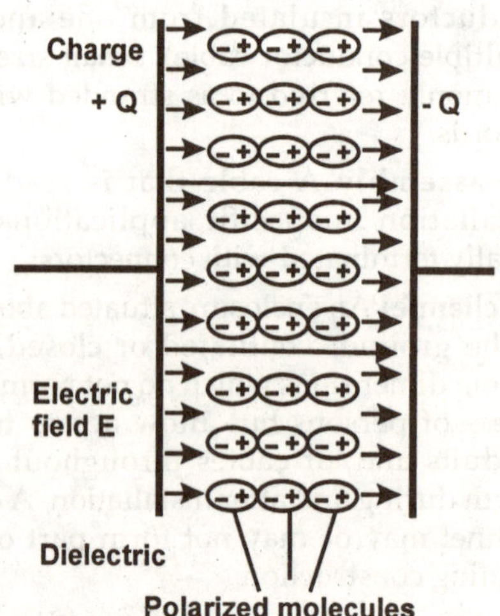

Fig. Capacitor

Capacitor filter This filter is used on extremely high-voltage, low-current power supplies and also where the ripple frequency is not critical.

Capacitor microphone Microphone whose operation depends on variations in capacitance caused by varying air pressure on the movable plate of a capacitor.

Capacitor motor A single-phase induction motor with a main winding arranged for direct connection to the power source, and auxiliary winding connected in series with a capacitor. There are three types of capacitor motors: capacitor start, in which the capacitor phase is in the circuit only during starting, permanent-split capacitor, which has the same capacitor and capacitor phase in the circuit for both starting and running; two-value capacitor motor, in which there are different values of capacitance for starting and running.

Capacitor start motor A type of single-phase, ac induction motor in which a starting winding and a capacitor are placed in series to start the motor. The values of XC and R are such that the main-winding and starting-winding currents are nearly 90 degrees apart and the starting torque is produced as in a two-phase motor.

Capacitor start The capacitor start single phase motor is basically the same as the split phase start, except that it has a capacitor in series with the starting winding. The addition of the capacitor provides a more ideal phase relation and results in greater starting torque with much less power input. As in the case of the split phase motor, this type can be reversed at rest, but not while running unless special starting and reversing switches are used. When properly equipped for reversing while running, the motor is much more suitable for this service than the split phase start as it provides greater reversing ability at less watts input.

Capacity The maximum load a generating unit, generating station, or other electrical apparatus is rated to carry by the manufacturer or can actually carry under existing service conditions.

Capacity charge An charge or assessment based on the amount of capacity being purchased.

Capacity factor The ratio of the electric energy produced by a generating unit to the electrical energy that could have been produced at continuous full-power operations.

Captive load Load which may be supplied by an Embedded Generator, in addition to the generator auxiliaries, which is within the Generating Company premises.

Carbon film resistor Device made by depositing a thin carbon film on a ceramic form.

Carbon microphone A microphone in which sound waves vary the resistance of a pile of carbon granules. May be single-button or double-button.

Carbon resistor Resistor of fixed value made by mixing carbon granules with a binder which is moulded and then baked.

Cardbus One of the different types of PCMCIA interfaces. CardBus implements the 32-bit PCI bus standard into the PCMCIA form factor.

Cardioid "Heart" shaped pickup pattern characteristic of some microphones which reduces sensitivity to sounds from the sides and back.

Carnot's cycle An ideal reversible four step cycle of operations for the working substance of a heat engine.

Carrier In a semiconductor, the mobile electrons or holes which carry charges are called carriers.

Carrier controlled approach A shipboard radar system used to guide aircraft to safe landings in poor visibility conditions.

Carrier frequency The frequency of an unmodulated transmitter output.

Carrier wave A continuous electromagnetic radiation, of constant amplitude and frequency, emitted by a transmitter. By modulation of the carrier wave, oscillations caused at the transmitting end are conveyed to the receiving end.

Cascade A method of connecting circuits in series so that the output of one is the input of the next.

Cascade connection An arrangement of two or more components or circuits such that the output of one is the input to the next.

Cascade control A control system in which one controller provides the setpoint for one or more other controllers.

Cascaded amplifier An amplifier with two or more stages arranged in a series configuration.

Cascode amplifier A high frequency amplifier made up of a common-source amplifier with a common-gate amplifier in its drain network.

Case temperature rating Maximum temperature the motor case can reach without the inside of the motor exceeding its internal temperature rating.

Catcher grid In a velocity-modulated tube, a grid on which the spaced electron groups induce a signal. The output of the tube is taken from the catcher grid.

Catenary Curve formed by a chain or string hanging from two fixed points.

Cathode 1. In an electron tube the electrode that is the source of current flow.

2. The general name for any negative electrode.

3. The negative terminal of a forward-biased semiconductor diode, which is the source of the electrons.

Cathode bias The method of biasing a vacuum tube in which the biasing resistor is placed in the common-cathode return circuit, thereby making the cathode more positive with respect to ground.

Cathode foil An aluminum electrolytic capacitor is typically composed of an anode foil, a cathode foil and separator material which are wound together and impregnated with an electrolyte. The cathode foil typically does not have an aluminum oxide layer acting as the dielectric. The aluminum foil (12.7 to 40 microns) is electrochemically etched to increase it's surface area, and therefore its capacitance. A thin layer of aluminum oxide naturally grows on the surface of the foil, but is only equivalent to approximately 0.5 to 1.5 volts.

Non-polar and bi-polar capacitors utilize two anode foils instead of anode and cathode foils. Each foil would thus be capable of supporting the full rated DC+AC voltage.

Cathode keying A system in which the cathode circuit is interrupted so that neither grid current nor plate current can flow.

Cathode modulator Voltage on the cathode is varied to produce the modulation envelope.

Cathode ray oscilloscope (CRO) An instrument based upon the cathode ray tube, which provides a visible image of one or more rapidly varying electrical quantities.

Cathode ray tube CRT An electron-beam tube in which the beam can be focused to a small cross-section on a luminescent screen and varied in both position and intensity to produce a visible pattern. Electric potentials applied to the deflection plates are used to control the position of the beam, and its movement across the screen, in a desired manner.

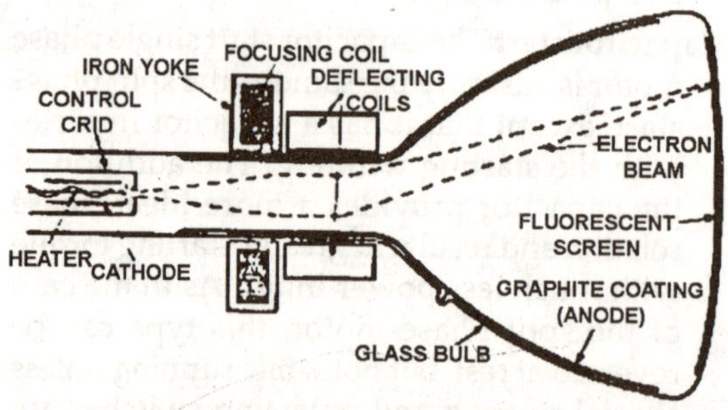

Fig. Cathode ray tube CRT

Cathode sputtering A process of producing thin film components.

Cation A positively charged ion (Na+, H+).

Caulking Technique for making airtight joints by applying a sealing material. A form of weatherstripping.

Cavitation The boiling of a liquid caused by a decrease in pressure rather than an increase in temperature.

Cavity resonator A space totally enclosed by a metallic conductor and supplied with energy in such a way that it becomes a source of electromagnetic oscillations. The size and shape of the enclosure determine the resonant frequency.

Cavity wall A wall built of two leaves, separated usually by a continuous gap. The two leaves are connected by ties at intervals. The inner layer may be double for floor bearing.

Cavity wavemeter An instrument used to measure microwave frequencies.

Ceiling plenum Space below the flooring and above the suspended ceiling that accommodates the mechanical and electrical equipment and that is used as part of the air distribution system.

Cell Single unit used to convert chemical energy into a DC electrical voltage.

Cellar A storey in a building whose floor line is below ground level at any entrance or exit, the ceiling of which is not more than 5ft above ground level whose primary function can be accommodation or storage.

Center feed method Connecting the center of an antenna to a transmission line which is then connected to the final (output) stage of the transmitter.

Center frequency Frequency to which an amplifier is tuned. The frequency half way between the cut-off frequencies of a tuned circuit.

Center of gravity (Mass Center) The center of gravity of a body is that point in the body through which passes the resultant of weights of its component particles for all orientations of the body with respect to a uniform gravitational field.

Center tap Midway connection between the two ends of a winding.

Center tapped rectifier Circuit that make use of a center tapped transformer and two diodes to provide full wave rectification.

Center tapped transformer A transformer with a connection at the electrical center of a winding.

Centi (c) Decimal sub-multiple prefix corresponding to one-hundredth or 10-2. This is not a preferred suffix.

Centimeter cube A unit of volume of large rectangular or square conductors. The cross-sectional area equals 1 square centimeter with a length of 1 centimeter.

Centimetre gram second system (CGS system) A decimal system which is an earlier form of the metric system.

Central processing unit (CPU) The unit of a computing system that includes circuits controlling the interpretation of instructions and their execution.

Central station This is the heart of a BEMS, and also the main communication channel for the operator. Here is contained the software and the main storage of data relating to the plant and buildings controlled.

Centralized intelligence Description of a system where algorithm processing is only possible at the central station. The outstations are dormant when not in contact with the central station.

Centralized system A BEMS in which all executive control takes place at the central station.

Centre of gravity The centre of gravity of a body is the fixed point through which the resultant force due to the Earth's attraction upon it always passes, irrespective of the position of the body.

Centrifugal cutout switch A centrifugally operated automatic mechanism used in conjunction with split phase and other types of single phase induction motors. Centrifugal cutout switches will open or disconnect the starting winding when the rotor has reached a pre-determined speed, and reconnect it when the motor speed falls below it. Without such a device, the starting winding would be susceptible to rapid overheating and subsequent burnout.

Centrifugal fan A fan in which the air is turned from parallel to the axis of rotation on entry to a direction tangential to the arc described by the tips of the rotating blades or vanes.

Centripetal force The radial force imposed by the restraining system, necessary to keep the body moving in a circular path. (Note: the centrifugal force and the centripetal force are equal and opposite).

Ceramic Polycrystalline ferroelectric materials which are used as the sensing units in piezoelectric accelerometers. There are many different grades, all of which can be made in various configurations to satisfy different design requirements.

Ceramic capacitor Generally a single layer capacitor that is flat and has a brown coating, Also have the name monoblock or monolithic in which the capacitor is made even smaller by creating multy-layers and coated in orange or blue paint.

Ceramic insulation High-temperature compositions of metal oxides used to insulate a pair of thermocouple wires The most common are Alumina (Al2O3), Beryllia (BeO), and Magnesia (MgO). Their application depends upon temperature and type of thermocouple. High-purity alumina is required for platinum alloy thermocouples. Ceramic insulators are available as single and multihole tubes or as beads.

Cermet A composite material consisting of a combination of ceramic and metallic meterials. The most common cermets are the cemented carbides, composed of an extremely hard ceramic (e.g. WC, TiC), bonded together by a ductile metal such as cobalt or nickel.

Chain reaction Any self-sustaining molecular or nuclear reaction, the products of which contribute to the propagation of the reaction.

Channel Common name for a complete amplifying stage in any audio amplifier. Most amplifiers are denominated as 1, 2, 4, 5, or 6 channel units. Each of these is a discrete audio amp on its own, capable of taking a small line signal input and amplifying it sufficiently to be heard on an appropriate speaker. Some amplifiers are capable of bridging two channels together, to form one channel of double the power of each separately. The manufacturer's instructions differ widely on how to accomplish this, and each must be followed exactly.

Channel crosstalk Coupling of a signal from one channel to another or any other output by conduction or radiation. Crosstalk is expressed in decibels (dB) at a specified load impedance and over a specific frequency range or ranges.

Character A letter, digit or other symbol that is used as the representation of data. A connected sequence of characters is called a character string.

Characteristic impedance The ratio of voltage to current at any given point on a transmission line represented by a value of impedance.

Charge Represents electrical energy. A material having an excess of electrons is said to have a negative charge. A material having a shortage of electrons is said to have a positive charge.

Charge controller A device which regulates the voltage of a battery. These may be used to assure a battery is not overcharged and/or overly discharged.

Charge coupled device (CCD) A semiconductor device, often used for sensing light, which operates by storing charge on capacitors and selectively moving that charge through the device by manipulating voltages on its electrodes.

Charge current Current that flows to charge a capacitor or battery when voltage is applied.

Charge cycle The period of time that a capacitor in an electrical circuit is storing a charge.

Charge rate The amount of energy per unit time that is being added to the battery, commonly expressed as the ratio of rated capacity to charge duration in hours.

Charge Sensitivity For accelerometers that are rated in terms of charge sensitivity, the output voltage (V)is proportional to the charge (Q) divided by the shunt capacitance (C). This type of accelerometer is characterized by a high output impedance. The sensitivity is given in terms of charge; picocoulombs per unit of acceleration (g).

Chassis The structure supporting or enclosing a power supply or other electrical circuit.

Chassis ground The voltage potential of the chassis or other reference point in a unit with a non-conducting chassis.

Chatter The rapid cycling on and off of a relay in a control process due to insufficient bandwidth in the controller.

Chebyshev filter A type of active filter characterized by high roll-off rates (40 db per decade per pole) and midband gain that is not constant.

Check bit A bit, such as a parity bit, derived from and appended to a bit string for later use in error detection and possibly error correction.

Check password A hacking program used for cracking VMS passwords.

Check sum A number that has been calculated as a function of some message, by adding up the bytes in the message. The [message] data is summed with out regard overflow. Perhaps this was what early checksums were. Today, however, although more sophisticated formulae are used, the term "checksum" is still used and interchangeable with the term CRC.

Chemical reaction The interaction of two or more substances resulting in chemical changes in them.

Chernobyl packet Also called Kamikaze Packet. A network packet that induces a broadcast storm and network meltdown. Typically an IP Ethernet datagram that passes through a gateway with both source and destination Ethernet and IP address set as the respective broadcast addresses for the subnetworks being gated between.

Chip A die (unpackaged semiconductor device) cut from a silicon wafer, incorporating semiconductor circuit elements such as a sensor, actuator, resistor, diode, transistor, and/or capacitor.

Choke A coil of low resistance and high inductance used in electrical circuits to pass low frequency or direct components while suppressing (or choking) the higher frequency undesirable alternating currents.

Choke coil An inductance device used in a circuit to present a high impedance to high frequencies without appreciably limiting the flow of direct current.

Choke joint A joint between two sections of waveguide that provides a good electrical connection without power losses or reflections.

Chop mode A vertical mode of operation for dual-trace oscilloscopes in which the display is switched between the two channels at some fixed rate much less than the sweep time.

Chopped waveform The standard surge waveform applied is suddenly made zero (chopped) at a predefined time to simulate the surge waveform with flashover.

Chord A combination of two or more notes played simultaneously

Chromatography The general name for a series of methods for separating mixtures by employing a system with a mobile phase and a stationary phase.

CIBSE Acronym for The Chartered Institution of Building Services Engineers.

Circuit An assembly of electrical equipment supplied from the same origin and protected against overcurrent by the same protective device(s).

Circuit breaker A mechanical switching device capable of making, carrying, and breaking currents under normal conditions. Also making, carrying for a specific time, and automatically breaking currents under specified abnormal circuit conditions, such as those of short circuit. It is usually required to operate infrequently although some types are suitable for frequent operation.

Circuit level gateway One form of a firewall. Validates TCP and UDP sessions before opening a connection. Creates a handshake, and once that takes place passes everything through until the session is ended.

Circuit protective conductor (CPC) A protective conductor connecting exposed conductive parts of equipment to the main earthing terminal.

Circular mil An area equal to that of a circle with a diameter of 0.001 inch. It is used for measuring the cross-sectional area of wires.

Circular mil foot A unit of volume of a conductor having a cross-sectional area of 1 circular mil and a length of 1 foot.

Circularly polarised light Light which can be resolved into two vibrations lying in planes at right angles, of equal amplitude and frequency and differing in phase by 90o.

Clamper A diode circuit used to change the DC level of a waveform without distorting the waveform.

Clamping level The voltage point at which a surge protector begins to limit surges.

Clapp oscillator A variation of the Colpitts oscillator. An added capacitor is used to eliminate the effects of stray capacitance on the operation of the basic Colpitts oscillator.

Clean power Electrical power which has been conditioned and/or regulated to remove electrical noise from the output power.

Clear To restore a device to a prescribed initial state, usually the zero state.

Cleaved coupled cavity Two or more aligned semiconductor lasers which through destructive and constructive interference are able to output light of a particular wavelength.'

Climate prevalent and predictable meteorological conditions of a geographical area; determined air temperature, solar radiation, humidity, wind parameters, clouding and precipitation data for the surrounding geographic region.

Clipboard Allows data to be copied within and between applications in Windows ie. when you use the "cut and paste" options in Windows.

Clipper A diode circuit used to eliminate part of a waveform .

Clipper chip A tamper-resistant VLSI chip designed by NSA for encrypting voice communications. It conforms to the Escrow Encryption Standard (EES) and implements the Skipjack encryption algorithm.

Clipping A signal that results from an amplifier that is either overloaded or underpowered relative to the signal Amplitude it being asked to generate. A clipped waveform is one in which the gently rounded peaks and valleys of the AC audio wave are instead sliced off or clipped, to yield what looks a lot like a square or alternating DC wave. When DC is applied to a speaker, the voice coil has no means of propelling itself relative to a constant magnetic field. Instead, it can only convert the incoming current to heat, and ultimately burns up. The effect of alternating DC on speakers is remarkable, irritating, painful, and short. If you are able to hear evident Distortion at high volume levels, or smell smoke, reduce the volume. It may already be too late for your speakers, but at least you may be able to save the amplifier.

Clock A square waveform used for synchronizing and timing of several circuits.

Closed circuit Circuit having a complete path for current flow.

Closed loop Describes a system where a measured output value is compared to a desired input value and corrected accordingly (e.g., a servomotor system).

Closed loop control A monitoring control system. This type of control system possess monitoring feedback, the deviation signal formed as a result of this feedback being used to control the action of a final control element in such a way as to tend to reduce the deviation to zero; cf. open-loop.

Closed loop gain Gain of an amplifier when a feedback path is present.

Closeness of control Total temperature variation from a desired set point of system. Expressed as "closeness of control" is ±2°C or a system bandwidth with 4°C, also referred to as amplitude of deviation.

Clutch A mechanical device for engaging and disengaging a motor often used when many starts and stops are required.

Clutter Confusing, unwanted echoes that interfere with the observation of desired signals on a radar indicator.

Coaxial cable A cable that has one conductor (shield) completely surrounding the other (center conductor), the two being coaxial and separated by an insulator. Standard industry types have a braided shield, or a semi-rigid copper or stainless steel shield material. Braided shield coaxial cable offers more physical flexibility but less shielding.

Coaxial line A type of transmission line that contains two concentric conductors.

Code Code is a combination of mark and space conditions representing symbols, figures, or letters.

Codons Organic bases in sets of three that form the genetic code.

Coefficient A number or other known factor or multiplier which measures some specified property of a given substance or algebraic expression.

Coefficient of coupling An expression of the extent to which two inductors are coupled by magnetic lines of force. This is expressed as a decimal or percentage of maximum possible coupling and represented by the letter K.

Coercive force The strength of the magnetic field to which a ferromagnetic material undergoing a hysteresis cycle must be subjected in order to demagnetise the material completely.

Coercivity The coercive force when the material is magnetised to saturation during the cycle.

Cogeneration The production of electricity and the utilization of waste heat, which could be used to produce steam for additional power generation (as in a combined cycle facility).

Cogging A term used to describe non-uniform angular velocity. It refers to rotation occurring in jerks or increments rather than smooth motion. When an armature coil enters the magnetic field produced by the field coils, it tends to speed up and slow down when leaving it. This effect becomes apparent at low speeds. The fewer the number of coils, the more noticeable it can be.

Cogging torque A measure of non-uniform velocity (e.g., jerkiness, momentary stalling, slipping).

Coherence A definite phase relationship between two energy waves, such as transmitted frequency and reference frequency.

Coherence function A frequency domain function computed to show the degree of a linear, noise-free relationship between a system's input and output. The value of the coherence function ranges between zero and one, where a value of zero indicates there is no causal relationship between the input and the output. A value of one indicates the existence of linear noise-free frequency response between the input and the output.

Coherent Radiation on one frequency.

Coherent oscillator In cw radar an oscillator that supplies phase references to provide coherent video from target returns.

Coil (Stator or Armature) The electrical conductors wound into the core slot, electrically insulated from the iron core. These coils are connected into circuits or windings which carry independent current. It is these coils that carry and produce the magnetic field when the current passes through them. There are two major types: "Mush" or "random" wound, round wire found in smaller and medium motors where coils are randomly laid in slot of stator core; and formed coils of square wire individually laid in, one on top of the other, to give an evenly stacked layered appearance.

Coincidence factor The ratio of the coincident maximum demand of two or more loads to the sum of their non-coincident maximum demands for a given period. The coincidence factor is the reciprocal of the diversity factor and is always less than or equal to one.

Coincidental demand Two or more demands that occur at the same time.

Coincidental peak load Two or more peak loads that occur at the same time.

Cold cathode tube A gas-filled electron tube that conducts without the use of filaments. Cold-cathode tubes are used as voltage regulators.

Cold junction Connection point between thermocouple metals and the electronic instrument.

Cold junction compensation A method of compensating for ambient temperature variations in thermocouple circuits.

Cold switching Closing the relay contacts before applying voltage and current, plus removing voltage and current before opening the contacts. (Contacts do not make or break current.)

Collector chamber Sealed box or other enclosure used to isolate a building component when conducting pressurization tests.

Collector characteristic curve A graph of collector voltage over collector current for a given base current.

Collector injection modulator The transistor equivalent of a plate modulator. Modulating voltage is applied to a collector circuit.

Collinear array An array with all the elements in a straight line. Maximum radiation is perpendicular to the axis of the elements.

Colour code The ANSI established colour code for thermocouple wires in the negative lead is always red. Colour Code for base metal thermocouples is yellow for Type K, black for Type J, purple for Type E and blue for Type T.

Colouration Listening term. A visual analogue. A "coloured" sound characteristic adds something not in the original sound. The colouration may be euphonically pleasant, but it is not as accurate as the original signal.

Colour rendering General expression for the effect of an illuminant on the colour appearance of objects in conscious or subconscious comparison with their colour appearance under a reference illuminant.

Colour temperature The temperature of a full radiator (radiation of all frequencies) which would emit visible radiation of the same spectral distribution as the radiation from the light source under consideration. (Unit kelvin, K)

Colour The sensation of colour is the result of the interpretation by the human central nervous system of the effect produced upon the eye by electromagnetic radiation of a particular wave length.

Colpitts oscillator An oscillator with a pair of tapped capacitors in the feedback network.

Comb Filter Effect This acoustical and electronic effect occurs when two signals interact in such a way as to produce an irregular spiked and choppy response pattern. When graphed on paper, this pattern looks like the teeth of a comb, hence the name. This effect is frequently the product of overlapping outputs from the various drivers in the system. It can develop in either a single speaker between the individual drivers, or between unmatched sets of speakers. This same effect can also be produced by wall reflections and other room anomalies. The usual remedy for this, if it becomes objectionable, (not every instance is even perceived as such) is to make sure the Crossover set points are appropriate for the drivers being used, or that the crossover is operating correctly. Obviously, one should also use only compatible speaker sets.

Combination array An array system that uses the characteristics of more than one array.

Combination circuit A series-parallel circuit.

Combination peaking A technique in which a combination of peaking coils in series and parallel (shunt) with the output signal path is used to improve high-frequency response.

Combined cycle Combines the gas turbine cycle together with a heat recovery steam cycle that extracts heat from the gas turbine exhaust flow to produce steam.

Combined cycle unit An electricity generating unit consisting of one or more gas (combustion) turbines combined with a steam turbine. The steam turbine utilises the waste exhaust heat from the combustion turbines. This process increases the efficiency of the electric generating unit.

Combined heat and power (CHP) A plant that generates electricity and supplies thermal energy, typically steam, to an industrial or other heating requirement.

Combustion or Burning A chemical reaction or complex chemical reaction, in which a substance combines with oxygen producing heat, light and flame.

Combustion turbine A type of generating unit normally fired by oil or natural gas. The combustion of the fuel produces expanding gases, which are forced through a turbine, thereby generating electricity.

Comfort measure of human acceptability of the physical environment

Commercial building A building whose primary purpose is to provide space for commercial activity rather than domestic. This includes offices, storage, plant, farm, public and some factory classifications.

Commercial operation Commercial operation occurs when control of the generator is turned over to the system dispatcher.

Commissioning The Start-up phase of a building that includes testing and adjusting HVAC, electrical, plumbing, and other systems to assure proper functioning and adherence to design criteria. Commissioning also includes the instruction of building representatives in the use of the building systems.

Common anode display A multisegment light emitting diode (LED) with a single positive voltage input connection. Separate cathode connections are provided for each individual segment.

Common base A transistor circuit in which the base electrode is the common element to both input and output circuits.

Common base amplifier A BJT circuit in which the base connection is common to both input and output.

Common Base Connection Same as ground base connection. A mode of operation in which the base is common to both the input and output circuits and is usually earthed. The emitter is used as the input terminal and the collector as the output terminal. (Grounded= grounded to AC signals).

Common base detector An amplifying detector in which detection occurs in the emitter-base junction and amplification occurs at the output of the collector junction.

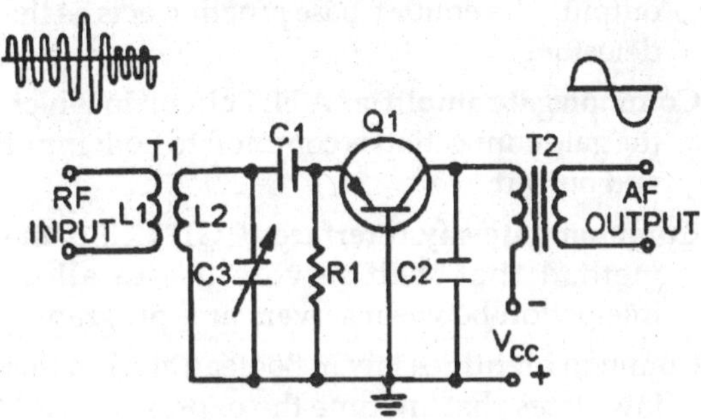

Fig. Common base detector

Common cathode display A multisegment light emitting diode (LED) with a single negative voltage input connection. Separate anode connections are provided for each individual segment.

Common collector A transistor circuit configuration in which the collector is the element common to both the input and the output circuits.

Common collector amplifier A BJT circuit in which the collector connection is common to both input and output.

Common Collector Connection Grounded collector connection. Also called the emitter-follower. A mode of operation in which the collector is common to both the input and the output circuits and is usually connected to one of the power rails.

Common drain amplifier A FET circuit in which the drain connection is common to both input and output.

Common emitter A circuit configuration in which the emitter is the element common to both the input and the output circuits.

Common emitter amplifier A BJT circuit in which the emitter connection is common to both input and output.

Common Emitter Connection Same as ground emitter connection. A mode of operation for a transistor in which the emitter is common to the input and output circuits. The base is the input terminal and the collector is the output terminal.

Common emitter detector Often used in receivers to supply detected and amplified output. The emitter-base junction acts as the detector.

Common gate amplifier A FET circuit in which the gate connection is common to both input and output.

Common gateway interface (CGI) CGI is the method that Web servers use to allow interaction between servers and programs.

Common identities law In Boolean algebra this law states that anytime the expression A(A + B) = AB or A + AB = A + B appears, it can immediately be simplified to AB without going through the process of using the distributive law, complementary law, or the law of union to simplify.

Common mode input isolation On a switching card, the isolation from signal high and low to guard (or shield) for a 3-pole circuit, or from signal high and low to chassis ground for a 2-pole circuit. Specified as resistance and capacitance.

Common mode line filter A device to filter noise signals on both power lines with respect to ground.

Common mode noise The component of the noise voltage that appears equally and in phase on conductors relative to a common reference.

Common mode rejection ratio (CMRR) The ratio of op-amp differential gain to common-mode gain. A measure of an op-amp's ability to reject common-mode signals such as noise.

Common mode tripping Automatic removal of two or more generating plant from the system owing to a cause that is common to both or all the generators.

Common mode voltage The voltage common to both sides of a differential circuit pair. The differential voltage across the circuit pair is the desired signal, whereas the common voltage signal is the unwanted signal which may have been coupled into the transmission pair.

Common source amplifier A FET circuit in which the source connection is common to both input and output.

Communication Transmission and reception of data among data processing equipment and related peripherals.

Communications module Controls the transmission between other controllers and between controllers and a central computer based on an established bus protocol.

Commutation angle, overlap angle The commutation period between two thyristors on the same side of the bridge is the angle by which one thyristor commutates to the next.

Commutation Controlling the currents or voltages in the motor phases in an effort to optimize motor performance; can be achieved mechanically or electrically.

Commutative law In Boolean algebra this law states that changing the order of the terms in an equation will not affect the value of the equation. Example: A + B = B + A; A

B = B

A.

Commutator A cylindrical device mounted on the armature shaft and consisting of a number of wedge-shaped copper segments arranged around the shaft (insulated from it and each other. The motor brushes ride on the periphery of the commutator and electrically connect and switch the armature coils to the power source.

Compact Disc (CD) The most popular format for conveying music and data currently available. It is among the first digital media to take over from the analogue formats of phonograph records and tapes; coming to the market in the early 1980's. Developed by Phillips, Sony, and Pioneer, it records information on the now familiar shiny discs by deforming the inner metal foil on the disc with tiny micro pits burned in by a laser. These pits taken together, form a binary digital code, which when converted to bits, then bytes, can recreate the original information, such as audio. It's superiority as a format, consists of the fact that the process gets around such problems as: noise, hiss, pops, transducer irregularities, and other audible problems that made analogue carriers a less than fully high fidelity mode. Dynamic range exceeds 100 decibels, a sufficient soft/loud difference to make the reproduction very lifelike. Frequency response is at the theoretical limits of human hearing and unwanted aural artifacts are generally below the threshold of perception. The only significant improvement is the DVD borne addition of multiple channels, to recreate the original sonic environment. In short, it's the best thing to come along since Mozart sat down at the piano.

Compact fluorescent lamp (CFL) A fluorescent lamp in compact form that may be conveniently used, in normal holders, in place of the lesser efficient incandescent lamps. The lamp life is significantly longer than incandescent lamps.

Comparing element (also 'error detector') A control element which compares the measured value and the desired value then sends an error signal (measured value minus desired value) to the part of the controller which determines control action. The control action implemented acts to reduce the magnitude of the error signal.

Comparitor An op-amp circuit that compares two inputs and provides a DC output indicating the polarity relationship between the inputs.

Compass The magnetic compass is used to obtain the direction of the earth's magnetic field at a point. In its simplest form consists of a magnetised needle pivoted at its centre so that it is free to move in a horizontal plane.

Compensated connector A connector made of thermocouple alloys used to connect thermocouple probes and wires.

Compensating alloys Alloys used to connect thermocouples to instrumentation. These alloys are selected to have similar thermal electric properties as the thermocouple alloys (however, only over a very limited temperature range).

Compensating loop Lead wire resistance compensation for RTD elements where an extra length of wire is run from the instrument to the RTD and back to the instrument, with no connection to the RTD.

Compensating windings Windings embedded in slots in pole pieces, connected in series with the armature, whose magnetic field opposes the armature field and cancels armature reaction.

Compensation An addition of specific materials or devices to counteract a known error.

Compensation control A process of automatically adjusting the control point of a given controller to compensate for changes in a second measured variable (e.g., outdoor air temperature). For example, the hot deck control point is normally reset upward as the outdoor air temperature decreases. Compensation control isone form of open-loop control.

Compiler A program that translates a high-level language, such as Basic, into machine language.

Complement A number or state that is the opposite of a specified number or state. The negative of a number is often represented by its complement.

Complement number A number that when added to another number gives a sum equal to the base of the number system of operation. For example, in the decimal number system, the complement of 1 is 9.

Complementary (Secondary) colours of light The colours of light produced when two of the primaries are mixed in overlapping beams of light. The complementary colours of light are magenta, yellow, and cyan.

Complementary law In Boolean algebra this law states that the logical addition of a quantity and its complement will result in 1 and the logical multiplication of a quantity and its complement will result in a product of 0.

Complementary MOS,CMOS A method of reducing the current drain of a digital circuit by combining n-channel and p-channel mosfets.

Complementary symmetry amplifier A class B amplifier using matched complementry transistors. Does not require a phase inverter for push-pull output.

Complementry transistors Two transistors, one NPN and one PNP having near identical charastics. N-channel and P-channel fets can also be complementry.

Complex function Any mathematically defined relationship given by the following expression:

$y(x) = a(x) + ib(x)$

Where: = the real variable

$a(x)$ = the real part of $y(x)$

$b(x)$ = the imaginary part of $y(x)$

Complex functions are usually expressed in terms of both their amplitude and phase.

Complex number Consists of two parts, real and imaginary. They obey the ordinary laws of algebra except that their real and imaginary parts must be equated separately.

Complex wave The resultant form of a number of sinusoidal waves that are summed together forming a periodic wave. Such waves may be analyzed in the frequency domain to readily determine their component parts.

Compliance The measurement in liters or cubic feet of the volume of air that is equal to the compliance, or maximum extension of a speaker's total suspension.

Compliance current The maximum output current of a constant voltage source. Also known as current limit.

Compliance voltage The maximum output voltage of a constant current source. Also known as voltage limit.

Component leakage The leakage of air through the building envelope which is directly attributable to flow through cracks around specific doors, windows or other components.

Component System This term is used in relation to speaker systems, to indicate a system in which separate mounting arrangements are provided for each component of the system. In a typical car system you might see a woofer in a box in the rear, midranges at the side and tweeters mounted on the dash panel. This compares to the typical integrated speaker enclosure in which all the Drivers are mounted in the same box.

Component video A three-channel video signal wherein the luminance, hue and colour saturation information are carried as R, G and B (Red, Green and Blue) signals or as one of several variations of colour difference signals.

Composite video A single video signal carrying combined luminance, chrominance and raster synchronizing information.

Compound motor A DC Motor with both a series connected winding as well as a shunt connected winding. Depending on whether the fields of the series winding and the shunt winding aid each other or oppose each other, they are called cumulative compound or differential compound.

Compound wound DC motors Designed with both a series and shunt field winding, the compound motor is used where the primary load requirement is heavy starting torque, and adjustable speed is not required. Also used for parallel operation. The load must tolerate a speed variation from full-load to no-load. Industrial machine applications include large planers, boring mills, punch presses, elevators, and small hoists.

Compound wound motors and generators Machines that have a series field in addition to a shunt field. Such machines have characteristics of both series- and shunt-wound machines.

Compression 1. An increase in density and pressure in a medium, such as air, caused intermittently by the passage of a sound wave.

2. The region in either air or material in which this occurs.

Compression Audio/Video Files A process of temporarily or permanently reducing audio data for more efficient storage or transmission. A temporary reduction in file size is called 'non-lossy' compression, and no information is lost. A permanent reduction in file size (such as with mp3 files) is called 'lossy' compression, and involves discarding (supposedly) unnecessary information which is irretrievably lost.

Compression Driver Compression drivers are usually dynamic; that is, with a magnet and interacting coil arrangement, and a small diaphragm as the main transducer. These are the motor parts, also known as the driver, of a compression horn tweeter or compression horn general-purpose speaker, such as those used for Public Address (PA) purposes. These drivers are usually coupled to the throat of an exponential horn. Such an arrangement enables this type of tweeter to have very high directional characteristics, which allows them to be especially effective in situations requiring a very wide sound field. In typical home and car stereo near-field applications, large horns can be a bit too narrowly directional to be practical. For this reason, compression horn systems are usually found only in special purpose speakers used in mid and wide field applications such as PA systems or the sound systems installed in large theaters. However, some specialized horns have small apertures and very shallow horns, and can be quite suitable for close spaces.

Compression waves Longitudinal waves that have been compressed (made more dense) as they move away from the source.

Compressor A type of dynamic range processor which reduces the gain of audio signals which are over an adjustable 'threshold' level, therefore reducing the dynamic range. Generally allows the operator control over threshold, ratio, attack and release times. Both analogueue and digital types are available.

Compromise An intrusion into a computer system where unauthorized disclosure, modification or destruction of sensitive information may have occurred.

Computer A data processor that can perform substantial computation, including numerous arithmetic or logic operations, without intervention by a human operator during the run.

Computer abuse The willful or negligent unauthorized activity that affects the availability, confidentiality, or integrity of computer resources. Computer abuse includes fraud, embezzlement, theft, malicious damage, unauthorized use, denial of service, and misappropriation.

Computer fraud Computer-related crimes involving deliberate misrepresentation or alteration of data in order to obtain something of value.

Computer ground A line for the ground connections to computers or microprocessor-based systems. It is isolated from safety ground.

Computer network A set of interconnected computer systems, terminals, and communications equipment.

Computer network attack Operations to disrupt, deny, degrade, or destroy information resident in computers and computer networks, or the computers and networks themselves.

Computer Oracle and Password System (COPS) A computer network monitoring system for Unix machines. Software tool for checking security on shell scripts and C programs. Checks for security weaknesses and provides warnings.

Computer program A schedule or plan that specifies actions which may or may not be taken, expressed in the form of a set of instructions suitable for execution by a computer.

Computer security Technological and managerial procedures applied to computer systems to ensure the availability, integrity and confidentiality of information managed by the computer system.

Computer security incident Any intrusion or attempted intrusion into an automated information system (AIS). Incidents can include probes of multiple computer systems.

Computer security intrusion Any event of unauthorized access or penetration to an automated information system (AIS).

Concentric Circles having the same centre.

Concert Pitch A standard for the tuning of musical instruments, internationally agreed in 1960, in which the note A above middle C has a frequency of 440 Hz.

Concurrent Pertaining to the occurrence of two or more events or activities within the same specified interval of time.

Condensation The precipitation of liquid from its vapour phase resulting from the lowering of temperature at constant pressure: especially the deposition of water from moist, warm air onto a relatively cold surface or between two surfaces such as within a cavity wall.

Condenser Microphone A mike that depends on an external power supply or internal battery to electrostatically charge capacitor plates, one of which is subjected to sonic motion. Also called a 'Capacitor' microphone.

Conditioned air Air that has been heated, cooled, humidified, or dehumidified to maintain an interior space within the "comfort zone." (Sometimes referred to as "tempered" air.)

Conductance (G) The ability to conduct electricity. Defined by G =Re (I/V) where G is the conductance in Siemens, I is the current in Amps, and V is the voltage in Volts.

Conducted noise Line noise which is generated from a power supply or other electrical components. There are several major standards as FCC, VCCI, CISPR, and EN.

Conduction Heat conduction involves the transfer of heat from one molecule to an adjacent one as an inelastic impact in the case of fluids, as oscillations in solid nonconductors of electricity, and as motions of electrons in conducting solids such as metals. Conduction is the only mechanism of heat transfer through an opaque solid. Some heat may be transfered through transparent solids such as glass, quartz and certain plastics, by radiation. In fluids the conduction process is supplemented by convection and if the fluid is transparent, by radiation.

Conduction band The unfilled energy levels into which electrons can be excited to become conductive electrons; a band that when partially occupied by mobile electrons, permits their net movement in a particular direction, producing the flow of electricity through the solid.

Conductivity Conductivity is the measure of conduction within a material. The conductivities of materials vary widely, being greatest for metals, less for nonmetals, still less for liquids and least for gases. Any material which has a low conductivity may be considered an insulator.

Conductor 1. A material with a large number of free electrons.

2. A material that easily permits electric current to flow.

Conductor loss Loss occurring in a conductor due to the flow of current. Also known as the I2 R loss and copper loss.

Conduit A tubular raceway for power or data cables. Both metallic conduit and non-metallic forms may be used.

Conduit box The metal container usually on the side of the motor where the stator (winding) leads are attached to leads going to the power supply.

Cone The cone-shaped diaphragm of a speaker. This is directly attached to the voice coil motor which actions produces the pulsation's of air that the ear detects as sound. Also useful for holding ice cream.

Cone excursion The distance that the driver cone can move (back and forth) without damage. Excursion is generally measured in millimeters (mm) or fractions of an inch.

Confidence level The range (with a specified value of uncertainty, usually expressed in percent) within which the true value of a measured quantity exists.

Confidentiality Assuring information will be kept secret, with access limited to appropriate persons.

Conformal coating An insulating layer that can be applied by spraying, dipping, or vapor deposition that covers and protects the components on a circuit board.

Conformity error For thermocouples and RTDs, the difference between the actual reading and the temperature shown in published tables for a specific voltage input.

Conical scanning Scanning in which the movement of the beam describes a cone, the axis of which coincides with that of the reflector.

Connected array Another term for driven array.

Connection head An enclosure attached to the end of a thermocouple which can be cast iron, aluminum or plastic within which the electrical connections are made.

Connector The part of a cable coupler or of an appliance coupler which is provided with female contacts and is intended to be attached to the end of the flexible cable remote from the supply.

Console The part of a computer used for communication between the human operator and the comnputer.

Constant concentration A Tracer gas method for measuring ventilation rates, whereby an automated system injects tracer gas at a rate required to maintain the concentration of tracer gas within a room or zone at a fixed, pre-determined level. The ventilation rate is proportional to the rate at which the tracer gas must be injected.

Constant current circuit Circuit used to maintain constant current to a load having resistance that changes.

Constant current power supply A power supply that regulates its output current, within specified limits, against in line voltage, load, ambient temperature, and time.

Constant Directivity (CD) Horn A horn-loaded high frequency driver that exhibits more or less constant distribution of high-frequency sound in the horizontal direction. This is done by using one of several special dual shaped horn designs created to solve the traditional problem of horn-loaded driver output varying with frequency. All CD horns exhibit a high frequency roll-off of approximately 6 dB/octave beginning somewhere in the 2 kHz to 4 kHz area.

Constant flow/emission A Tracer gas method for measuring ventilation rates whereby tracer gas is continually emitted at a uniform rate. The equilibrium concentration of tracer gas in air is then measured.

Constant H.P. A designation for variable or adjustable speed motors used for loads requiring the same amount of H.P. regardless of their motor speed during normal operation.

Constant speed A DC motor which changes speed only slightly from a no load to a full load condition. In AC motors, these are synchronous motors.

Constant torque Refers to loads whose H.P. requirements change linearly with changing speeds. Horsepower varies with the speed, i.e.- 2/1 HP at 1800/900 RPM. (Seen on some 2-speed motors). Possible applications include conveyors, some crushers, or constant-displacement pumps.

Constant voltage power supply A power supply that regulates its output voltage, within specified limits, against in line voltage, load, ambient temperature, and time.

Contact Current carrying part of a switch, relay or connector.

Contact bounce The intermittent and usually undesired opening of mechanical relay contacts during closure, or closing of contacts during opening. Contact bounce period depends upon the type of relay and varies from .5mS for small reed relays to 10-20mS for larger solenoid types. Solid-state or mercury wetted contacts (Hg) do not have a contact bounce characteristic.

Contact life The maximum number of expected closures before failure. Life is dependent on the switched voltage, current, and power. Failure is usually when the contact resistance exceeds an end of life value. Typical failure mode is non-closure of the contact as opposed to a contact sticking closed.

Contact potential A voltage produced between contact terminals due to the temperature gradient across the relay contacts, and the reed-to-terminal junctions of dissimilar metals. (The temperature gradient is typically caused by the power dissipated by the energized coil.) Also known as contact offset voltage, thermal EMF, and thermal offset. This is a major consideration when measuring voltages in the microvolt range. There are special low thermal relay contacts available to address this need. Special contacts are not required if the relay is closed for a short period of time where the coil has no time to vary the temperature of the contact or connecting materials (welds or leads).

Contact rating The voltage, current, and power capacities of relay contacts under specified environmental conditions.

Contact resistance The resistance in ohms or milliohms across closed contacts.

Contactor An electro-mechanical device that is operated by an electric coil and allows automatic or remote operation to repeatedly establish or interrupt an electrical power circuit. A contactor provides no overload protection as required for motor loads.

Contaminant An unwanted airborne constituent that may reduce the acceptability of the air (quality) and may be detrimental to the health of building occupants.

Continuity An uninterrupted, complete path for current flow.

Continuous action The action of an element, regulator, or automatic control system whose output is a continuous function of it's input signal.

Continuous spectrum A frequency spectrum that is characterized by non-periodic data The spectrum is continuous in the frequency domain and is characterized by an infinite number of frequency components.

Continuous stall current Amount of current applied to the motor to achieve the continuous stall torque.

Continuous stall torque Maximum amount of torque a motor can provide at zero speed without exceeding its thermal capacity.

Continuous wave keying The on-off keying of a carrier.

Contract price Price marketed on a contract basis for one or more years.

Control The control is the ability of the system to respond to the changing requirements imposed upon it by the fluctuation of outside conditions.

Control action The action generated by the controller and fed to the correcting unit, i.e., the relationship between the input signal and the output signal of a control element; cf. Control mode.

Control agent The medium in which the manipulated variable exists. In a steam heating system, the control agent is the steam, and the manipulated variable is the flow of steam; cf. Controlled medium.

Control character A character whose occurrence in a particular context starts, modifies or stops an operation that effects the recording, processing, transmission or interpretation of data.

Control differential transmitter (CDX) A type of synchro that transmits angular information equal to the algebraic sum or difference of the electrical input supplied to its stator, the mechanical input supplied to its stator, and the mechanical input supplied to its rotor. The output is an electrical voltage taken from the rotor windings.

Control element A general term for a constituent part of a control system.

Control field In a protocol data unit (PDU), the field that (a) contains data interpreted by the receiving destination logical-link controller (LLC) and (b) may be the field immediately following the destination service access point (DSAP) and source service access point (SSAP) address fields of the PDU.

Control function Generally, this term is used for the operations carried out by an automatic control system.

Control grid modulator Uses a variation of grid bias to vary the instantaneous plate voltage and current. The modulating signal is applied to the control grid.

Control grid The electrode of a vacuum tube, other than a diode, upon which a signal voltage is impressed to regulate the plate current.

Control law This defines the control algorithm which represents the logic or "control action" of a controller. For example, the control law implemented may be PID control.

Control loop Generally, this term means any control network consisting of the control elements required for automatic control.

Control mode The output form or type of control action used by a temperature controller to control temperature, i.e., on/off, time proportioning, PID.

Control parameter A variable used in the control algorithm e.g. setpoint, proportional band.

Control point The temperature at which a system is to be maintained.

Control point adjustment The procedure of changing the operating point of a local loop controller from a remote location.

Control range The change between the initial and the potential value of the controlled condition.

Control synchro systems Synchro systems that contain control synchros and are used to control large amounts of power with a high degree of accuracy. The electrical outputs of these systems control servosystems, which in turn generate the required power to move heavy loads.

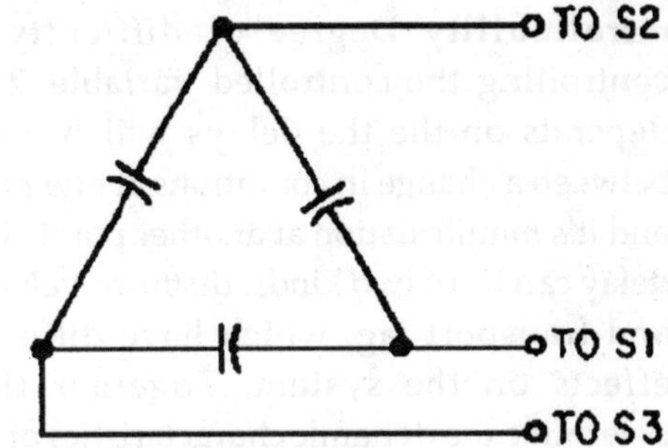

Fig. Control synchro systems

Control system A group of components systematically organized to perform a specific control purpose. These systems are categorized as either closed- or open-loop systems. The main difference between the two is that the closed-loop system contains some form of feedback.

Control transformer (CT) A type of synchro that compares two signals: the electrical

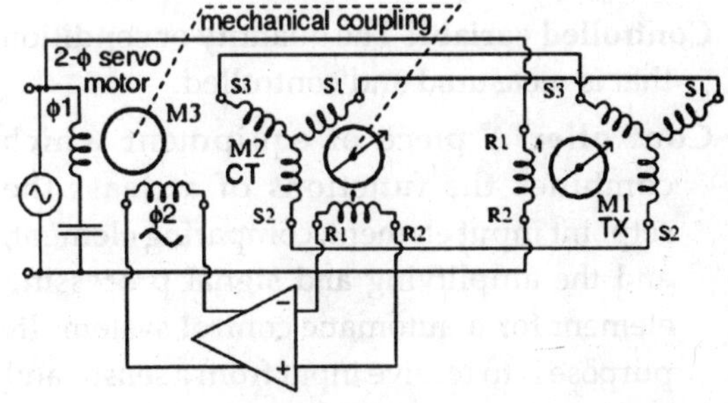

Fig. Control transformer

signal applied to its stator and the mechanical signal applied to its rotor. The output is an electrical voltage, which is taken from the rotor winding and is used to control a power-amplifying device. The phase and amplitude of the output voltage depends on the angular position of the rotor with respect to the magnetic field of the stator.

Control transmitter (CX) A type of synchro that converts a mechanical input, which is the angular position of its rotor, into an electrical output signal. The output is taken from the stator windings and is used to drive either a CDX or CT.

Controllability Degree of difficulty in controlling the controlled variable. This depends on the the delays which occur between a change in conditions at one point and it's manifestation at another point. This delay can be of two kinds: distance-velocity and transport lag, which have different effects on the system. Together they determine the dynamic characteristics of the plant.

Controlled medium The medium in which the controlled variable exists. In a space temperature control system, the controlled variable is the space temperature and the controlled medium is the the air within the space; cf. Controlled agent.

Controlled sequence Equipment operating order established upon a correlated set of environment data conditions.

Controlled variable The quantity or condition that is measured and controlled.

Controller A piece of equipment which combines the functions of at least the setpoint input element, comparing element, and the amplifying and signal processing element for a automatic control system. Its purpose is to receive input from a sensor and then derive the proper correction output which is then sent to the actuator.

Convection The essential process in the case of convection is the flow of a fluid over a solid surface, accompanied by a transfer of heat between the surface and the fluid. The movement of the fluid may be due to changes in its density caused by changes in its temperature, by natural convection; or it can be created by mechanical means, by forced convection.

Conventional controllers All controllers with the exception of microprocessor-based controllers. Thus the term 'conventional controllers' describes the following analogueue controllers:- pneumatic controllers, hydraulic controllers, fluidic controllers, electrical controllers, and electronic (solid-state) controllers.

Conventional current flow During the early period of work on electric circuits it was thought that current flowed from positive to negative.

Conversion A process where a signal is changed from an analogue to digital (A-D) representation, or digital to analogue (D-A).

Conversion rate The rate at which sampled analogue data is converted to digital data or digital data is converted to analogue data.

Converter In communications, equipment that changes the audio output of a receiver to dc pulses. These pulses are fed to a tty to indicate marks and spaces.

Convolution The convolution of two signals consists of time-reversing one of the signals, shifting it, and multiplying it point by point with the second signal,and integrating the product. It is used to characterise physical systems.

Cookie A cookie is a mechanism that allows your movements through the Web to be remembered and stored by your computer (usually when you browse the Web there is no record kept about the sites you have visited or the route you have taken - but with cookies switched on, this information is kept). Cookies are used to run certain Web services as they can record your preferences when using particular Web sites or take you back to a particular point on the Web at a later visit.

Cookie cutter tuner A mechanical magnetron tuning device that changes the frequency by changing the capacitance of the anode cavities.

Cooling The transfer of energy from a body of solid, liquid or gas by the existence of a temperature gradient from that body to its surroundings which are at a lower temperature, and may also be solid, liquid or gas. This process is the opposite of heating.

Co-ordinated Usually refers to characteristics which are co-ordinated to give optimum performance.

Copolymer A polymer that consists of two or more dissimilar monomer units in combination in its molecular chains. Also a polymer formed from the polymerization of more than one type of monomer.

Copper loss Power lost in transformers, generators, connecting wires and other parts of a circuit due to current flow through the resistance of copper conductors.

Cordwood module A method of increasing the number of discrete components in a given space. Resembles wood stacked for a fireplace.

Core The iron portion of the stator and rotor; made up of cylindrical laminated electric steel. The stator and rotor cores are concentric separated by an air gap, with the rotor core being the smaller of the two and inside to the stator core.

Core loss The Loss occurring in a magnetic core due to alternating magnetisation. It is the sum of the hysteresis loss and the eddy current loss.

Coriolis force A result of centripetal force on a mass moving with a velocity radially outward in a rotating plane.

Corner reflector antenna A half-wave antenna with a reflector consisting of two flat metal surfaces meeting at an angle behind the radiator.

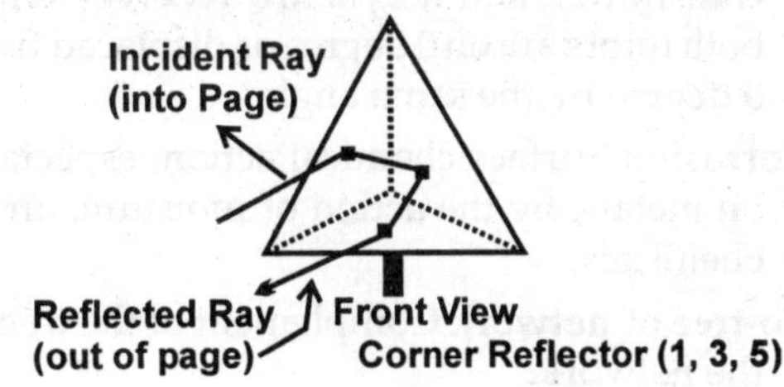

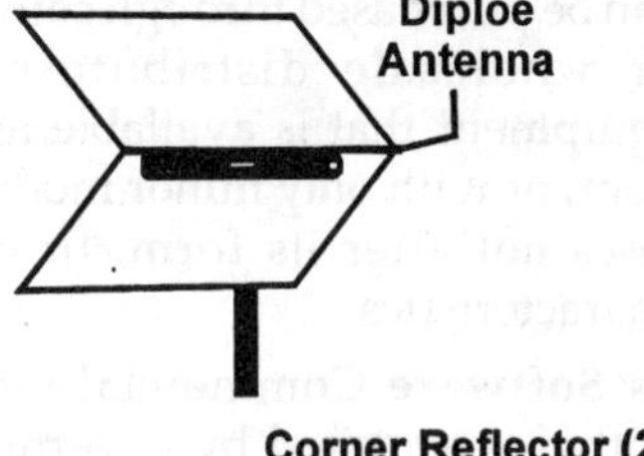

Corner Reflector

Fig. Corner reflector antenna

Corona discharge Bluish white luminous discharge which appears surrounding a conductor surface at a field exceeding corona inception, but not being sufficient to cause sparking or flash over.

Corona inception Inception of the ionisation of the air on the surface of a conductor, caused by the voltage gradient exceeding a critical value, but not being sufficient to cause sparking or flash over.

Corona The discharge of electricity from a conductor with a high potential.

Corona loss power loss due to corona.

Correction (Balancing) plane A plane perpendicular to the shaft axis of a rotor in which correction for unbalance is made.

Corrective action Control action that results in a change of the manipulated variable. Initiated when the controlled variable deviates from the setpoint.

Corrective maintenance Includes location and repair of equipment failures.

Correspondence The term given to the positions of the rotors of a synchro transmitter and a synchro receiver when both rotors are on 0 degree or displaced from 0 degree by the same angle.

Corrosion Surface chemical action, especially on metals, by the action of moisture, air or chemicals.

Co-tree of network Complement of the tree of the network.

Cots Commercial off-the-shelf equipment that can be purchased through commercial retail or wholesale distributors as is (i.e., equipment that is available as a cataloged item) or with only minor modifications that does not alter its form, fit or functional characteristics.

Cots Software Commercial Off the Shelf - Software acquired by government contract through a commercial vendor. This software is a standard product, not developed by a vendor for a particular government project.

Coulomb (C) SI unit of electric charge. One coulomb is equal to the amount of charge accumulated in one second by a current of one ampere.

Coulomb sensitivity Charge/unit acceleration, expressed in Pc/g (charge sensitivity).

Coulomb's law Also called the LAW OF ELECTRIC CHARGES or the LAW OF ELECTROSTATIC ATTRACTION. Coulomb's Law states that charged bodies attract or repel each other with a force that is directly proportional to the product of their individual charges and inversely proportional to the square of the distance between them.

Counter electromotive force (CEMF) The voltage generated within a coil by a moving magnetic field cutting across the coil itself. This voltage is in opposition (counter) to the moving field that created it. Counter emf is present in every motor, generator, transformer, or other inductance winding whenever an alternating current flows.

Counter firing A method for reducing harmonic distortion at low frequencies that involves the use of a secondary coil in a woofer or mid bass driver to cancel non-fundamental wave elements not found in the original signal. The signal from the coil is used by the amplifier to refine it's damping control over the cone motion. This proprietary servo-feedback technology was originally developed by Velodyne in the early '90s.

Counter weight A weight added to a body so as to reduce a calculated unbalance at a desired place.

Counter/Timer A circuit that counts pulses or measures pulse duration.

Countermeasures Action, device, procedure, technique, or other measure that reduces the vulnerability of an automated information system. Countermeasures that are aimed at specific threats and vulnerabilities involve more sophisticated techniques as well as activities traditionally perceived as security.

Counterpoise A network of wire connected to a quarter-wave antenna at one end. The network provides the equivalent of an additional one-fourth wavelength.

Counts The number of time intervals counted by the dual-slope A/D converter and displayed as the reading of the panel meter, before addition of the decimal point.

Coupler Device that fastens the output shaft of the motor assembly to the shaft of the load.

Coupling The process of transferring energy from one point in a circuit to another point, or from one circuit to another.

Coupling capacitor A capacitor used to transmit an ac signal from one node to another.

Coupling coefficient The coupling coefficient of a pair of coils is a measure of the magnetic coupling between two coils.

Coupling device A coupling coil that connects the transmitter to the feeder.

Covalent bond The way some atoms complete their valence shells by sharing valence electrons with neighbouring atoms.

Crack A popular hacking tool used to decode encrypted passwords. System administrators also use Crack to assess weak passwords by novice users in order to enhance the security of the AIS.

Crack length The total length of the narrow gaps found around doors and windows etc, through which ventilation air may pass.

Crack/crackage Small gaps around doors, windows and other parts of a building envelope through which ventilation air may pass.

Cracker One who breaks security on an AIS.

Cracking The act of breaking into a computer system.

Crash A sudden, usually drastic failure of a computer system.

Crawlspace A shallow space in a building, usually under the floor, which provides access to pipes, wires and other equipment.

Creep The time-dependent permanent deformation that occurs under stress; for most materials it is important only at elevated temperatures.

Creepage distance Shortest path along the surface of insulating material between two conductive parts.

Crest (TOP) The peak of the positive alternation (maximum value above the line) of a wave.

Crest factor The ratio of the peak value to the root-mean-square (rms) value of a waveform.

Critical angle The maximum angle at which radio waves can be transmitted and still be refracted back to earth.

Critical damping Critical damping of a measuring instrument causes the equilibrium deflection to be reached in the shortest possible time, with the oscillations of the needle being quickly damped out. (Note: Under damped instruments have their needles oscillating for some time, while over damped instruments take a long time to reach the final equilibrium deflection)

Critical frequency The maximum frequency at which a radio wave can be transmitted vertically and still be refracted back to earth.

Critical mass The minimum amount of fissile material required in a nuclear reactor to sustain a chain reaction.

Critical speed The rotational speed of the rotor or rotating element at which resonance occurs in the system. The shaft speed at which at least one of the "critical" or natural frequencies of a shaft is excited.

Critically damped Describes a system where the response to an input change is achieved in the minimum possible time.

Cross bonding A method of connecting the sheaths of single core cables in a three phase system in order to reduce the circulating currents flowing in the sheaths.

Cross contamination (of air or masses) The contamination of one stream of air by pollutants in another, due to air movement between the two streams (or masses).

Cross regulation In a multiple output power supply, the percent voltage change at one output caused by the load change on another output.

Cross sectional area The area of a "slice" of an object. When applied to electrical conductors it is usually expressed in circular mils.

Cross sensitivity The influence of one measurand on the sensitivity of a sensor, another measurand.

Cross Talk 1. Undesired capacitive, inductive, or conductive coupling from one circuit, part of a circuit, or channel, to another.

2. Any phenomenon by which a signal transmitted on one circuit or channel of a transmission system creates an undesired effect in another circuit or channel. Note: In telecommunications, cross talk is usually distinguishable as speech or signaling tones.

Cross ventilation Air enters on one side of a room and leaves on a different side of the same room. Airflow between the entry and exit provides ventilation. Also used for flow between rooms, where the inlet is in one room and the outlet is in another.

Crossed field amplifier A high-power electron tube that converts dc to microwave power by a combination of crossed electric and magnetic fields.

Crosslinked polymer A polymer in which adjacent linear molecular chains are joined at various positions by covalent bonds.

Crossover A device or passive circuit used in systems with separate tweeter and/or midrange Drivers. It Rolls Off frequencies above and below certain points in the range, to allow the sound to be tailored for the specific driver to which it is sent. Most speakers have crossovers that consist of passive elements such as capacitors, coils, and resistors to separate the various frequencies. In a bi-amped or multi-amped system, the crossover is an active device that feeds the various frequency bands to the inputs of the amplifiers that operate the individual drivers.

Crossover distortion Distortion caused by both devices in a class B amplifier being cut-off at the same time.

Crossover Frequencies The frequencies at which a passive or electronic crossover network divides the audio signals, which are then routed to the appropriate amplifiers or speakers.

Crossover Network A unit which divides the audio spectrum into two or more frequency bands.

Crossover point The frequency point at which sound is transferred from one driver to another. The loudness of the two drivers is the same at the crossover point.

Crossover Slope The rate at which a crossover circuit attenuates the blocked frequencies. Slope is expressed as decibels per octave. A 6dB per octave crossover reduces signal amplitude level by 6dB in every octave starting at the crossover point. This means that every time the frequency of the audio signal is changed by a factor of 2 (one octave), the level of the audio signal is attenuated by 6dB. For example, if a low-pass crossover is set at 60Hz with a 6dB slope, you'll see a drop in level of 6dB at 120Hz. With slopes of 12dB and higher, the output beyond the crossover point will be reduced to below the level of audibility.

Crosspoint switch A switch which, when closed, connects the signal on an input bus to one or more output buses. Also referred to as a matrix switch or switching array.

Crosstalk The coupling of a signal from one input to another (or from one channel to another or to the output) by conduction or radiation. Crosstalk is expressed in decibels at a specified load and up to a specific frequency.

Crosstalk/Crosstalk isolation Unwanted interference in an output resulting from other input and output signals, measured in dB below the nominal signal level, and is expressed in decibels (dB) at a specified load impedance and over a specific frequency range or ranges. Also referred to as All Hostile or Hostile Crosstalk.

Crowbar Circuit used to protect the output of a souce from a short circuited load. Load current is limited to a value the source can deliver without damage.

Cryogenic Empire Magnetics cryogenic motors and related products are rated for an ambient temperature of 20° K though motors rated for an ambient temperature of 4° K have been provided on a custom basis.

Cryptanalysis 1) The analysis of a cryptographic system and/or its inputs and outputs to derive confidential variables and/or sensitive data including cleartext. 2) Operations performed in converting encrypted messages to plain text without initial knowledge of the crypto-algorithm and/or key employed in the encryption.

Cryptographic hash function A process that computes a value (referred to as a hashword) from a particular data unit in a manner that, when a hashword is protected, manipulation of the data is detectable.

Cryptography The art of science concerning the principles, means, and methods for rendering plain text unintelligible and for converting encrypted messages into intelligible form.

Cryptology The science which deals with hidden, disguised, or encrypted communications.

Crystal A natural substance, such as quartz or tourmaline, that is capable of producing a voltage when under physical stress or of producing physical movement when a voltage is applied.

Crystal controlled oscillator Oscillator that uses a quartz crystal in its feedback path to maintain a stable output frequency.

Crystal furnace A device for artificially growing cylindrical crystals to be used in the production of semiconductor substrates.

Crystal microphone A microphone that uses the piezoelectric effect of crystalline matter to generate a voltage from sound waves.

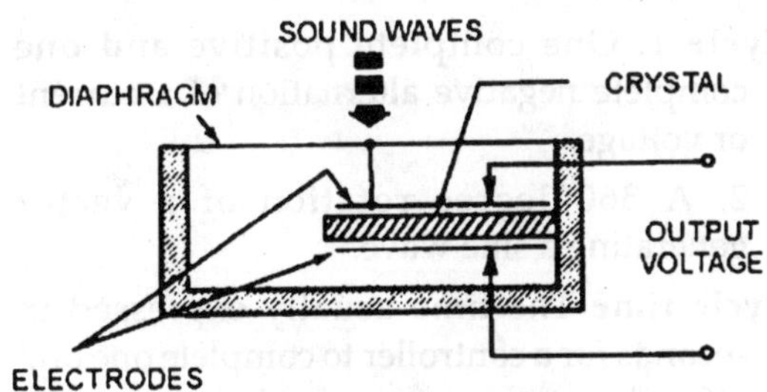

Fig. Crystal microphone

Crystal oven A closed oven maintained at a constant temperature in which a crystal and its holder are enclosed to reduce frequency drift.

Crystal structure For crystalline materials, the manner in which atoms or ions are arrayed in space. It is defined in terms of the unit cell geometry and the atom positions within the unit cell.

CSMA/CD Acronym for Carrier Sense, Multiple Access with Collision Detection. It is a method used to control access to shared transmission medium, such as a coaxial cable bus to which a number of stations are connected.

Cumulative uncertainty Cumulative uncertainty will arise in successive stages of the calibration chain of a measuring instrument. this is usually compounded into a single value supplied by the calibration laboratory.

Cumulo nimbus cloud Thunder cloud from which lightning strikes originate.

Cure point The temperature at which a normally magnetic material goes through a magnetic transformation and becomes non-magnetic.

Curie Measure of the activity of a radioactive substance: It is defined in terms of the rate of decay of a quantity of a radioactive isotope.

Curie temperature (also Curie point) (Tc) The temperature above which a ferromagnetic or ferrimagnetic material becomes paramagnetic. For iron the Curie point is 760oC and for nickel 356oC.

Current (A) Measure of rate of flow of electric charge: A one-ampere current is a flow of 1 C of charge per second.

Current amplifier Amplifier to increase signal current.

Current at peak torque Amount of current required to produce peak torque.

Current balance Instrument for the determination of an electric current in absolute electromagnetic units.

Current carrying capacity of a conductor The maximum current which can be carried by a conductor under specified conditions without its steady state temperature exceeding a specified value.

Current divider Parallel network designed to divide the total current of a circuit .

Current feed method Same as center-feed method.

Current feedback Feedback configuration where a portion of the output current is fed back to the amplifier input.

Current limiting resistor Resistor in the path of current flow to control the amount of current drawn by a device.

Current mirror Term used to describe the fact that DC current through the base circuit of a class B amplifier is approximately equal to the DC collector current.

Current proportioning An output form of a temperature controller which provides a current proportional to the amount of control required. Normally is a 4 to 20 milliamp current proportioning band.

Current rating This is the maximum current, which the fuse will carry for an indefinite period without undue deterioration of the fuse element.

Current regulator A circuit that provides a constant current output.

Current sink Transistor output configured so that the load is wired from the (+) side of the power source to the output and the transistor makes the circuit sink to common.

Current source Transistor output configured in such a way that the load is wired from the output of the sensor to the common side of the power source so that when the transistor turns on voltage is sourced to the load.

Current standing wave ratio (ISWR) The ratio of maximum to minimum current along a transmission line.

Current surge limiting The circuitry necessary to protect relay contacts from excessive and possibly damaging current caused by capacitive loads.

Current transformer An instrument transformer specifically designed to give an accurate current ratio for measurement and/ or control purposes. They are always connected in series with the circuit (like an ammeter) and hence should never be allowed to have their secondary to be on open circuit to avoid saturation

Curve fitting Curve fitting is the process of computing the coefficients of a function to approximate the values of a given data set within that function. The approximation is called a "fit". A mathematical function, such as a least squares regression, is used to judge the accuracy of the fit.

Cutoff Condition when an active device is biased such that output current is near zero or beyond zero.

Cutoff frequency The frequency at which the attenuation of a waveguide increases sharply and below which a traveling wave in a given mode cannot be maintained. A frequency with a half wavelength that is greater than the wide dimension of a waveguide.

Cutoff Frequency Filters The frequency at which a signal falls off by 3 dB (the half power point) from it's maximum value. Also referred to as the -3 dB points, or the corner frequencies

CW Demodulator A circuit that detects the presence of RF oscillations and converts them into a useful form.

Cyberspace Describes the world of connected computers and the society that gathers around them. Commonly known as the INTERNET.

Cycle 1. One complete positive and one complete negative alternation of a current or voltage.

2. A 360-degree rotation of a vector generating a sine wave.

Cycle time The time usually expressed in seconds for a controller to complete one on/ off cycle.

Cycles per second (Hertz) One complete reverse of flow of alternating current per rate of time. (A measure of frequency.) 60 HZ (cycles per second) A.C. power is common throughout the U.S. and 50 HZ is more common in some foreign countries.

Cycling A periodic change in the controlled variable from one value to another. Uncontrolled cycling is called hunting.

Cyclotron A type of particle accelerator in which an ion introduced at the center is accelerated in an expanding spiral path by use of alternating electrical fields in the presence of a magnetic field.

Cylindrical parabolic reflector A parabolically shaped reflector that resembles part of a cylinder.

D

D Flange A special end shield with holes for through bolts in the flange and is primarily used for mounting the motor on gear boxes or bulkheads. Standardized for frames 143T through 445T. "D" flanges are not threaded and the bolt holes extend beyond the motor frame.

DAC Abbreviation for "digital to analogue converter." .

Daily peak The maximum amount of power or energy or service demanded in one day from a service.

Daisy chaining The serial control connection of two or more mainframes in a master/ slave(s) configuration. Also, some switching modules or cards can be daisy-chained to yield more inputs. This term is also used in reference to control panels daisy chaining (looping) from control panel to control panel to the final destination, the switching system.

Damped wave A sinusoidal wave in which the amplitude steadily decreases with time. Often associated with energy loss.

Damper Part of the suspension connected to the bottom of the speaker cone at the voice coil that centers the voice coil in the magnetic gap. It is sometimes referred to as the spider

Damping (Damping factor, etc.) Refers to the ability of an audio component to "stop" after the signal ends. For example, if a drum is struck with a mallet, the sound will reach a peak level and then decay in a certain amount of time to no sound. An audio component that allows the decay to drag on too long has poor damping, and less definition than it should. An audio component that is overdamped does not allow the initial energy to reach the full peak and cuts the decay short. "Boomy" or "muddy" sound is often the result of underdamped systems. "Lifeless" sound may be the result of an overdamped system.

Damping Factor The ratio of rated load impedance to the internal impedance of an amplifier. The higher the value, the more efficiently an amplifier can control unwanted movement of the speaker coil. A high damping factor is crucial for large speakers that reproduce bass. Usually the higher the number, the better, although it is debatable if anything over 50 is audible. Damping factor is calculated by dividing the load (speaker) impedance by the output impedance of the amplifier. Thus, a given amplifier's damping factor will decrease as the speaker's impedance decreases. This means an amp optimized at 4 ohms will provide tighter bass than at 2 ohms.

Damping ratio Ratio of actual damping to critical damping; if the damping ratio is less than one a system is said to be underdamped and if it is greater than one a system is said to be overdamped.

Danger Risk of injury to persons (and livestock where expected to be present) from: (i) fire, electric shock and bums arising from the use of electrical energy, and (ii) mechanical movement of electrically controlled equipment, in. So far as such danger is intended to he prevented by electrical emergency switching or by electrical switching for mechanical maintenance, of non electrical parts of such equipment.

Dark side hacker A criminal or malicious hacker.

Darlington pair An amplifier consisting of two bipolar junction transistors with their collectors connected together and the emitter of one connected to the base of the other. Circuit has an extremely high current gain and input impedance.

Darpa Defense Advanced Research Projects Agency.

Data A representation of facts, concepts, or instructions in a formalised manner suitable for communication, interpretation, or processing by humans or by automatic means.

Data Base A large amount of data stored in a well-organized manner. A data base management system (DBMS) is a program that allows access to the information.

Data driven attack A form of attack that is encoded in innocuous seeming data which is executed by a user or a process to implement an attack. A data driven attack is a concern for firewalls, since it may get through the firewall in data form and launch an attack against a system behind the firewall.

Data encryption standard (DES) 1. An unclassified crypto algorithm adopted by the National Bureau of Standards for public use.

2. A cryptographic algorithm for the protection of unclassified data, published in Federal Information Processing Standard (FIPS) 46. The DES, which was approved by the National Institute of Standards and Technology (NIST), is intended for public and government use.

Day economization A control scheme which permits heating or cooling plant to be turned off completely if the anticipated fall or rise in internal temperature does not exceed pre-selected limits within a period, usually 1 hour.

Daylight factor distribution This is included to indicate artificial lighting needs.

DC (Direct current) A current that flows only in one direction in an electric circuit. It may be continuous or discontinuous and it may be constant or varying.

DC Converter A circuit or device that changes a dc input to a DC output of different value and/or polarity.

DC load line A graph representing all possible combinations of voltage and current for a given load resistor in an amplifier.

DC Motor A motor using either generated or rectified D.C. power (see Motor definition). A DC motor is usually used when variable speed operation is required

DC offset The change in input voltage required to produce a zero output voltage when no signal is applied to an amplifier.

DC potentiometer A potentiometer in which the supply is a battery and the balance is under d.c. Conditions.

DC power supply Any source of DC power for electrical equipment.

DC Pressurization Building airtightness levels can be measured by using a fan, temporarily installed in the building envelope (a blower door) to pressurize the building. Air flow through the fan creates an internal, uniform, static pressure within the building. The aim of this type of measurement is to relate the pressure differential across the envelope to the air flow rate required to produce it. Generally the higher the flow rate required to produce a given pressure difference, the less airtight the building. (See blower door; internal fan pressurization; external fan pressurization)

Dead band 1. For chart records the minimum change of input signal required to cause a deflection in the pen position.

2. For temperature controllers the temperature band where heat is turned off upon rising temperature and turned on upon falling temperature expressed in degrees. The area where no heating (or cooling) takes place.

Dead short Short circuit having zero resistance.

Dead time The time interval between a change in a signal and the initiation of a perceptible response to that change.

Dead volume The volume of the pressure port of a transducer at room temperature and ambient barometric pressure.

Debug To find and correct mistakes in a program.

Debye shielding The Debye length in front of a sensing electrode depends on the ionic strength of the electrolyte used. In a 0.001N NaCl the Debye length measures 96.5 Å, while for a 1.0 N solution it is reduced to 3.0 Å. An adsorbed protein can stick out from the surface for as much as 50 to 100 Å. As a result, the charges which could contribute to the surface potential will be shielded in a 1.0 N solution. To make more sensitive measurements a solution of low ionic strength should be used.

Decade A frequency factor of ten.

Decay method (tracer gas) A tracer gas method for measuring the ventilation rate whereby a quantity of tracer gas is released and the decrease in concentration is measured as a function of time.

Decay The time of reduction of the level of a signal immediately after its cessation.

Deceleration Rate of decrease in velocity with respect to time.

Decentralized intelligence A system where data processing is carried out at outstations as well as at the central station.

Deci (d) Decimal sub-multiple prefix corresponding to one-tenth or 10-1. This is not a preferred suffix.

Decimal Pertaining to the number representation system with a radix of ten.

Decimal digit In decimal notation, one of the characters 0 through 9.

Decimal notation A fixed radix notation where the radix is ten.

Decimal numeral A decimal representation of a number.

Decimal point The radix point in decimal representation.

Decipol The decipol attempts to quantify the concentration of odour as perceived by humans. The decipol represents the perception of odour measured by the "pol" unit. To obtain a usable unit it has been suggested that one tenth of the pol unit is used, the "decipol". The perceived air pollution is defined as that concentration of human bioeffluents that would cause the same dissatisfaction as the actual air pollution concentration.

Deck In HVAC terminology, the air discharge of the hot or cold coil in a duct serving a conditioned space.

Decoupling capacitor A capacitor used to transfer unwanted signals out of a circuit; for example, coupling an unwanted signal to ground. Also called a BYPASS CAPACITOR.

Deemphasis In FM transmission, the process of restoring (after detection) the amplitude-vs.frequency characteristics of the signal.

Deep cycle battery A battery designed to regularly discharge 80 percent of its capacity before recharging.

Default The value(s) or option(s) that are assumed during operation when not specified.

Defibrillation The use of electric shock to stop abnormally fast heart rhythms. Electrical current is used to restore the heart's natural pacemaker function which resumes a normal heartbeat. The shock is administered through electrodes placed on the chest wall (external defibrillation) or in the heart (internal defibrillation).

Defibrillator The machine or device that produces the electric shock current for defibrillation

Definite purpose motor A definite purpose motor is any motor design, listed and offered in standard ratings with standard operating characteristics with special mechanical features for use under service conditions other than usual or for use on a particular type of application.

Deflection Movement of an indicating needle.

Deflection coils In a cathode-ray tube, coils used to bend an electron beam a desired amount.

Deflection plates Two pairs of parallel electrodes, one pair set forward of the other and at right angles to each other, parallel to the axis of the electron stream within an electrostatic cathode-ray tube.

Defrost circuits(s) As heat pumps extract heat from the external atmosphere even at very low temperatures itis inevitable the external coil freezes with ice. The coil therefore has to be warmed periodically in order to remove the ice, this is achieved by running the refrigeration circuit in reverse for a brief period of time. This cycle is referred to as the defrost cycle and therefore unlike a condensing unit (i.e. cooling only unit)heat pump units form water externally. Therefore, consideration has to be given to the removal of the water formed, by the provision of an external tray or other device.

Degeneration The process whereby a part of the output signal of an amplifying device is returned to its input circuit in such a manner that it tends to cancel part of the input.

Degenerative feedback Also called negative feedback. A portion of the output of an amplifier is inverted and connected back to the input. This controls the gain of the amplifier and reduces distortion and noise.

Degradation A term used to describe the deteriorative processes that occur with polymeric materials, including swelling, dissolution, and chain scission.

Degree The increments in a temperature scale, or the increments of rotation of a dial. The location of a reference point in electric or phase in a cycle, in mechanical or electrical cyclic scales. (One cycle is equal to 360 degrees).

Degree day(s) The concept of degree days or **accumulated temperature difference** allows the requirement for heating in a building to be assessed.

Degree of freedom The number of axes about which a gyro is free to precess.

Dehumidification The process of reducing the moisture content of the air; serves to increase the cooling power of the air and can contribute to occupant comfort.

Deionization potential The potential at which ionization of the gas within a gas-filled tube ceases and conduction stops; also referred to as extinction potential.

Deionization time In a spark gap, the time required for ionized gas to return to its neutral state after the spark is removed.

Deka (da) Decimal multiple prefix corresponding to ten or 10. This is not a preferred suffix.

Delay A signal processing device or circuit used to delay one or more of the output signals by a controllable amount. This feature is used to correct for loudspeaker drivers that are mounted such that their points of apparent sound origin (not necessarily their voice coils) are not physically aligned. Good delay circuits are frequency independent, meaning the specified delay is equal for all audio frequencies (constant group delay). Delay circuits based on digital sampling techniques are inherently frequency independent and thus preferred.

Delay angle or control angle The control angle for rectification (also known as the ignition angle) is the angle by which firing is delayed beyond the natural take over for the next thyristor.

Delay line A transmission line, or equivalent device, used to delay a signal.

Delay time The time for collector current to reach 10% of its maximum value in a BJT switching circuit.

Delimiter 1. A character used to indicate the beginning and end of a character string, i.e., a symbol stream, such as words, groups of words, or frames.

2. A flag that separates and organizes items of data.

Delocalized (electrons) Electrons that are no longer bound to a given atomic nucleus and are highly mobile.

Delta A three-phase connection in which windings are connected end-to-end, forming a closed loop that resembles the Greek letter delta. A separate phase wire is then connected to each of the three junctions.

Delta connection A method of connecting three elements of a three-phase electrical system in a closed triangle or delta, and with the three phases being taken from the corners of the triangle.

Demag current Current at which the motor magnets will begin to permanently demagnetize; usually equal to the peak current.

Demagnetisation The process of removing the magnetic properties from a material.

Demand The rate at which electric energy is delivered to or by a system, part of a system, or a piece of equipment. It is expressed in kilowatts, kilovoltamperes or other suitable unit at a given instant or averaged over any designated period of time. The primary source of "Demand" is the power-consuming equipment of the customers.

Demand charge The sum to be paid by a large electricity consumer for its peak usage level.

Demand control (also 'Demand Limiting', 'Load Limiting', 'Load Control' An energy management technique used to monitor a facility's energy use in order to limit the peak demand by automatically shutting down selected equipment, on a priority basis,for short periods of time. Demand limits are pre-programmed into the demand control softwarefor this purpose. Demand control is most often applied to electrical usage, and sometimessteam plant. Unlike most other BEMS techniques, monetary benefits are not a direct result of energy savings since electricity usage is often merely postponed, not eliminated. Thebenefits are reduced demand charges to the customer and alleviated peak demand for utilities.

Demand controlled ventilation (DCV) A ventilation strategy where the airflow rate is governed by a chosen pollutant concentration level. This level is measured by air quality sensors located within the room or zone. When the pollutant concentration level rises above a preset level, the sensors activate the ventilation system. As the occupants leave the room the pollutant concentration levels are reduced and ventilation is also reduced. Common pollutants are usually occupant dependent, such as, carbon dioxide, humidity or temperature.

Demand factor The ratio of the maximum demand of a system, or part of a system, to the total connected load of a system or the part of the system under consideration.

Demodulation The removal of intelligence from a transmission medium. The recovery, from a modulated carrier, of a signal having substantially the same characteristics as the original modulating signal.

Demodulator A circuit used in servo-systems to convert an ac signal to a dc signal. The magnitude of the dc output is determined by the magnitude of the ac input signal, and its polarity is determined by whether the ac input signal is in or out of phase with the ac reference voltage.

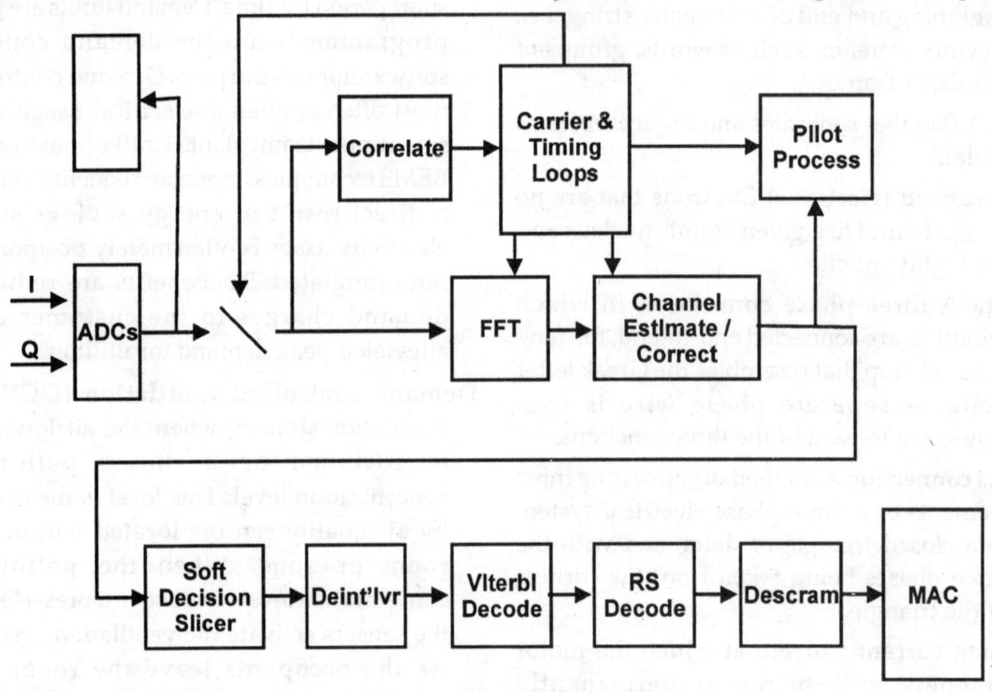

Fig. Demodulator

Demon dialer A program which repeatedly calls the same telephone number. This is benign and legitimate for access to a BBS or malicious when used as a denial of service attack.

Demorgan's theorem A theorem which states that the inversion of a series of AND applications is equal to the same series of inverted OR applications, or the inversion of a series of OR applications is equal to the same series of inverted AND applications.

Denaturation The breaking down of the three-dimensional structure of a protein resulting in the loss of its function.

Denial of service Action(s) which prevent any part of an AIS from functioning in accordance with its intended purpose.

Density Mass per unit volume of a substance usually expressed in lbs/ft3 or grams/cm3.

Deoxyribonucleic acid (DNA) A huge nucleotide polymer having a double helical structure with complementary bases on two strands. Its major functions are protein synthesis and the storage and transport of genetic information.

Dependable capacity The system's ability to carry the electric power for the time interval and period specified.

Depletion mode In a FET, an operating mode where reverse gate-source voltage is used to deplete the channel of free carriers. This reduces the size of the channel and increases its resistance.

Depletion mode MOSFET A MOSFET designed to operate in either depletion mode or enhancement mode.

Depletion region The region in a semiconductor where essentially all free electrons and holes have been swept out by the electrostatic field which exists there.

Depot level maintenance (SM&R CODE D) Supports SM&R Code I and SM&R Code O activities through extensive shop facilities and equipment and highly skilled personnel.

Depreciation, straight line Straight-line depreciation takes the cost of the asset less the estimated salvage value and allocates the cost in equal amounts over the asset's estimated useful life.

Depressurization A measurement technique used to evaluate the airtightness of a building or component. The air inside the room or building is extracted by the use of a fan, creating a lower pressure inside, than outside the room or building.

Depth of discharge The amount of energy withdrawn from a battery or cell expressed as a percentage of its rated capacity.

Derating Reduction in rated power output as a function of elevated ambient temperature and altitude.

Derating factor A value that tells how much to reduce the power rating of a device for each degree above the reference temperature.

Deregulation The elimination of regulation from a previously regulated industry or sector of an industry.

Derf The act of exploiting a terminal which someone else has absent mindedly left logged on.

Derivative action The action of a controller in which the output signal is proportional to the direction and the rate of change of deviation of the input signal. This means that the derivative action term 'looks' to the future, by examining the rate of change of the error. It is the controller's ' accelerator' and 'brake', and is used in addition to proportional action, and possibly integral action, to improve on a controller's response to sudden or very large load changes. It cannot be used by itself since it does not respond to a constant error.

Derivative action time (DAT) In a controller having proportional + integral action, the time interval in which thepart of the signal due to proportional action increases by an amount equal to the part of the output signal due to derivative action, when the derivative action is changing at a constant rate.

Derivative The derivative function senses the rate of rise or fall of the system temperature and automatically adjusts the cycle time of the controller to minimize overshoot or undershoot.

Derivative Kick Electronic 'noise' can cause sudden changes in the sensor signal input to the controller resulting in sudden error changes in the controller output. This is especially true in the derivative part of a P+I+D controller. where this sudden rate of change causes the derivative action term to change dramatically. A change in set-point can also produce derivative kick and, to a lesser extent, proportional kick.

Derived units Units of physical measurement, other than the fundamental units, but derived from these.

Design To plan and delineate with an end in mind and subject to constraints.

Design current (of a circuit) The magnitude of the current (rms value for a.c.) To be carried by the circuit in normal service.

Designated agent An agent that acts on behalf of a transmission provider, consumer or transmission consumer as required under the tariff.

Destination The equipment connected to the output of a routing switcher, crosspoint switch or switching array. Used when defining the size of a switching array, the user must specify how many sources and destination there are in the system.

Destructive Interference (phase cancellation) A phenomenon that occurs when speakers are 180 degrees out of phase, i.e., what one speaker is trying to produce, the other speaker is fighting to cancel. One speaker's wave is in the positive phase (rarefaction), while the other speaker's wave is in the negative phase (compression).

Detachable Face Occasionally referred to as 'Removable Panel' or 'Theft Deterrent

Faceplate', or some variation. This is a physical method for foiling receiver thieves.

Detection The separation of low-frequency (audio) intelligence from the high-frequency carrier.

Detector An instrument to detect the unbalance in a bridge circuit.

Detent torque Torque that is present in a non-energized motor.

Deutsche industrial norm (DIN) A set of technical, scientific and dimensional standards developed in Germany. Many DIN standards have worldwide recognition.

Deviation The difference between the value of the controlled variable and the value at which it is being controlled.

Deviation signal The difference between the setpoint and the measured value.

Device A unit of an electrical system that is intended to carry but not utilize electric energy.

Devitrification The process in which a glass (noncrystalline or vitreous solid) transforms to a crystalline solid.

Diac A two terminal bidirectional thyristor. Has a symmetrical switching mode.

Dialectric constant Peoperty of a material that determines how much electrostatic energy can be stored per unit volume when unit voltage is applied.

Dialectric strength The maximum voltage an insulating material can withstand without breaking down.

Diamagnetism A weak form of induced or nonpermanent magnetism for which the magnetic susceptibility is negative. A type of magnetism associated with paired electrons, that causes a substance to be repelled from the inducing magnetic field.

Diaphragm This term describes the sound-producing element in a tweeter, or Horn. This is the surface that produces the sound you actually hear. The motor that drives it can be any of several technologies including Piezo, conventional dynamic, or ribbon types. Diaphragms do not produce low and low midrange frequencies well, so they are not usually found in that application.

Dibit A group of two bits. Note: The four possible states for a dibit are 00, 01, 10, and 11.

Die bonding Process of mounting a chip to a package.

Die Cast (basket) A type of speaker basket or frame that is cast as a single piece of relatively thick, rigid metal. This contrasts with a Stamped frame that is shaped by pressure, much like a car body fender. Cast metal is heavier and more rigid, and thus less likely to "ring" at certain frequencies, and will hold its shape somewhat longer against the pull of gravity. This is mainly advantageous in the larger woofers of 12" or greater. Smaller drivers will likely not benefit perceptibly from being cast.

Dielectric A substance in which an electric field may be maintained with zero or near-zero power dissipation, i.e., the electrical conductivity is zero or near zero. An insulator; a term applied to the insulating material between the plates of a capacitor.

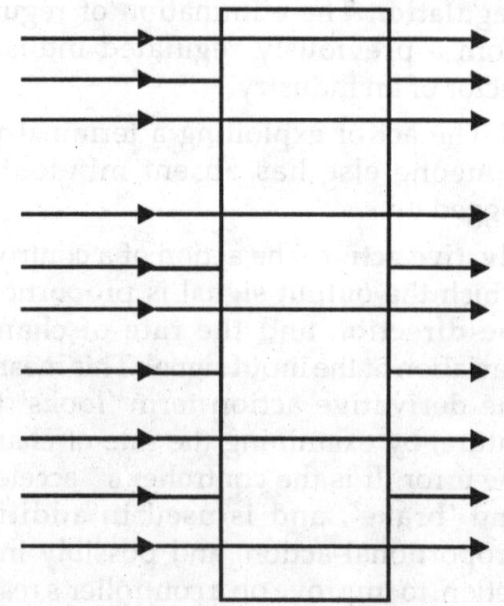

Fig. Dielectric

Dielectric (breakdown) strength The magnitude of an electric field necessary to cause significant current passage through a dielectric material.

Dielectric adsorption Also called dielectric hysteresis or dielectric soak. A characteristic of dielectrics which determines the length of time a capacitor takes to deliver the total amount of its stored energy. It manifests itself as the reappearance of a potential on the electrodes after the capacitor has been discharged. Its magnitude depends on the charge and discharge time of the capacitor.

Adsorption mostly occurs in imperfect dielectrics, primarily solid dielectrics, such as glass, wax, Bakelite, etc. Positive and negative charges are separated and then accumulated at certain regions within the volume of the dielectric, usually manifesting itself as a gradually decreasing current with time after the application of a fixed direct voltage. Its magnitude depends on the nature, composition and purity of the material. Moisture present in the dielectric reduces the insulation resistance of the dielectric and thus tends to suppress absorption, but it increases the leakage current.

Adsorption occurs from the penetration of the electrical charge into the mass of the dielectric for an appreciable time, following the almost instantaneous charge when a capacitor is connected to a continuous voltage through a very low resistance. The current caused by the absorption is steadily decreasing and when the dielectric is saturated with the charge, only leakage current is flowing in the circuit.

The leakage current is steady and ordinarily of an exceedingly small value. Upon short-circuiting, the free charge is released almost instantaneously, but if the capacitor exhibits absorption and is left open-circuited for a while, it may again be discharged (once or several times) because the residual charge slowly seeps out of the dielectric. The residual charges are of a smaller magnitude than the free charge.

Adsorption effects are accompanied by heat dissipation (only part of the absorbed charge is recoverable) and constitute the main portion of losses in solid dielectrics observed when these are subjected to alternating (or fluctuating) potentials.

The apparent capacitance of a capacitor may be strongly affected by absorption effects. A capacitor measured at low frequencies may appear to have a much greater capacitance than at high frequencies.

Dielectric constant (å) The ratio of the permittivity of a medium to that of a vacuum. Also called the relative dielectric constant or relative permittivity.

Dielectric displacement The magnitude of charge per unit area of capacitor plate.

Dielectric field The space between and around charged bodies in which their influence is felt. Also called electric field of force or an electrostatic field.

Dielectric heating A form of heating in which electrically insulating material is heated by being subjected to an alternating electric field. Results from energy being lost by the field to electrons within the atoms and molecules of the material.

Dielectric hysteresis loss Power loss of a capacitor because of the changes in orientation of electron orbits in the dielectric; the changes in orientation are caused by rapid reversal in polarity of line voltage. The higher the frequency, the greater the loss.

Dielectric leakage Power loss of a capacitor because of the leakage of current through the dielectric. Also relates to leakage resistance; the higher the leakage resistance, the lower the dielectric leakage.

Dielectric losses The losses resulting from the heating effect on the dielectric material between conductors.

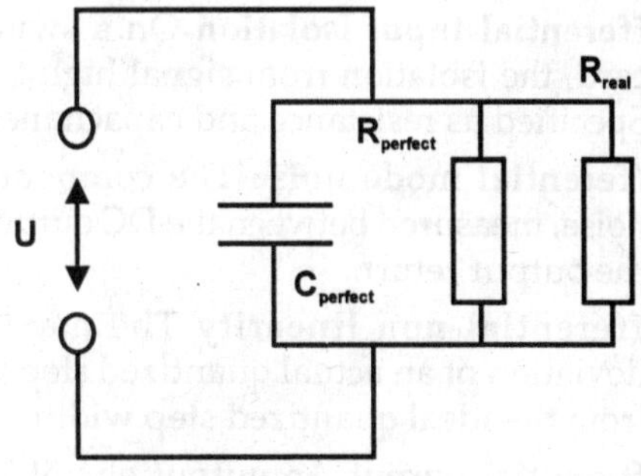

Fig. Dielectric losses

Dielectric strength The ability of an insulator to withstand a potential difference without breaking down (usually expressed in terms of voltage).

Diesel oil The oil left after petrol and kerosene have been distilled from crude petroleum. Used as a fuel in diesel engines.

Difference amplifier A device that amplifies the difference between two inputs. It rejects any signals common to the two input.

Difference of potential A voltage between two points.

Differential For an on/off controller, it refers to the temperature difference between the temperature at which the controller turns heat off and the temperature at which the heat is turned back on. It is expressed in degrees.

Differential amplifier An amplifier in which the output is in proportion to the differences between voltages applied to its two inputs.

Differential gain Unwanted variations in a video signal's chrominance subcarrier's amplitude that result from changes in the signal's DC level, usually specified between 10% and 90% of full scale. Expressed in a percentage, or a fraction of a percentage.

Differential input An input circuit that actively responds to the difference between two terminals rather than the difference between one terminal and ground. Often associated with balanced input circuitry, but also may be used with an unbalanced source. The opposite is the single-ended or unbalanced input.

Differential input isolation On a switching card, the isolation from signal high to low. Specified as resistance and capacitance.

Differential mode noise The component of noise, measured between the DC output and the output return.

Differential non linearity The maximum deviation of an actual quantized step width from the ideal quantized step width.

Differential output An output circuit where the output voltage appears between two active output terminals rather than between one terminal and ground. Normally associated with balanced circuitry.

Differential phase Unwanted variations in a subcarrier's phase as a result of changes in the chrominance signal's DC level, usually specified between 10% and 90% of full scale. Expressed in degrees, or fractions or a degree.

Differential pressure The difference in static pressure between two identical pressure taps at the same elevation located in two different locations in a primary device.

Differentiator An op amp whose output is proportional to the rate of change of the input signal.

Diffraction The bending of waves (as light or RF) when the waves are met with some form of obstruction.

Diffuse reflection Diffusion by reflection in which, on the macroscopic scale, there is no regular reflection.

Diffuse transmission Transmission in which, on the macroscopic scale, there is no regular transmission.

Diffused lighting Lighting in which the light on the working plane or on an object is not incident predominantly from a particular direction.

Diffuser A commercial device like a lens or grillwork that diffuses, or scatters sound

Diffusers and grilles Components of the ventilation system that distribute and diffuse air to promote air circulation in the occupied space. Diffusers supply air and grilles return air.

Diffusion 1. The scattering of reflected light waves from an object, such as white paper. 2. Controlled application of impurity atoms to a semiconductor substrate.

Diffusion coefficient The constant of proportionality between the diffusion flux and the concentration gradient in Fick's first law. Its magnitude is indicative of the rate of atomic diffusion.

Digit A measure of the display span of a panel meter. By convention, a full digit can assume any value from 0 through 9, a 1/2-digit will display a 1 and overload at 2, a 3/4-digit will display digits up to 3 and overload at 4, etc. For example, a meter with a display span of ±3999 counts is said to be a 3-3/4 digit meter.

Digital Relating to devices or circuits that have outputs of only two discrete levels. Examples: 0 or 1, high or low, on or off, true or false etc.

Digital computer 1. A computer in which discrete representation of data is used.

2. A computer that operates on discrete data by performing arithmetic and logic processes on these data.

Digital electronics The branch of electronics dealing with information in binary form.

Digital I/O. Abbreviation for digital input/output.

Digital lines/Ports/Bits/Channels. In hardware, a digital line is physical hardware connection to a pin with a digital signal. A digital port is a physical grouping of digital lines. In software, a digital bit (1 or 0) is a logical representation of a digital line. A digital channel is a logical grouping of digital bits.

Digital meter Show a discrete reading, in the form of a decimal number, for a given input quantity.

Digital output An output signal which represents the size of an input in the form of a series of discrete quantities.

Digital signal A discrete-time signal with quantized amplitude. If the discrete-time signal can assume a continuous range of values, then it is called a 'sampled-data signal'.

Digital subscriber line (DSL) In Integrated Services Digital Networks (ISDN), equipment that provides full-duplex service on a single twisted metallic pair at a rate sufficient to support ISDN basic access and additional framing, timing recovery, and operational functions.

Digital Technology Human beings directly process information in an analogue format. That is, our senses directly convey information to the brain in the form of constantly varying amplitudes and frequencies that approximate the impact of outside stimuli upon the receptors (eyes and ears, etc.) of the organs involved.

Digital technology is a method of translating this same information in the form of packets (bytes and bits) of numerically equivalent values. This enables input transducers (microphones, cameras, chemical sensors, etc.) to produce differential waveforms that can be reduced, down to the smallest differences, to a number. This number is transmitted to a receiver with a digital to analogue converter that allows the original analogue information to be reconstructed for the benefit of our human brains. This is of course an egregiously simplified explanation of the most significant technology of the present time. We invite you to explore more this topic deeply, not just for audio/video, but to understand the new world we have entered.

Digital trigger An event that occurs at a user-selected point on a digital input signal. The polarity and sensitivity of the digital trigger can often be programmed.

Digitize To convert an analogue signal into a digital signal carrying equivalent information.

Dilution ventilation dilution of contaminated air with uncontaminated air in a general area, room, or building for the purpose of health hazard or nuisance control.

Dimensionless number Dimensionless grouping of important coefficients for the current process.

Dimensions of unit The dimensions of a physical quantity are the powers to which the fundamental units expressing that quantity are raised.

Dimer A molecule formed by the joining of two identical monomers.

Diode An electron tube containing two electrodes: a cathode and a plate.

2. A two element, solid-state device made of either germanium or silicon; it is primarily used as a switching device.

Diode detector A demodulator that uses one or more diodes to provide a rectified output with an average value that is proportional to the original modulation.

Diode drop The forward voltage developed across a diode when it is operating.

DIP Abbreviation for "dual in line package." .

Dip or through hole mounting A power supply that has pin terminals that are used for both connection and soldering. Holes in the PCB are required.

Dipole A common type of half-wave antenna made from a straight piece of wire cut in half. Each half operates at a quarter wavelength of the output.

Direct contact Contact of persons or livestock with live parts.

Direct coupled amplifier (DC amplifier) An amplifier in which the output of one stage is coupled to the input of the next without the use of a capacitor. This type of amplifier will amplify Direct Currents and low frequency waveforms.

Direct coupling Where the output of an amplifier is connected directly to the input of another amplifier or to a load. Also known as DC coupling because DC signals are not blocked.

Direct current The type of current normally supplied by batteries. It is called Direct Current (DC) because the flow of electrons (electricity) is always in the same directions; from the negative (-) pole of the power source to the positive (+) pole.

Direct digital control (DDC) A control loop in which a microprocessor-based controller directly controls equipment based on sensor inputs and setpoint parameters, i.e. the plant is under the direct control of software (either in an outstation or central station) and not through the intermediary of some non-programmable controller. The programmed control sequence determines the output to the equipment.

Direct energy conversion Production of electricity from an energy source without transferring the energy to a working fluid or steam. For example, photovoltaic cells transform light directly into electricity. Direct conversion systems have no moving parts and usually produce direct current.

Direct expansion equipment 'Direct expansion,' 'DX,' 'refrigeration' or 'split' units are all generic terms used to identify the same equipment. It is accepted that the terms refer to two or more units, one usually positioned externally and one or more usually positioned internally. The units are connected together by site installed refrigeration pipework which is charged with a refrigerant. The external unit may take one of three forms:

1. The heat pump - which consists of a fan, compressor, coil and reversing valve, and rejects unwanted heat to atmosphere during the cooling cycle and extracts heat from the atmosphere during the heating cycle.

2. The condensing unit - which is as described above but does not have a reversing valve and therefore cools only.

3. The condenser which consists of a fan and coil (as the compressor is contained in the indoor unit); the condenser is used less often than (1) and (2).

The indoor units consist of fan coil units or air handling units which may be located in the atmosphere being air conditioned or remotely in a plantroom. Some manufacturers produce 'external' units that may be located internally and in the case of these units ductwork is usually connected to atmosphere to reject heat or extract heat. DX systems are in direct contrast to hydraulic systems or chilled water systems. With these systems cooling is achieved by circulating chilled water with a hydraulic pump

Direct lighting Lighting by means of luminaires with a light distribution such that 90 to 100 per cent of the emitted luminous flux reaches the working plane direct, assuming that this plane is unbounded.

Direct sound Sound that arrives at the listeners ear first. Sound reaching the listening location without reflections, i.e., sound that travels in the most direct path from the source to the listener.

Directional antenna An antenna that radiates most effectively in only one direction.

Directional coupler A device that samples the energy travelling in a waveguide in one direction only.

Directivity The ability of an antenna to radiate or receive more energy in some directions than in others. The degree of sharpness of the antenna beam.

Directly heated cathode A wire, or filament, designed to emit the electrons that flow from cathode to plate. The filament is designed so that a current is passed through it; the current heats the filament to the point where electrons are emitted.

Director The parasitic element of an array that reinforces energy coming from the driver element.

Directory A list of files or other directories on a computer at an Internet site.

Discharge Electrical discharge can occur by the release of the electric charge stored in a capacitor through an external circuit. It can also occur by the breakdown of gaseous dielectrics within solid dielectrics on the application of a field.

Discharge coefficient A dimensionless coefficient relating the mean flow rate through an opening to an area and the corresponding pressure difference across the opening.

Discharge current The surge current that flows through the surge diverter during spark over or operation.

Discharge lamp Lamp in which the light is produced, directly or indirectly, by an electric discharge through a gas, a metal vapor, or a mixture of several gases and vapors.

Discharge rate The rate , usually expressed in amperes, at which electrical current is taken from a cell or battery.

Discharge time constant The time required for the output-voltage from a sensor or system to discharge 37% of its original value in response to a zero rise time step function input. This parameter determines a low frequency response.

Disconnecting means A device or group of devices, or other means whereby all the ungrounded conductors of a circuit can be disconnected simultaneously from their source of supply.

Disconnector A mechanical switching device which, in the open position, complies with the requirements specified for isolation. A disconnector is otherwise known as an isolator.

Discontinuous control The controller produces a maximum or minimum output signal at a upper and lower pre-set limits of the measured variable, in order to maintain this measured variable between the limits (differential). Although these may be an optimum desired value, this is never maintained. The measured value is always increasing or decreasing. There are basically two types of discontinuous control:- 1. Step control;

2. Float control. Discontinuous control is an inexpensive and relatively simple form of control. However, when the desired value has to be constantly maintained, some form of continuous control must be used.

Discount/interest Rate The discount rate is used to determine the present value of future or past cash flows. The rate accounts for inflation and the potential earning power of money.

Discrete A term used for circuit made up of individual components.

Discrete component Package containing only a single component as opposed to an integrated circuit containg many components in a single package.

Discrete device A class of electronic components that contain one active element, such as a transistor or diode. However, hybrids, optoelectronic devices, and intelligent discretes may contain more than one active element.

Discrete Output Devices A separate active unit in an amplifier, capable of performing a single essential function within the output circuit. There are three basic types of output devices found on car audio amplifiers – integrated circuits, bipolar transistors, or MOSFETs. Integrated Circuits (or IC)are found only on relatively low-wattage (20 watts RMS per channel or less) amplifiers and receivers. An IC incorporates many functional devices and thus is not considered a discrete output device. Most cannot handle more than 25 watts RMS.

Discrimination The ability to discriminate. The characteristics of protective devices must be such that a fault on one circuit will not disconnect another circuit.

Discrimination control A control mode in which sensor signals from a number of sensors are fed into the controller which then decides on which sensor value to use when comparing a sensed value with the desired value. For example, in some multi-zone air-heating systems, the sensed temperature 'selected' is the one indicating the zone with the maximum heating requirements. In this way the heat energy input to the system supply air is kept to a minimum.

Discriminator A circuit in which amplitude variations are derived in response to phase or frequency variations. The part of an FM receiver that extracts the desired signal from an incoming FM wave by changing frequency variations into amplitude variations.

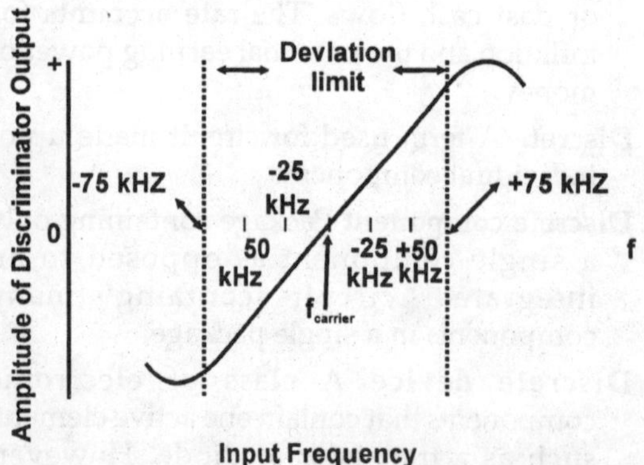

Fig. Discriminator

Disk A memory device which uses a magnetic media for the storage of information. Disk, as a term, has expanded into other areas often used to describe the shape of the storage media, that is: floppy disk, compact disk, laser disk, or hard disk.

Disk operating system (DOS) Program used to control the transfer of information to and from a disk, such as MS DOS.

Dislocation A linear crystalline defect around which there is atomic misalignment. Plastic deformation corresponds to the motion of dislocations in response to an applied shear stress. Edge, screw, and mixed dislocations are possible.

Dispersion The extent to which a sound emitter yields acoustic radiation over a given area. This is a particular concern in tweeters and midranges whose portion of the audio spectrum has a much more directional character than woofers. Many Horn tweeters, while very energetic, have a more limited area of dispersion within which their effect can be fully perceived. Generally, Dome tweeters can be heard over a much wider area, all other factors being equal. But each particular example must be assessed individually for this factor. Some radiator designs are better than others.

Displacement The measured distance traveled by a point from its position at rest. Peak to peak displacement is the total measured movement of a vibrating point between its positive and negative extremes. Measurement units expressed as inches or millinches.

Displacement current The current that flows through a capacitor in response to the rate of voltage change across it: displacement current = C * dV/dt

Displacement flow ventilation With displacement ventilation, air is introduced into the air conditioned space at low level and at low velocity. Displacement air distribution has gained in popularity, mainly due to comfort and cleanliness considerations.

Display The visual representation of a signal on a screen.

Dissipation Release of electrical energy in the form of heat.

Dissipation constant The ratio for a thermistor which relates a change in internal power dissipation to a resultant change of body temperature.

Dissociation constant (K) A value which quantitatively expresses the extent to which a substance dissociates in solution. The smaller the value of K, the less dissociation of the species in solution. This value varies with temperature, ionic strength, and the nature of the solvent.

Distance/velocity lag The dead time between an alteration in the value of a signal and its manifestation unchanged at a later part of the system, arising solely from the finite speed of propagation of the signal. For example, if a flow detector is located at a distance of 10 metres from a mixing valve and the position of the valve is suddenly changed, then if the water velocity is 0.5 m/s, it will take 20 seconds for the new temperature front to arrive at the detector; thus the distance/velocity lag equals 20 seconds. As long as the distance/velocity lag has not elapsed, the controller is neither in a position to counteract the effect of a disturbance nor to correct that of any action it may have initiated.

Distilled water Water that has been purified through a process of evaporation and condensation.

Distortion Any departure from a true and accurate reproduction of the original waveform. It can include Noise, Clipping Distortion, Harmonic, and Intermodulation Distortion. These last two forms are fairly common in loudspeaker reproduction and can be reduced but not entirely eliminated in the existing technology. It would be fair to say that modern amplifier design fairly eliminates nearly all forms of inherent perceived distortion, leaving only that caused by inappropriate user settings and overloading.

Distortion is the name given to anything that alters a pure input signal in any way other than changing its size. The most common forms of distortion are unwanted components or artifacts added to the original signal, including random and hum-related noise. Distortion measures a system's linearity - or nonlinearity. Anything unwanted added to the input signal changes its shape (skews, flattens, spikes, alters symmetry or asymmetry). A spectral analysis of the output shows these unwanted components. If a circuit is perfect, it does not add distortion of any kind. The spectrum of the output shows only the original signal - nothing else - no added components, no added noise - nothing but the original signal.

It's rather amusing to see amplifier manufacturers making great claims about the advantage of the extra .001 % Distortion they've wrung out of their products, while most speakers are considered very good if they can keep such distortions below 5 %. It's true that the reduction of any distortion anywhere is a positive contribution to the goal of high fidelity, but the disparity between the two technologies in this regard points up the largely subjective nature of many such claimed advantages.

Distributed capacitance Any capacitance other than that within a capacitor. For example, the capacitance between adjacent turns of wire in a coil.

Distributed constants. The constants of inductance, capacitance, and resistance in a transmission line. They are spread along the entire length of the line and cannot be distinguished separately.

Distributed generation A distributed generation system involves small amounts of generation located on a distribution system for the purpose of meeting local peak loads and/or displacing the need to build additional local distribution lines.

Distributed inductance Any inductance other than that within an inductor. Example inductance in any conductor.

Distributed power Distributed power or power distribution is a method of achiving a highly effient and dense design. Usually 12V, 24V, or 48V are used for the intermediate bus voltages to distribute power safely. But 300VDC is also used in certain applications.

Distribution Outside a building distribution refers to the process of delivering power from the transmission system to the premises. Inside the building, distribution is the process of using feeders and circuits to provide power to devices.

Distribution board An assembly containing switching or protective devices (e.g. Fuses, circuit breakers, residual current operated devices) associated with one or more outgoing circuits fed from one or more incoming circuits, together with terminals for the neutral and protective circuit conductors. It may also include signalling and other control devices. Means of isolation may be included in the board or may be provided separately.

Distribution line This is a line or system for distributing power from a transmission system to a consumer. It is any line operating at less than 69,000 volt.

Distribution system The portion of an electric system (after the transmission system) that is dedicated to delivering electric energy to an end user.

Distributive law In Boolean algebra the law which states that if a group of terms connected by like operators contains the same variable, the variable may be removed from the terms and associated with them by the appropriate sign of operation (for example, A(B + C) = AB + AC).

Distributor A person who distributes electricity to consumers using electrical lines and equipment that he owns or operates.

Diversity exchange Exchange of capacity or energy between systems that have peak loads occurring at different times.

Diversity factor The ratio of the sum of the non-coincident maximum demands of two or more loads to their coincident maximum demand for the same period. Diversity factor The ratio of the sum of the maximum power demands of the subdivisions, or parts of a system, to the maximum demand of the whole system or of part of the system under consideration.

Diversity Tuner An FM tuning method which employs two antennas. The tuner can switch between the two antennas in order to attain better reception.

DMA (Direct memory access) channels ISA bus PCs offer eight parallel channels for DMA mode data transfers. A number of these are reserved for exclusive use by the computer. The remainder are available for use by user-supplied I/O options, such as plug-in data acquisition cards. Also called DMA levels.

DMM An electronic instrument that measures voltage, current, resistance, or other electrical parameters by converting the analogue signal to digital information and display. The typical five-function DMM measures DC volts, DC amps, AC volts, AC amps, and resistance.

DNA Probes A DNA or nucleic acid probe is a short strand of DNA that locates and binds to its complementary sequence in samples containing single strands of DNA or RNA enabling identification of specific sequences. Nucleic acid probe assays exploit the fundamental hybridization reaction that occurs spontaneously between two complementary DNA:DNA or DNA:RNA strands. As in immunoassays, detection of the hybrid requires that the probe be labeled. Various direct and indirect methods have been devised for the detection of the hybrid. Direct labeling involves attaching the label directly to the probe sequence; indirect labeling binds an antibody to the DNA:DNA or DNA:RNA hybrid. As in immunoassays, non-isotopically-labeled probes are preferred over radio-labeled probes primarily because of radiation hazards, disposal problems, and short reagent shelf life. In addition, the factors determining the detection limits of hybridization assays based on labeled probes are similar to those in immunoassays. Therefore, the development of a simple, inexpensive and sensitive direct detection system which eliminates the use of labels is highly desirable.

DNA Sequencing There are two main classical methods for sequencing DNA: The first method, developed by Allan Maxam and Walter Gilbert, involves chemicals used to cleave the DNA at certain positions, generating a set of fragments that differ by one nucleotide. The second method, developed by Fred Sanger and Alan Coulson, involves enzymatic synthesis of DNA strands that terminate in a modified nucleotide. Analysis of fragments is similar for both methods and involves gel electrophoresis and autoradiography or fluorescence. The enzymatic method has largely replaced the chemical method as the technique of choice, although there are some situations where chemical sequencing can provide data more easily than the enzymatic method.

DNS Spoofing Assuming the DNS name of another system by either corrupting the name service cache of a victim system, or by compromising a domain name server for a valid domain.

Dolby Digital Dolby's name for its format for the digital soundtrack system for motion picture playback. Utilizes their AC-3 method of digital compression. The signal is optically printed between the sprocket holes. Introduced to Home Theater on laser disc and DVD and CD. Dolby Digital may use any number of primary audio delivery and reproduction channels, from 1 to 5, and may include a separate bass-only effects channel. The designation "5.1" describes the complete channel format. Surround decoder systems with Dolby Digital automatically contain Dolby Pro Logic processing to ensure full compatibility with the many existing program soundtracks made with Dolby Surround encoding.

Dolby The most common shared logo or feature between all brands is the Dolby labs' 'double D' symbol. Named after audio engineer, Ray Dolby, it represents a number of sound processing technologies that are incorp-orated in a diverse number of products since the early 1970's.

Domain theory A theory of magnetism based upon the electron-spin principle. Spinning electrons have a magnetic field. If more electrons spin in one direction than another, the atom is magnetized.

Dome Tweeter Tweeters come in several different types; cones, horns and domes being the most popular. Dome types are heavily favored in many standard applications. They are efficient, and have low Distortion and wide dispersion. There is a choice between hard and soft dome materials, but all have relatively low mass and high power handling capabilities.

This same design in a larger form is quite effective as a midrange Driver also, and for many of the same reasons.

Dominant mode The easiest mode to produce in a waveguide, and the most efficient mode in terms of energy transfer.

Donor An impurity that can make a semiconductor material an N-type by donating extra "free" electrons to the conduction band.

Donor atoms Pentavalent atoms that give up electrons to the conduction band in an N type semiconductor material.

Doorknob tube An electron tube that is similar to the acorn tube but larger. The doorknob tube is designed to operate, at high power, in the uhf frequencies.

Dopant An impurity element that is deliberately added to a semiconductor.

Doping The process of adding impurities to semiconductor crystals to increase the number of free charges that can be moved by an external, applied voltage. Doping produces N-type or P-type material.

Doppler effect 1. The apparent change in frequency or pitch when a sound source moves either toward or away from a listener. 2. In radar, the change in frequency of a received signal caused by the relative motion between the radar and the target.

Doppler frequency The difference between transmitted and reflected frequencies; caused by the Doppler effect.

Dot convention Standard used with transformer symbols to indicate whether the secondary voltage is in phase or out of phase with the primary voltage.

Dot notation A notation used to denote similar ends of mutually coupled coils.

Double exponential waveform A waveform obtained by taking the difference of two exponential waveforms. Usually they have time constants which are highly different to give a waveform with fast rise time and relatively slow overall decay.

Double insulation Insulation comprising both basic insulation and supplementary insulation.

Double layer, electric In electrochemistry "electrical double layer" refers to the electrical double layer at the electrode-solution interface. In other words, there is a separation of charges at the interface between two phases, usually a solid and liquid phase where the solid is the electrode and the liquid is the electrolyte.

The charge layers consist of an excess or deficit of electrons on the metal side and of an excess or deficit of ions of the opposite charge on the solution side. In addition to these free charge carriers, there are oriented molecules with dipole moments which also contribute to the overall electrical moment of the layer.

A further contributor is the posibility of a subsidiary dipole layer being formed by ions which are "specifically adsorbed" (i.e. by chemical bonding) to the metal, and whose charge is neutralized (partly or wholly) by ions of opposite sign held by electrostatic attraction nearby in the solution.

Double moding In a transmitter output tube, the abrupt and random change from one frequency to another.

Double negative law In Boolean algebra, the law which states that the complement of a complement is the equivalent of the original term.

Double precision The degree of accuracy that requires two computer words to represent a number. Numbers are stored with 17 digits of accuracy and printed with up to 16 digits.

Double receiver A fine and coarse synchro receiver enclosed in a common housing with a two-shaft output (one shaft inside the other).

Double Stacked (magnet) Two magnets overlaying each other to increase the overall magnetic field. In some designs for home theater applications the polarity is reversed which provides the effect of "shielding," inhibiting the size of the magnetic field radiation, thus allowing the speakers to be used next to a television without creating Distortions on the picture tube. This is achieved without a detrimental effect on the normal power handling capabilities of the speaker.

Doublet Another name for the dipole antenna.

Doubling up This is a type of two-equipment installation where one unit can be substituted for another in the event of failure.

Down link The frequency used to transmit an amplified signal from a satellite or other craft back to earth.

Downdraught An airstream with a significant downward directional component of velocity. Often occurs adjacent to cold surfaces. It may be generated artificially by air curtains, air doors etc.

Download The process of obtaining a data file or program from a distant location by means of a communications channel, such as a telephone line.

Downtime The time to locate a fault and then repair it.

Draught Excessive air movement in an occupied enclosure causing discomfort.

Draughtproofing The action of filling the gaps around doors and windows, in order to prevent outside cold air leaking into the building, causing draughts. **Dress** The arrangement of signal leads and wiring for optimum circuit operation, cosmetic appeal, and protective routing.

Drift Gradual departure of the instrument output from the calibrated value. An undesired slow change of the output signal.

Drift space In an electron, a region free of external fields in which relative electron position depends on velocity.

Drift velocity The average velocity of a carrier that is moving under the influence of an electric field in a conductor, semiconductor, or electron tube.

Drip proof guarded A drip-proof machine with ventilating openings guarded (with screens) as in a guarded motor.

Drip proof motor An open motor in which the ventilating openings are so constructed that drops of liquid or solid particles falling on it, at any angle not greater than 15 degrees from the vertical, cannot enter either directly or by striking and running along a horizontal or inwardly inclined surface.

Drive Amplifier that converts step and direction input to motor currents and voltages.

Driven array An array in which all of the elements are driven.

Driven element The element of an antenna connected directly to the transmission line.

Driver An alternate term for: speaker, transducer, or radiator. Properly speaking, the term speaker should refer to an entire sound producing system with whatever combination of woofer, midrange and tweeter; in whatever enclosure type it is housed.

Droop A common occurrence in time-proportional controllers. It refers to the difference in temperature between the set point and where the system temperature actually stabilizes due to the time-proportioning action of the controller.

Dropping resistor Resistor whose value has been chosen to drop or develop a given voltage.

Drum type armature An efficient, popular type of armature designed so that the entire length of the winding is cutting the field at all times. Most wound armatures are of this type.

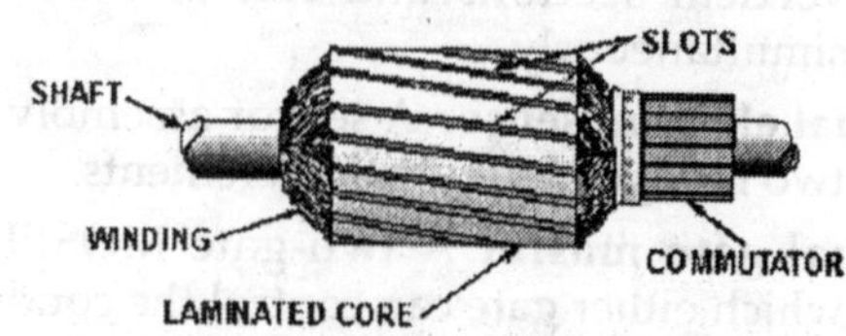

Fig. Drum type armature

Dry air system Provides dehumidified air for electronic equipment that is moisture critical.

Dry bulb temperature The temperature indicated by a dry temperature sensing element (such as the bulb of mercury in a glass thermometer) shielded from the effects of radiation.

Dry cell An electrical cell in which the electrolyte is not a liquid. In most dry cells the electrolyte is in the form of a paste.

Dry circuit switching Switching below specified levels of voltage and current to minimize any physical and electrical changes in the contact junction.

Dry location A location not normally subject to dampness or wetness. A location classified as dry may be temporarily subject to dampness or wetness, as in the case of a building under construction.

Dry reed relay A glass enclosed, hermetically sealed, magnetically actuated contact. No mercury or other wetting material is used. Typical atmosphere inside the glass enclosure is nitrogen.

Dual Channel or Dual Voice Coil (speaker) A woofer with two voice coils mounted to a common cone, which can be connected to separate amplifiers, to produce a common bass output. Since bass has a non-directional character, this still permits the optimum reproduction of the stereo image via other speakers. Care should be taken in making connection, to observe proper polarities, however. Failure to do so can result in the quick extinction of the Driver if the amplifiers are pulling the cone in different directions at once.

Dual channel oscilloscope An oscilloscope that has two independent input connectors and vertical sections and can display them simultaneously.

Dual element sensor A sensor assembly with two independent sensing elements.

Dual gate mosfet A two-gate MOSFET in which either gate can control the conductor independently, a fact which makes this MOSFET very versatile.

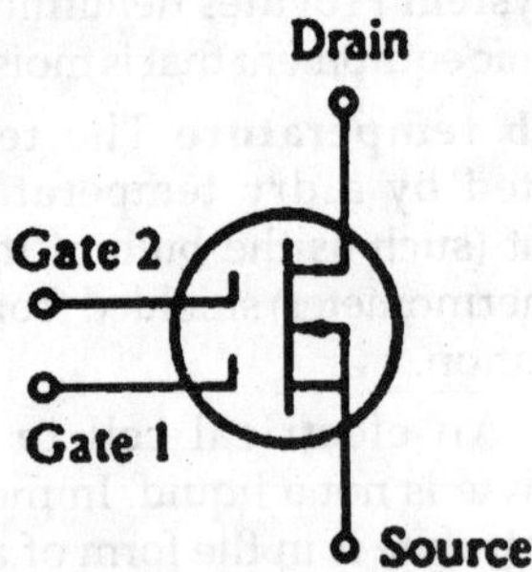

Fig. Dual gate mosfet

Dual in line package (DIP) IC package having two parallel rows of performed leads.

Dual slope A/D converter An analogue-to-digital converter which integrates the signal for a specific time, then counts time intervals for a reference voltage to bring the integrated signal back to zero. Such converters provide high resolution at low cost, excellent normal-mode noise rejection, and minimal dependence on circuit elements.

Dual torque Dual speed motor whose torque varies with speed (as the speed changes the horsepower remains constant).

Dual trace oscilloscope Oscilloscope that can simultaneously display two signals.

Dual voltage Some motors can operate on two different voltages, depending upon how it is built and connected. The voltages are either multiples of two or the 3 of one another.

Duality principle The duality principle establishes an analoguey between similar variables when governed by analogueous differential equations. For example, series circuits and parallel circuits are analogueous if resistance is replaced by conductance, conductance by resistance, inductance by capacitance or vice versa, current by voltage or vice versa. There is also an analoguey between electrical and magnetic circuits.

Dub A copy or the process of making a copy of a recording on another storage device.

Duct A closed passageway formed underground or in a structure and intended to receive one or more cables which may be drawn in.

Ductile 1. The propagation of radio waves within an atmospheric duct.

2. Easily drawn out (as to form filaments or wires).

Ductility A measure of a material's ability to undergo plastic deformation before fracture; expressed as percent elongation (%EL) or percent area reduction (%AR) from a tensile test.

Ducting Trapping of an RF wave between two layers of the earth's atmosphere or between an atmospheric layer and the earth.

Dummy load A dissipative but non-radiating device that has the impedance characteristics of an antenna or transmission line.

Duplex Pertaining to simultaneous two-way independent data communication transmission in both direction. Same as "full duplex".

Duplex circuit A circuit that permits transmission in both directions.

Duplex transmission Simultaneous independent transfer of data in two directions; cf. Half-duplex.

Duplex wire A pair of wires insulated from each other and with an outer jacket of insulation around the inner insulated pair.

Duplexer A radar device that switches the antenna from the transmitter to the receiver and vice versa.

Duration adjust type signal (DAT signal) This scheme is used to modulate an intermediate device by sending it a train of on/off signals where the on to off ratio varies, according to proportional duty cycle. A common example is DAT control of a thyrsitor which in turn varies the amount of energy supplied to an electric heating element; cf. Position Adjust Type Signal.

Duration The time interval between the first and last instants at which the instantaneous amplitude reaches a stated fraction of the peak pulse amplitude.

Dust Cover/Cap A rigid cardboard or plastic dome placed over the opening to the voice coil cylinder in a dynamic cone driver. The main purpose is to prevent dust from falling into the voice coil-magnet gap and causing problems. In some units it may have a slight ancillary function in extending - slightly - the midrange/ high response. Occasionally the dome may be inverted in such a manner as to extend the inner surface of the woofer and provide a very slight improvement in the smoothness of low frequency to midrange roll-off.

Dustproof Constructed or protected so that dust will not interfere with its successful operation.

Duty cycle The relationship between the operating and rest times or repeatable operation at different loads. A motor which can continue to operate within the temperature limits of its insulation system, after it has reached normal operating (equilibrium) temperature is considered to have a continuous duty (CONT.) rating. One which never reaches equilibrium temperature, but is permitted to cool down between operations is operating under intermittent duty (INT.) conditions such as a crane and hoist motor which are often rated 15 or 30 min. duty.

Duty cycling A control method which alternates or cycles the sequence of plant.

Duty ratio The ratio of pulse width to repetition period. Also known as Duty Cycle.

Dynamic Relating to conditions that are changing or in motion.

Dynamic (Two Plane) Balancing Machine A dynamic balancing machine is a centrifugal balancing machine that furnishes information for performing two-plane balancing.

Dynamic braking Braking that can be enacted while the motor is in motion.

Dynamic calibration Calibration in which the input varies over a specific length of time and the output is recorded vs. Time.

Dynamic characteristics A description of an instrument's behavior between the time a measured quantity changes value and the time the instrument obtains a steady response.

Dynamic data exchange (DDE) A Microsoft Windows standard mechanism for communication between programs. It allows your application to send and share data with other applications such as spreadsheets.

Dynamic error The error that occurs when the output does not precisely follow the transient response of the measured quantity.

Dynamic link library (DLL) A software module in Microsoft Windows containing executable code and data that can be called or used by Windows applications or other DLLs. DLL functions and data are loaded and linked at run time when they are referenced by a Windows application or other DLLs.

Dynamic load A load that rapidly changes from one level to another. To be properly specified, both the total change and the rate of change must be stated.

Dynamic microphone A device in which sound waves move a coil of fine wire that is mounted on the back of a diaphragm and located in the magnetic field of a permanent magnet.

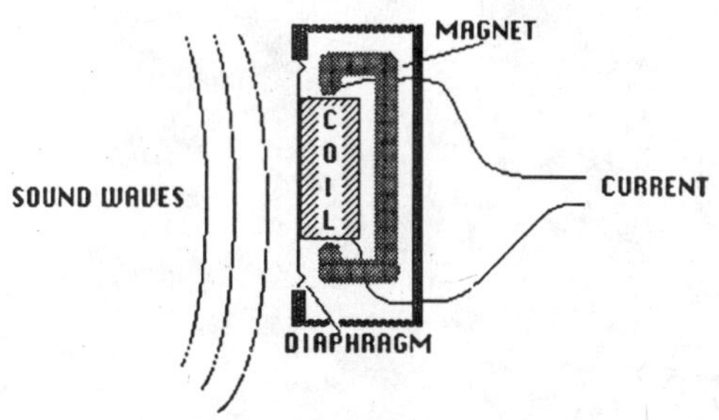

Fig. Dynamic microphone

Dynamic pressure The difference in pressure levels from static pressure to stagnation pressure caused by an increase in velocity. Dynamic pressure increases by the square of the velocity.

Dynamic RAM (DRAM)A type of semiconductor memory that stores data as capacitor charges that need to be refreshed periodically.

Dynamic range The ratio of the loudest (undistorted) signal to that of the quietest (discernible) signal in a unit or system as expressed in decibels (dB). Dynamic range is another way of stating the maximum S/N ratio. With reference to signal processing equipment, the maximum output signal is restricted by the size of the power supplies, i.e., it cannot swing more voltage than is available. While the minimum output signal is determined by the noise floor of the unit, i.e., it cannot put out a discernible signal smaller than the noise. Professional-grade analogue signal processing equipment can output maximum levels of +26 dBu, with the best noise floors being down around -94 dBu. This gives a maximum dynamic range of 120 dB - pretty impressive numbers, which coincide nicely with the 120 dB dynamic range of normal human hearing (from just audible to uncomfortably loud).

Dynamic Range Suppression A signal compression technique which raises the level of lower passages without affecting overall volume. Especially useful with high noise levels, such as a moving vehicle.

Dynamic unbalance Dynamic unbalance is that condition in which the central principal axis is not coincident with the shaft axis.

Dynamo Device for converting mechanical energy into electrical energy. The mechanical energy of rotation is converted into electrical energy in the form of a current in the armature.

Dynamometer A device which loads the motor to measure output torque and speed accurately by providing a calibrated dynamic load. Helpful in testing motors for nameplate information and an effective device in measuring efficiency.

Dynamometer instrument This instrument is also a moving coil instrument except that in this case, the permanent magnet is replaced by a pair of fixed coils to give the fixed field.

Dyne Unit of force in the c.g.s. System. 1 dyne = 10-5 N

E core Laminated form in the shape of the letter "E", onto which inductors and transformers are wound.

E field Electric field that exists when a difference in electrical potential causes a stress in the dielectric between two points.

E transformer A special form of differential transformer employing an E-shaped core. The secondaries of the transformer are wound on the outer legs of the E, and the primary is wound on the center leg. An output voltage is developed across the secondary coils when its armature is displaced from its neutral position. This device is used as an error detector in servo-systems that have limited load movements.

E type T junction A waveguide junction in which the junction arm extends from the main waveguide in the same direction as the E-field in the waveguide.

Earphone (earpiece) A device for tuning electrical energy into sound waves - fits in your ear.

Earth An electrical connection to the earth, which represents 0 volts or 'ground potential' by way of a metal or conductive rod.

Earth electrode A conductor or group of conductors in intimate contact with, and providing an electrical connection to, Earth.

Earth electrode resistance The resistance of an earth electrode to Earth.

Earth fault current A fault current which flows to Earth.

Earth fault loop impedance The impedance of the earth fault current loop starting and ending at the point of earth fault. The earth fault loop comprises the following, starting at the point of fault: the circuit protective conductor, and the consumer's earthing terminal and earthing conductor, and for TN systems, the metallic return path, and for TT and IT systems, the earth return path, and the path through the earthed neutral point of the transformer, and the transformer winding, and the phase conductor from the transformer to the point of fault.

Earth ground A metal rod, usually copper, that provides an electrical path to the earth, to prevent or reduce the risk of electric shock.

Earth leakage circuit breaker (ELCB) The elcb is designed to protect both equipment and users from fault currents between the live and earth conductors by detecting the rise in voltage of the frame earth connection with respect to a reference earth.

Earth leakage Flow of current from a live conductor to earth in an unintended path through the insulation.

Earthed concentric wiring A wiring system in which one or more insulated conductors are completely surrounded throughout their length by a conductor, for example a metallic sheath, which acts as a PEN conductor.

Earthed equipotential zone A zone within which exposed conductive parts and extraneous conductive parts are maintained at substantially the same potential by bonding, such that, under fault conditions, the differences in potential between simultaneously accessible exposed and extraneous conductive parts will not cause electric shock.

Earthing Connection of the exposed conductive p" of an installation to the main earthing terminal of that installation. Earthing conductor. A protective conductor connecting the main earthing terminal of an installation to an earth electrode or to other means of earthing.

EAU Estimated annual usage is the amount of a specific part number a customer requires in a one year period.

Echo Effect produced when sound is reflected from a surface sufficiently far away for the reflected sound to be separately distinguishable.

Echo box A resonant cavity device that is used to check the overall performance of a radar system. It receives a portion of the transmitted pulse and retransmits it back to the receiver as a slowly decaying transient.

ECL Emitter Coupled Logic. Where transistors are held in the turned-on state to increase the speed of the gate.

ECL Logic Abbreviation for Emitter Coupled Logic, a very high speed digital technology.

Eclipse A condition in which the satellite is not in view or in direct line of sight with the sun. This happens when the earth is between them.

Economic dispatch The distribution of total generation requirements among alternative sources for optimum system economy with consideration to both incremental generating costs and incremental transmission losses.

Economic efficiency A term that refers to the optimal production and consumption of goods and services. This generally occurs when prices of products and services reflect their marginal costs. Economic efficiency gains can be achieved through cost reduction, but it is better to think of the concept as actions that promote an increase in overall net value (which includes, but is not limited to, cost reductions).

Economic energy Energy produced and substituted for the traditional but less economical source of energy. Economic energy is usually sold without capacity and is priced at variable costs plus administration costs.

Economiser control An energy management function whichaims to minimise energy consumption by the use of 'free-cooling'. Internal heat generation in a building may require the HVAC to provide cooling, even though the air temperatures are lower than the thermostat setpoint. Under this condition, it is possible to introduce outdoor air into the building to provideall or part of the cooling normally accomplished by refrigeration equipment. To use this'free-cooling', the economiser measures the dry-bulb temperature of the return air and the outdoor air, and selects an appropriate amount of the cooler air for the building conditioning by adjusting outdoors, return, and exhaust dampers.

Eddy current Localized currents induced in an iron core by alternating magnetic flux. These

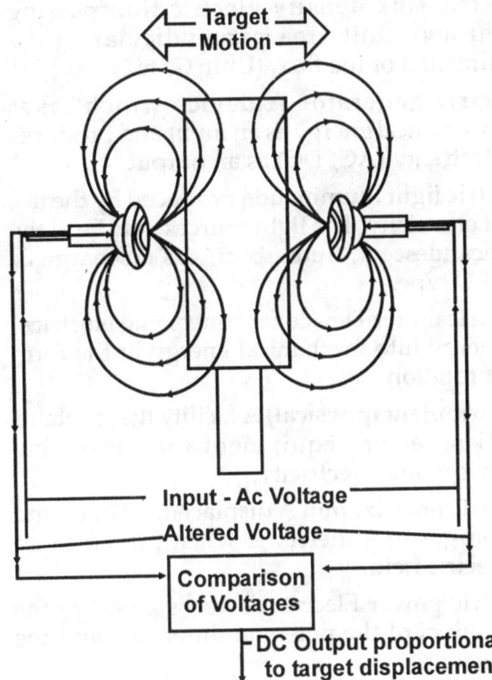

Fig. Eddy current

currents translate into losses (heat) and their minimization is an important factor in lamination design.

Eddy current loss power loss in magnetic materials due to eddy currents. This loss is proportional to the square of the thickness and hence can be reduced by the use of laminations.

Edison effect Also called Richardson effect. The phenomenon wherein electrons emitted from a heated element within a vacuum tube will flow to a second element that is connected to a positive potential.

EDP Ethylene diamine pyrocatechol.

Effective dead time In order to emphasize the essential difference between transfer lag and exponential lag, the term 'effective dead time' is used to to denote the time interval between the change of a signal to an element or system and the build-up of the response to a specific proportion, say until 5% of the final change has taken place.

Effective earthing Effective earthing avoids having dangerous potentials on the equipment even during electrical faults and also ensures the proper operation of electrical protection equipment during fault conditions. A system is said to be effectively earthed if the factor of earthing does not exceed 80%, and non-effectively earthed if it does.

Effective Piston Area (Sd) (measured in square inches or meters) The active radiating area of a speaker cone, including that part of the surround which displaces sufficient air to produce a measurable acoustic output.

Effective temperature (ET, ET*, ET*-DISC) A temperature index that accounts for radiative and latent heat transfers. ET* represents the new effective temperature which evolves with time rather than being steady-state.

Effective value The effective value of a periodic waveform (current or voltage) is the equivalent constant waveform that delivers the same average power to a resistor as the periodic waveform.

Efficiency The ability of an audio system to convert electrical energy (watts) into mechanical energy (Decibels of acoustical energy). This ratio is usually given as the amount of energy measured in Decibels at a distance of one meter from the input of one watt of electrical energy. In most speakers, the greater the efficiency rating, the louder the unit will play in response to the same setting of the volume control, in comparison to less efficient types. The overall efficiency for most speakers systems is under 20 percent. Typical speakers can be rated at anywhere from 85 to 110 dB. Keep in mind, of course, that efficiency is only one parameter of a speaker's overall quality.

Efficiency Bandwidth Product (EBP) A parameter that helps determine the suitability of a driver for a sealed or ported enclosure. An EBP of less than 50 indicates the driver should be used in a sealed box, 50

- 90 indicates flexible design options including bass reflex, over 90 indicates the need for a ported enclosure. EBP = Fs / Qes

Efficiency of a machine The ratio of the output energy to the input energy, usually expressed as a percentage. The efficiency of a machine can never exceed unity or 100%.

Elastic deformation A nonpermanent deformation that totally recovers upon release of an applied stress.

Elastic modulus or modulus of elasticity The ratio of the stress to the strain in a given material.

Elasticity The ability of a substance to return to its original state.

Elasticity of demand The ratio of the percentage change in the quantity demanded of a good to the percentage change in price.

Elastomer A polymeric material that may experience large and reversible elastic deformations.

Electric (E) field The field of force that is produced as a result of a voltage charge on a conductor or antenna.

Electric capacity The ability of a power plant to produce a given output of electric energy at an instant in time.

Electric charge Charge is an electrical property of the atomic particles of which matter is made. The elementary particle called the electron is negatively charged while the proton is equally positively charged so that normal matter is electrically neutral. (Unit: coulomb or C)

Electric circuit An interconnection of electrical elements.

Electric current An electric current flows through a conductor when there is an overall movement of charge through it and is measured as the time rate of change of charge. (Unit: ampere or A)

Electric distribution company The company that owns the power lines and equipment necessary to deliver purchased electricity to the consumer.

Electric field [V/m] In simplest form, the potential difference between two points divided by the distance between the two.

Electric flux density Electric flux passing through unit area perpendicular to the direction of the flux. (Unit: C/m2)

Electric generator A device which takes mechanical energy as an input and produces electricity (AC/DC) as an output.

Electric light Illumination produced by the use of electricity. The light sources may be of the incandescent, fluorescent, gas discharge or LED type.

Electric motor Device for converting electrical energy into mechanical energy in the form of rotation.

Electric plant (physical) A facility that contains all necessary equipment for converting energy into electricity.

Electric polarization A displacement of bound charges in a dielectric when placed in an electric field.

Electric power Electric power is given by the product of the potential difference and the current.

Electric shock A dangerous physiological effect resulting from the passing of an electric current through a human body or livestock. Injury to the skin or internal organs that results from exposure to an electrical current. Electric shock occurs when the body becomes a part of an electric circuit. The electrical current must enter the body at one point and leave at another. The human body is a good conductor of electricity. Direct contact with electrical current can be potentially fatal. While some electrical shocks may appear not to be serious, there still may be serious internal damage, especially to the heart and brain.

Electric system All of the elements needed to distribute electrical power. It includes overhead and underground lines, poles, transformers, and other equipment.

Electric welding In electrical welding, a very high electric current produces the heat needed to melt the material and join two metals together.

Electrical analogueue The analoguey that exists between electrical flow and heat flow may be used to construct electrical analogueue devices for the study of complex heat flow phenomena. The technique, although having had its use as a research tool, has little application in a design context.

Electrical breakdown Condition in which, particularly with high electric field, a nominal insulator becomes electrically conducting.

Electrical charge Symbol Q, q. Electric energy stored on or in an object. The negative charge is caused by an excess of electrons; the positive charge is caused by a deficiency of electrons.

Electrical chemical The action of converting chemical energy into electrical energy.

Electrical conductivity The ability of a material to carry an electric current; it is the reciprocal of resistivity with units of ohm.

Electrical control A control system that operates on line or low voltage and uses a mechanical means, such as a temperature-sensitive bimetal, to perform control functions, such as actuating a switch or positioning a potentiometer. The controller signal usually operates or positions an electric actuator, or may switch an electrical load either directly or through a relay.

Electrical degree One cycle in a rotating electric machine is accomplished when the rotating field moves from one pole to the next pole of the same polarity. There are 360 electrical degrees in this time period. (i.e. For each pair of poles there are 360 electrical degrees. In a machine with more than one pair of poles, one electrical cycle is completed for each pair of poles in the mechanical cycle; or the electrical degrees per revolution is obtained by multiplying the number of pairs of poles by 360.)

Electrical equipment Any item for such purposes as generation, conversion, transmission, distribution or utilisation of electrical energy, such as machines, transformers, apparatus_ measuring instruments, protective devices, wiring systems, accessories, appliances and luminaires.

Electrical heating The heating characteristic of an electric current is used extensively in industrial and domestic heating applications. Electric heating can be obtained from (a) resistance heating, (b) induction heating, (c) eddy current heating (d) dielectric heating, and (e) electric arc heating.

Electrical installation (abbr: Installation) An assembly of associated electrical equipment supplied from a common origin to fulfil a specific purpose and having certain co-ordinated characteristics.

Electrical interference Electrical noise induced upon the signal wires that obscures the wanted information signal.

Electrical lock A synchro zeroing method. This method is used only when the rotors of the synchros to be zeroed are free to turn and their leads are accessible.

Electrical power system. Provides the necessary input power.

Electrical resistance The measure of the difficulty of electric current to pass through a given material; its unit is the ohm.

Electrical symbols Graphic symbols used to illustrate the various electrical or electronic components of a circuit.

Electrical unbalance In a 3 phase supply, where the voltages of the three different phases are not exactly the same. Measured in % of unbalance.

Electrical zero A standard synchro position, with a definite set of stator voltages, that is used as the reference point for alignment of all synchro units.

Electrically independent earth electrodes Earth electrodes located at such a distance from one another that the maximum current likely to flow through one of them does not significantly affect the potential of the other(s).

Electricity Science states that certain particles possess a force field or charge. The charge possessed by an electron is negative while the charge possessed by a proton is positive. Electricity can be divided into two groups, static and dynamic. Static electricty deals with charges at rest and dynamic electricity deals with charges in motion.

Electroacoustic transducer Device that produces an energy transfer from electric to acoustic (sound) or from acoustic to electric. Examples include a microphone, earphones and loudspeakers.

Electrochemical breakdown In a practical insulation ions may arise from dissociation of impurities or from slight ionisations of the insulating material itself. When these ions reach the electrodes, reactions occur in accordance with Faraday's law of electrolysis, but on a much smaller scale. The products of the electrode reaction may be chemically or electrically harmful and in some cases can lead to rapid failure of the insulation.

Electrode The terminal at which electricity passes from one medium into another, such as in an electrical cell where the current leaves or returns to the electrolyte.

Electrode potential (E) The difference in potential established between an electrode and a solution when the electrode is immersed in the solution.

Electrodynamic The interaction of magnetism and electrical current.

Electrodynamic meter movement A meter movement using fixed field coils and a moving coil; usually used in ammeters and wattmeters.

Electrodynamometer An instrument dependant on the interaction of the

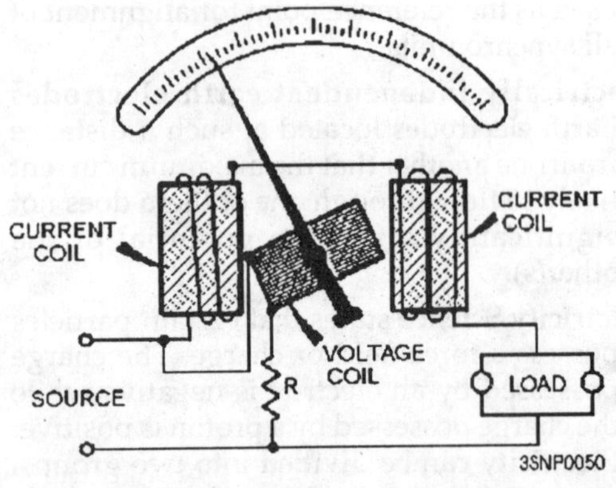

Fog. Electrodynamometer

electromagnetic fields of fixed and movable coils. It can measure current, voltage or power in both d.c. And a.c.

Electroluminescence In electrical engineering: the emission of visible light by a p-n junction across which a forward-biased voltage is applied. In electrochemistry: emission of light by a molecule which is being reduced or oxidized on a biased electrode. If the exciting cause is a photon, rather than an electron, the process is called photoluminescence.

Electrolysis Electric current passing through an electrolyte which produces chemical changes in it.

Electrolyte A conducting medium in which the flow of electric current takes place by migration of ions. Commonly an electrolyte is defined as a substance that when dissolved in a specific solvent, usually polar, like water or ethylene glycol, produces an ionically conducting solution.

Typical electrolytes are solutions of acids, bases, salts, and molten salts. Molten salts are a special class of electrolytes in which the ions and the solvent are the same species.

In an electrolytic capacitor the electrolyte constitutes the second electrode, or plate, separated from the anode, or positive plate by the barrier layer of oxide formed on the anode surface.

The electrolyte impregnated in the element (typically the wound anode foil, paper separator, and cathode foil of an electrolytic capacitor) is referred to as driving or working electrolyte and performs the following two functions:

1. It impregnates and adheres to the surfaces of the anode and cathode foils to realize 100 % of their capacitance (essentially a cathode action).

2. Repairs defects in the anode oxide film.

Electrolyte driving An electrolyte is broadly defined as any substance that passes electric current by the transport of ions. Electrolytes may be liquid solutions, fused salts or ionically conducting solids. An electrochemical process is a reaction that occurs across an interface, such as that between an electrolyte and an electrode. No electrochemical process can occur in isolation, that is a complete circuit must exist. Two electrodes separated by an electrolyte and connected by an external circuit (e.g., a wire) form a complete electrochemical cell. Thus, an electrolytic capacitor in circuit is an electrochemical cell.

Some compounds possess the ability to dissociate or ionize into positively and negatively charged species in solution. A common example of this is ordinary table salt, sodium chloride. Salts, however, are a subclass of all ionizing compounds because they exist as discrete ions in the solid state. Many molecular compounds, such as adipic acid dissociate in solution.

The anode oxide is formed on etched aluminum foil in an electrochemical process involving a so called formation electrolyte. Generally, these are aqueous solutions. Formation electrolytes are optimized to form aluminum oxide at maximum efficiency and highest oxide quality. Being aqueous, their use in finished capacitors is usually incompatible with the requirements of modern capacitors. The most obvious being the wide temperature range over which they are specified, e.g., -40°C to greater than 105°C. At lower temperatures aqueous systems solidify, resulting in loss of capacitance and excessive dissipation factors and ESR's. An aqueous electrolyte vaporizes in the upper temperature regime, accelerating the occurrence of the typical open circuit end-of-lifetime failure mode.

Thus modern capacitor electrolytes are based on such non-aqueous solvents as:

ethylene glycol

gamma-butyrolactone (GBL)

dimethylformamide (DMF)

N-methylpyrrolidinone (NMP)

other solvents

mixtures of the above

Usually, the concentration of water is maintained between 2 to 10 weight percent in order to balance physical and electrical properties. As operating temperatures of up to 150°C are specified, however, the trend towards completely "anhydrous" electrolytes has intensified. Nevertheless, the importance of water for proper oxide maintenance has been demonstrated. It is significant that paper contains approximately 8% w/w moisture at 73°F and 48% relative humidity because it is hygroscopic.

Electrolytes are characterized by their specific resistivity. For the sake of brevity, resistivity will be taken to mean specific resistivity. A more physically meaningful term is the equivalent conductivity, which considers the concentration number of ions generated by each compound in solution.

Electrolyte/Insulator/Silicon (EIS) Structures at the heart of a broad family of potentiometric silicon sensors. The best-known member of the family is the ion-sensitive field effect transistor, known as the ISFET or CHEMFET and the light-addressable potentiometric sensor LAPS The principle of operation of devices using such structures is as follows. A potential with respect to a reference electrode is generated at the interface between the liquid solution and the insulator. The surface potential (ø0) is determined by that ionic species which has the fastest exchange rate (io) with the membrane covering the insulator. If no intentional membrane is deposited on an oxide covered insulator that species will be H+. Surface potential changes in turn change the Si flat-band voltage VFB. The flat-band voltage is the potential one needs to apply to the Si in order to have the bands flat throughout the semiconductor. The flat band voltage of an EIS structure has been shown to be given by: VFB = EREF - ÖSi/q - ø0 - Qins/Cinss where VFB stands for the flat-band voltage of the structure, EREF for the reference electrode potential, ÖSi for the work function of silicon, ø0 for the surface potential at the insulator/electrolyte interface, Qins for the charge at the insulator/silicon interface and Cins for the insulator capacitance. At least two terms in the above equation are not known with a precision greater than a few hundred millivolts. This is true for EREF as well as for Qins/Cins which can vary from device to device by several hundred millivolts. For a given EIS sensor, these inaccurately known quantities are constant, and variations in flat-band voltage can be equated to variations of the surface potential.

Electrolytic A type of capacitor that has a liquid or paste between the plates to increase its capacitance.

Electrolytic capacitor A capacitor is typically made up of two (2) parallel plates, the electrodes, with a dielectric between them. The amount of capacitance is directly proportional to the surface area of the electrode and indirectly proportional to the dielectric thickness.

An electrolytic capacitor consists of a series combination of two capacitors (e.g. foil electrodes or plates) at least one of which is a valve metal, separated by an electrolyte, and between which a dielectric film is formed adjacent to the surface of one or both of the electrodes.

The high volumetric efficiency of an electrolytic capacitor is due to its enhanced plate surface area and a very thin dielectric layer. The dielectric is an oxide and has a high dielectric strength which is electrochemically deposited in very thin layers, usually on the order of hundreds to tens of thousands of Ångstrom units.

Electromagnet A coil of wire usually wound on a soft iron or steel core. When current is passed through the coil a magnetic field is generated. The core provides an easy path for the magnetic lines of force. This concentrates the field in the core.

Electromagnetic Relating to a magnetic field created by an electric current.

Electromagnetic communication Use of an electromagnetic wave to pass information between two points. Also called wireless communication.

Electromagnetic compatibility (EMC) The ability of equipment or a system to function as designed in its electromagnetic environment without introducing intolerable electromagnetic disturbances to that environment, or being affected by electromagnetic disturbances in it.

Electromagnetic damping Electromagnetic damping is produced by the induced effects when the coil moves in the magnetic field and a closed path is provided for the currents to flow.

Electromagnetic field Electric and magnetic force field that surrounds a moving electric charge.

Electromagnetic induction Voltage produced in a coil due to relative motion between the coil and magnetic lines of force.

Electromagnetic interference (EMI) Electrical and magnetic noise imposed on a system. There are many possible causes, such as switching AC power on inside the sine wave. EMI can interfere with the operation of controls and other devices.

Electromagnetic radiation (Waves) A series of energy waves that travel in a vacuum at the speed of 3 x 108 m/s; includes radio waves, microwaves, visible light, infrared, and ultraviolet light, x-rays, and gamma rays.

Electromagnetic relay A relay controlled by electromagnetic means, to open and close electric contacts.

Electromagnetic spectrum The range of frequencies over which electromagnetic radiations are propagated.

Electromagnetic wave Wave taht consists of both electric and magnetic variation.

Electromagnetism Relates to the magnetic field generated around a conductor when current is passed through it.

Electromechanical effects A relay that uses an electromagnet to move an armature thereby actuating current.

Electromechanical transducer Device that transforms electrical energy into mechanical energy (electric motor) or mechanical energy into electrical energy (generator).

Electrometer A highly refined DC multimeter. In comparison with a digital multimeter, an electrometer is characterized by higher input resistance and greater sensitivity. It can also have functions not generally available on DMMs (e.g., measuring electrical charge, sourcing voltage).

Electromotive force (Emf) series A series of chemical elements arranged in order of their electromotive force. The electromotive force is the greatest potential difference that can be generated by a particular source of electric current. In practice this potential may be observable only when the source is not supplying current, because of its internal resistance.

Electron Smallest sub atomic particle of negative charge that orbits the nucleus of an atom.

Electron emission The escape of electrons from certain materials.

Electron energy level In quantum mechanics, an energy which is allowed for an electron.

Electron flow The direction in which electrons flow. This is from negative to positive - as electrons are negatively charged.

Electron gun The source of electrons in a cathode ray tube. Consists of a cathode emitter of electrons, an anode with an aperture through which the beam of electrons can pass, and one or more focussing and control electrodes.

Electron lens A system of electric or magnetic fields used to focus a beam of electrons in a manner analogueous to an optical lens.

Electron state (level) One of a set of discrete, quantized energies that are allowed for electrons. In the atomic case each state is specified by 4 quantum numbers.

Electron volt Unit of energy used in dealing with subatomic particles. It is the increase in energy or the work done on an electron when passing through a potential rise of 1 volt. 1 ev = 1.602 x 10-19 J

Electron, free An electron not bound to the crystalline lattice, hence, free to conduct electricity.

Electronegative Describing elements that tend to gain electrons and form negative ions. The halogens are typical electronegative elements.

Electronic attack (EA) That division of EW involving the use of electromagnetic, directed energy, or antiradiation weapons to attack personnel, facilities, or equipment with the intent of degrading, neutralizing, or destroying enemy combat capability. EA includes actions taken to prevent or reduce an enemy's effective use of the electromagnetic spectrum, such as jamming and electromagnetic deception and employment of weapons that use either electromagnetic or directed energy as their primary destructive mechanism (lasers, radio frequency, particle beams).

Electronic ballast A ballast which uses semi-conductor components to increase the frequency and control the operation of a fluorescent lamp. Fluorescent system efficiency is increased due to the higher frequency.

Electronic control A control circuit that operates on low voltage and uses solid-state components to amplify input signals and perform control functions, such as operating a relay or providing an output signal to position an actuator. The controller usually furnishes fixed control routines based on the logic of the solid-state components.

Electronic journal A journal which is available electronically. It may be available in electronic format only, or it may have a print equivalent. Some are available free of charge, others require payment or a subscription.

Electronic protection (EP) That division of EW involving actions taken to protect personnel, facilities, and equipment from any effects of friendly or enemy employment of EW that degrade, neutralize, or destroy friendly combat capability.

Electronic warfare (EW) Any military action involving the use of electromagnetic and directed energy to control the electromagnetic spectrum or to attack the enemy. The three major subdivisions within electronic warfare are electronic attack, electronic protection, and electronic warfare support.

Electronic warfare support (ES) That division of EW involving actions tasked by, or under direct control of, an operational commander to search for, intercept, identify, and locate sources of intentional and unintentional radiated electromagnetic energy for the purpose of immediate threat recognition. Thus, electronic warfare support provides information required for immediate decisions involving EW operations and other tactical actions such as threat avoidance, targeting and homing. ES data can be used to produce signals intelligence.

Electrostatic Relating to an electric field created by an electric charge.

Electrostatic discharge (ESD) The flow of current that results when objects having a static charge come into close proximity, causing a discharge between them.

Electrostatic Driver The Electrostatic Driver generates a motive force for its diaphragm by the interaction of electric rather than magnetic fields as is the case for the dynamic driver. The electrostatic driver is basically "hung" on a frame. It does NOT have a box enclosure, so a great deal of "colouration" (frequency response fluctuations caused by the enclosure reflections) is avoided. A large diaphragm of lightweight material is placed between two perforated (acoustically transparent) electrodes. The diaphragm is electrically polarized at a few thousand volts relative to the electrodes, which maintain a large electrostatic field. The audio signal is applied to the two electrodes in a push-pull fashion. Under these conditions, the diaphragm will vibrate in step with the audio drive signal and produce sound. Because it is a true push-pull driver (i.e. its diaphragm is driven from both front and rear), it operates in a linear fashion.

In an electrostatic driver, the driving force is uniform over the entire diaphragm surface (note that for the Dynamic Driver, the diaphragm is driven over a small portion of its overall surface) as a result, Electrostatic Drivers do not suffer drastically from "breakup" effect as dynamic drivers. Gross distortion typically results only if the driving amplifier clips into the speaker, or when, in an attempt to play the driver at a higher sound level than its design permits, its step-up transformer may reach a point of saturation. Due to the diaphragm of the electrostatic driver being of low mass (lightweight), its transient response is excellent and reproduction of subtle, low-level musical details is exceptional.

Electrostatic field Force field produced by static electrical charges.

Electrostatic generator A machine designed for the continuous separation of electric charge. An example is the Van de Graaf Generator.

Electrostatic meter These basically work on the principle that the force (or torque) of attraction is proportional to the product of the charges and the force is proportional to the square of the voltage. Thus this meter reads the mean square value and hence is calibrated to read the root mean square value. The electrostatic meter is basically a voltmeter.

Electrostatic shield A conductive screen that shunts induced electrical energy to ground.

Element A substance, in chemistry, that cannot be divided into simpler substances by any means ordinarily available. A part of an antenna that can be either an active radiator or a parasitic radiator.

Elevation angle The angle between the horizontal plane and the line of sight to a target or object.

Ellipsoid A solid figure traced out by an ellipse rotating about one of its axes.

Emergency lighting Lighting provided for use when the supply to the normal lighting fails.

Emergency power An independent reserve source of electric power which, upon failure or outage of the normal power source, provides stand-by electric power.

Emergency stopping Emergency switching intended to stop an operation.

Emergency switching An operation intended to remove, as quickly as possible, danger, which may have occurred unexpectedly.

EMF (Electromotive force) The force that causes electricity to flow between two points with different electrical charges or when there is a difference of potential between the two points. The unit of measurement is volts.

EMI (Electromagnetic Interference) Any electronic disturbance that interrupts, obstructs, or otherwise impairs the performance of electronic equipment.

Emission Any pollution discharge from a source.

Emissivity The ratio of energy emitted by an object to the energy emitted by a blackbody at the same temperature. The emissivity of an object depends upon its material and surface texture; a polished metal surface can have an emissivity around 0.2 and a piece of wood can have an emissivity around 0.95.

Emitter The semiconductor region from which charge carriers are injected into the base of a bipolar junction transistor.

Emitter feedback Coupling from the emitter output to the base input of a bipolar junction transistor.

Emitter follower A common collector amplifier. Has a high current gain, high input impedance and low output impedance.

Emitter injection modulator The transistor equivalent of the cathode modulator. The gain is varied by changing the voltage on the emitter.

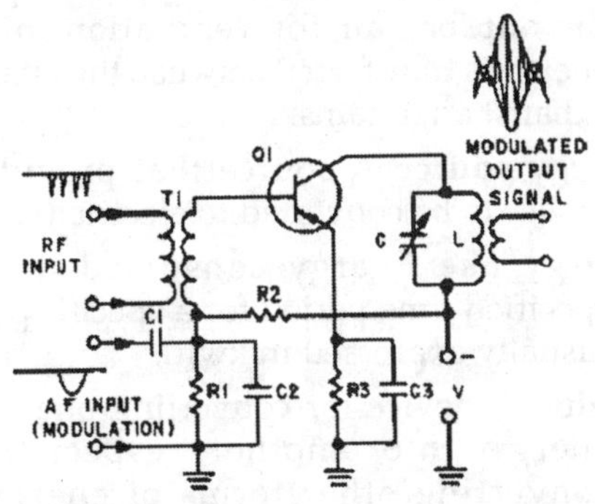

Fig. Emitter injection modulator

Empirical Based upon the results of experiment and observation only.

Emulation Real time simulation. An emulator consists of a real time simulation of the building and plant together with a hardware interface that is used to connect the simulator to a BEMS. The outputs from the control system are read by the hardware interface and used as the boundary values for the simulation of the building and plant. The simulated outputs of the sensors are transmitted through the hardware interface to the control system, which then responds by producing a new set of outputs.

Enamel A synthetic compound of cellulose acetate (wood pulp and magnesium). Used to insulate wire in meters, relays, and motor windings.

Encapsulated Imbedded in solid material or enclosed in glass or metal.

Encapsulated winding A motor which has its winding structure completely coated with an insulating resin (such as epoxy). This construction type is designed for exposure to more severe atmospheric conditions than the normal varnished winding.

Encapsulating security payload (ESP) A mechanism to provide confidentiality and integrity protection to IP datagrams.

Enclosure or Box A cabinet in which the various Drivers are housed. This arrangement is absolutely essential if bass response is desired from the woofer, which by itself in the open air, will produce very little low frequency response. This is so because of an effect called phase cancellation. This means that the sound wave coming from the back of the speaker at low frequencies is identical in form and intensity to the one coming from the front of the cone. When they meet, as in the open air, they are exactly 180 degrees out of phase and thus cancel each other out. An enclosure either prevents this from occurring or modifies the interaction so that the backwave actually reinforces the front wave. Tweeters and midranges are less severly effected by the phase cancellation effects, and are not in need of enclosures.

Enclosure Volume The total amount of internal airspace of an enclosure. Each woofer has an optimal airspace volume that helps it meet it's bass potential. This resonance cavity specification usually measured in cubic feet, includes the net driver and vent volumes. To find the volume of a cube multiply HxWxD in inches. divide the result be 1725 (the number of inches in a cubic foot) to get the volume in cubic feet.

Encode To use a code to represent individual characters in a message. Here are a number of Encoding Terms.

Encoder Feedback device that utilizes an optical source and sensor to provide velocity and position information in the form of a digital signal; not readily adaptable to different environments.

End feed method A method in which one end of an antenna is connected through a capacitor to the final output stage of a transmitter.

End fire array An array in which the direction of radiation is parallel to the axis of the array.

End play Amount of axial displacement resulting from the application of a load equal to the stated maximum axial load.

End point (Potentiometric) The apparent equivalence point of a titration at which a relatively large potential change is observed.

End use The specific purpose for which electric is consumed (i.e. Heating, cooling, cooking, etc.).

Endothermic Absorbs heat. A process is said to be endothermic when it absorbs heat.

Endshield The part of the motor housing which supports the bearing and acts as a protective guard to the electrical and rotating parts inside the motor. This part is frequently called the "end bracket" or "end bell.".

Energized Being electrically connected to a voltage source so the device is activated.

Energy Capacity for performing work or to cause heat flow. Like work itself, it is measured in Joules.

Energy audit A review of the customer's electricity and/or gas usage often including recommendations to alter the customer's electric demand or reduce energy usage. An audit usually includes a visit to the customer's facility.

Energy balance The arithmetic balancing of energy inputs versus outputs of an object or processing equipment; it is positive if energy is released, and negative if energy is absorbed.

Energy charge The amount of money owed by an electric consumer for kilowatt-hours consumed.

Energy conservation The deliberate design of a building or process to reduce its energy usage, or to increase its energy efficiency.

Energy consumption This represents running costs which can be broken down to indicate the principal causal factors. Issues such as: - larger windows; - heat gains from lights; are inter-related and affect energy consumption. For example, the electrical power savings which result from from enhanced daylight utilisation can significantly outweigh the higher heating energy consumption.

Energy costs Costs, such as for fuel, that are related to and vary with energy production or consumption.

Energy efficiency The efficient use of energy with minimum waste.

Energy efficiency programs Programs that reduce consumption.

Energy meter Instrument to measure energy, usually a house service meter.

Energy recovery ventilation system A device or combination of devices applied to provide the outdoor air for ventilation in which energy is transferred between the intake and exhaust airstreams.

Energy source A source that provides the power to be converted to electricity.

Energy use Energy consumed during a specified time period for a specific purpose (usually expressed in kwh).

Engine A device for converting one form of energy into another, especially for converting other forms of energy into mechanical (kinetic) energy.

Engineering notation A floating point system in which numbers are expressed as products consisting of a number greater than one multiplied by an appropriate power of ten that is some multiple of three.

Enhancement mode MOSFET A field effect transistor in which there are no charge carriers in the channel when the gate source voltage is zero.

Enthalpy (H) A property of a system equal to E + PV, where E is the internal energy of the system, P is the pressure of the system, and V is the volume of the system. At constant pressure the change in enthalpy equals the energy flow as heat.

Entrance cable/service entrance conductor Cable running outside of a consumer's house into the meter. This cable is owned by the consumer and its maintenance is the consumer's responsibility.

Environmental attributes Environmental attributes quantity the impact of various options on the environment. These attributes include particulate emissions, SO2 or Nox, and thermal discharge (air and water).

Environmental conditions All conditions in which a transducer may be exposed during shipping, storage, handling, and operation.

Environmental factors Conditions other than indoor air contaminants that cause stress, comfort, and/or health problems (e.g., humidity extremes, drafts, lack of air circulation, noise, and over-crowding).

Ep Ip curve The characteristic curve of an electron tube used to graphically depict the relationship between plate voltage (Ep) and plate current (Ip).

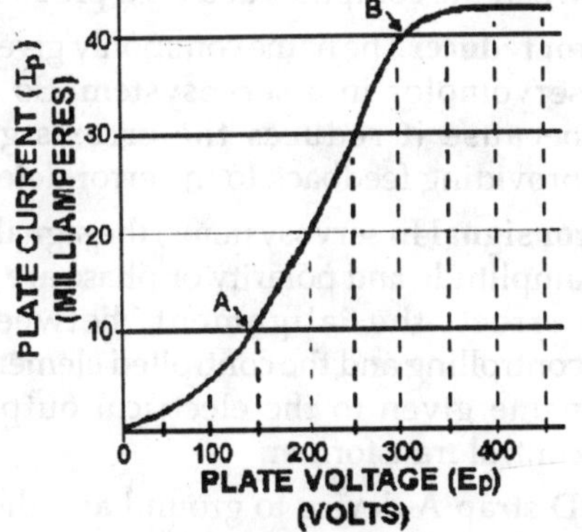

Fig. Ep Ip curve

Ephemeris A table showing the pre-calculated position of a satellite at any given time.

Epitaxial or epi A single-crystal semiconductor layer grown upon a single-crystal substrate having the same crystallographic characteristics as the substrate material.

Epitaxial process A method of depositing a thin, uniformly doped crystalline region (layer) on a substrate.

EQ (equalizer) A class of electronic filter circuits designed to augment or adjust electronic or acoustic systems. Equalizers can be fixed or adjustable, active or passive. Most consumer audio equallizers divide the spectrum into 3 to 12 bands, allowing each section to be either increased or decreased in amplitude without changing the response of the rest.

Equal percentage valve This type of valve is designed to produce equal percentage change in flow for equal increase change in valve lift.

Equalised run time (ERT) A facility which can be provided by a microprocessor based step controller. Rather than load in one directionj and unload in the other (last on - first off), ERT loads and unloads to equalise the time in which each step is made, or is in the on position by adopting a first on - first off schedule. In this way, the run times for each plant item in multiple compresor and boiler installations are equalised; cf. Step control and Intelligent step control

Equalization Selective amplification (signal restoration) applied to a signal in order to compensate for high frequency attenuation and other distortions encountered in long lengths of cable.

Equalizing charge A controlled overcharge of an already full battery to restore all the individual cells within the battery to the same state of charge.

Equatorial orbit An orbit that occurs when the plane of a satellite coincides with the plane of the earth at the equator.

Equilateral Having all the sides equal in length. Equilateral triangle is one which has all three sides equal.

Equilibrium State of balance between opposing forces or effects.

Equilibrium constant The product of the concentrations (or activities) of the substances produced at equilibrium in a chemical reaction divided by the product of concentrations of the reacting substances, each concentration raised to that power which is the coefficient of the substance in the chemical equation.

Equipment A general term including material, fittings, devices, appliances, luminaires (fixtures), apparatus, and the like used as a part of, or in connection with, an electrical installation.

Equipotential bonding Electrical connection maintaining various exposed conductive parts and extraneous conductive parts at substantially the same potential.

Equipotential lines and surfaces Lines and surfaces having the same electric potential.

Equitransference Equal diffusion rates of the positively and negatively charged ions of an electrolyte across a liquid junction without charge separation.

Equivalent conductance (l) Equivalent conductance of an electrolyte is defined as the conductance of a volume of solution containing one equivalent weight of dissolved substances when placed between two parallel electrodes 1 cm apart, and large enough to contain between them all of the solution. L is never determined directly, but is calculated from the specific conductance (Ls). If C is the concentration of a solution in gram equivalents per liter, then the concentration of a solution in gram equivalents per liter, then the concentration per cubic centimeter is C/1000, and the volume containing one equivalent of the solute, is, therefore, 1000/C.

Equivalent leakage area (ELA) The equivalent amount of orifice area that would pass the same quantity of air as would pass collectively through a building envelope or component at a specified reference pressure difference.

Equivalent resistance (Req) A resistance that represents the total ohmic values of a circuit component or group of circuit components. Usually drawn as a single resistor in a simplified circuit.

Erg Unit of work or energy in the c.g.s. System of units. 1 erg = 10-7 J

Ergonomics The accessibility and ease of controlling a system in normal operational mode with a minimum of motion and thought required.

Erosion In a surface discharge, if the products of decomposition are volatile and there is no residual conducting carbon on the surface, the process is simply one of pitting and is known as erosion. Erosion occurs in organic materials.

Error The difference between the value indicated by the transducer and the true value of the measurand being sensed. Usually expressed in percent of full scale output.

Error band The allowable deviations to output from a specific reference norm. Usually expressed as a percentage of full scale.

Error detector The component in a servo-system that determines when the load has deviated from its ordered position, velocity, and so forth.

Error message A textual message displayed when the computer detects a problem.

Error reducer The name commonly given to the servomotor in a servosystem. So named because it reduces the error signal by providing feedback to the error detector.

Error signal In servosystems, the signal whose amplitude and polarity or phase are used to correct the alignment between the controlling and the controlled elements. The name given to the electrical output of a control transformer.

ESD strap A device to ground and discharge static electricity from a person.

ESP(Electronic Shock Protection). An electronic circuit that stores the audio data stream from a CD or MD in a memory buffer. If the laser pick-up mistracks, audio still flows from the buffer preventing an interruption.

Estimating summer time temperature - the CIBSE method The CIBSE method of determining the peak inside environmental temperature.

Etched foil The heater type that utilizes a resistance foil as the element with a print and etch method of manufacture.

Eutectic alloy An alloy that changes directly from a solid to a liquid with no plastic or semi-liquid state.

Eutectic solder An alloy of 63 percent tin and 37 percent lead. Melts at 361° F.

Eutectic temperature The lowest possible melting point of a mixture of alloys.

Evaporation Conversion of a liquid into vapour, without necessarily reaching the boiling point.

Even symmetry or even function A function has even symmetry when its plot is symmetrical about the vertical axis. F(t) = f(-t)

Event A programmable ON/OFF output signal. Events can control peripheral equipment or processes, or act as input for another control or control loop.

Event initiated programs Computer programs initiated on the occurrence of an input/output operation or an alarm condition.

Exa (E) Decimal multiple prefix corresponding to 1018 .

Excitation The addition of energy to a nucleus, an atom or a molecule transferring it from its ground state to a higher energy level. The excitation is the difference in energy between the ground state and the excited state.

Excitation voltage The supply voltage required to activate a circuit.

Exciting current The current that flows in the primary winding of a transformer when the secondary is open-circuited; it produces a magnetic flux field. Also called magnetizing current.

Excursion The back-and-forth travel of a cone in a dynamic Driver. How loud a speaker can play depends on how much air it can move without overheating. How much air can be moved is determined by the surface area of the cone as it moves back and forth (Xmax), the Enclosure resonance, and the suspension compliance of the motor system.

Excursion Limited SPL The maximum sound pressure level the system can produce given an input signal equal to the rated excursion limited power handling.

Exfiltration air leakage outward through cracks and intersices and through ceilings, floors, and walls of a space or building.

Exhaust air Air removed from a space and not reused therein.

Exhaust ventilation Mechanical removal of air from a portion of a building (e.g., piece of equipment, room, or general area).

Exothermic Gives off heat. A process is said to be exothermic when it releases heat.

Expander A type of dynamic range processor which reduces the gain of audio signals which are under an adjustable 'threshold' level, therefore increasing the dynamic range. Generally allows the operator control over threshold, ratio, attack, release and 'hold' times. Both analogue and digital types are available.

Expansion factor Correction factor for the change in density between two pressure measurement areas in a constricted flow.

Expert system A formal definition is: 'The embodiment in a computer of a knowledge-based component from an expert that offers intelligent advice or takes an intelligent decision and, in addition, is able to justify its own reasoning'. This means that an expert system is a computer program that learns, deduces, diagnoses, and advises. The style adopted to attain these characteristics is rule-based programming. The rules are based on logic and standard computer statements.

Exposed conductive part A conductive part of equipment which can be touched and which is not a live part but which may become live under fault conditions.

Exposed junction A form of construction of a thermocouple probe where the hot or measuring junction protrudes beyond the sheath material so as to be fully exposed to the medium being measured. This form of construction usually gives the fastest response time.

Expression A validated series of variables, constants, and functions that can be connected by operating symbols to describe a desired computation.

Extended Pole Piece A1986 Extended pole pieces on the magnet assembly allow for more voice coil travel, and thus lower Frequency Response, and less chance of "bottoming out"

Extended temperature Empire Magnetics extended temperature (HT) motors and related products have a case temperature rating of -40° C to 155° C.

External fan pressurization A blower door is fitted to a building, induced air flow through the fan creates an artificial, uniform static pressure within the building. Internal and external pressure taps are made and a manometer is used to measure the air flow required to produce a given pressure difference. The higher the flow rate required to produce a given pressure difference, the less airtight the building.

External trigger An analogue or digital hardware event from an external source that starts an operation.

Externally excited meter A term used to describe meters that get their power from the circuit to which they are connected.

Externally synchronized radar A radar system in which timing pulses are generated by a master oscillator external to the transmitter.

Externally ventilated A motor using an external cooling system. This is required in applications where the motor's own fan will not provide sufficient cooling; this is true for certain duty cycle applications, slow speed motors, also in environments with extreme dirt. Often a duct with an external blower is used to bring clean air into the motor's air-intake.

Extract air Air that is removed from a building or space. A proportion is often used for recirculation and added to incoming air. Alternatively it is all exhausted to the outdoors, sometimes via an air to air heat exchanger or a heat pump.

Extract ventilation A mechanical ventilation system in which air is extracted from a space or spaces, so creating an internal negative pressure. Supply air is drawn through adventitious or intentional openings. Such a system allows heat to be recovered, using an exhaust air heat pump.

Extraneous conductive part A conductive part liable to introduce a potential, generally earth potential, and not forming part of the electrical installation.

Extrapolation Filling in values or terms of a series on either side of the known values thus extending the range of values.

Extremely high frequency The band of frequencies from 30 gigahertz to 300 gigahertz.

Extremely low frequency The band of frequencies up to 300 hertz.

Extrinsic A semiconductor in which impurities have been added to create certain charge carrier concentrations.

Extrinsic semiconductor A semiconductor in which the carrier density results mainly from the presence of impurities or other imperfections, as opposed to an intrinsic semiconductor in which the electrical properties are characteristics of the ideal crystal.

Eye pattern An oscilloscope display in which a pseudorandom digital data signal from a receiver is repetitively sampled and applied to the vertical input, while the data rate is used to trigger the horizontal sweep.

F

F is the coefficient of friction or "friction/skin factor"

F Type connector A threaded medium performance coaxial signal connector typically used in consumer applications (TV's and VCR's). This connector is typically usable as high as 1GHz. It is inexpensive since the pin of the connector is actually the center conductor of the coaxial cable.

F3 (measured in Hz) The frequency at which the acoustic power output from a system has fallen to one-half its reference value. Known as the systems 3dB down point. F3: determined by the frequency at which the output is 3dB lower than the level at 100Hz. This frequency was chosen because it is a typical crossover point. In the case of a Bandpass system, F3 is determined by the frequency at which the output is 3dB lower than the level at the middle of the pass band.

Fab For "fabrication", a term referring to the making of semiconductor devices such as microprocessors.

Fabric protection control A control function, the aim of which is to prevent condensation occurring.

Face centered cubic A crystal structure found in common elemental metals also FCC. Within the cubic unit cell, atoms are located at all corner and face-centered positions.

Facsimile The method for transmitting and receiving still images. These images can be maps, photographs, and handwritten or printed text.

Factor Any of the elements, quantities, or symbols that, when multiplied together, form a product.

Factor of earthing This is the ratio of the highest r.m.s. Phase-to-earth power frequency voltage on a sound phase during an earth fault to the r.m.s. Phase-to-phase power frequency voltage which would be obtained at the selected location without the fault. This ratio characterises, in general terms, the earthing conditions of a system as viewed from the selected fault location.

Fader Rather like the right to left balance control, however the fader moves the sonic position between the front and back speakers.

Fading Variations in signal strength by atmospheric conditions.

Fahrenheittemperature scale in which the melting point of ice is taken as 32 of and the boiling point of water under standard atmospheric pressure (760 torr) as 212 of. A Fahrenheit degree is 1/180 of the difference between these two temperatures.

Failsafe In terms of relay technology, when power is lost, the relay contacts fall back to a default position.

Fall time The time required for a signal to change from a large percentage (usually 90%) to a small percentage (usually 10%) of its peak-to-peak amplitude.

False negative Occurs when an actual intrusive action has occurred but the system allows it to pass as non-intrusive behavior.

False positive Occurs when the system classifies an action as anomalous (a possible intrusion) when it is a legitimate action.

Fan A mechanical device employing rotating aerofoil blades or vanes to continuously move air from one place to another.

Fan coil units and air handling units These may be manufactured for the direct expansion market or for the hydraulic market. They generally consist of a filter, fan and a cooling and/or heating coil. The difference between a fan coil unit and an air handling unit has become increasingly blurred. A fan coilunit generally is a small unit quite often located in the conditioned environment, they may also be located above a false ceiling and connected to a very small amount of ductwork. Air handling units are generally always intended to be connected to distribution ductwork, and the fan is sized to facilitate this. Large fan coil units have fans capable of providing up to 150 Pascals static pressure (~0.6 inches water gauge).

Farad (F) SI unit of electric capacitance. One farad is defined as the ability to store one coulomb of charge per volt of potential difference between the two conductors.

Faraday Quantity of electricity required to liberate or deposit 1 gram-equivalent of an ion. 1 Faraday = 96,490 coulomb.

Faraday rotation The rotation of the plane of polarization of electromagnetic energy when it passes a substance influenced by a magnetic field that has a component in the direction of propagation.

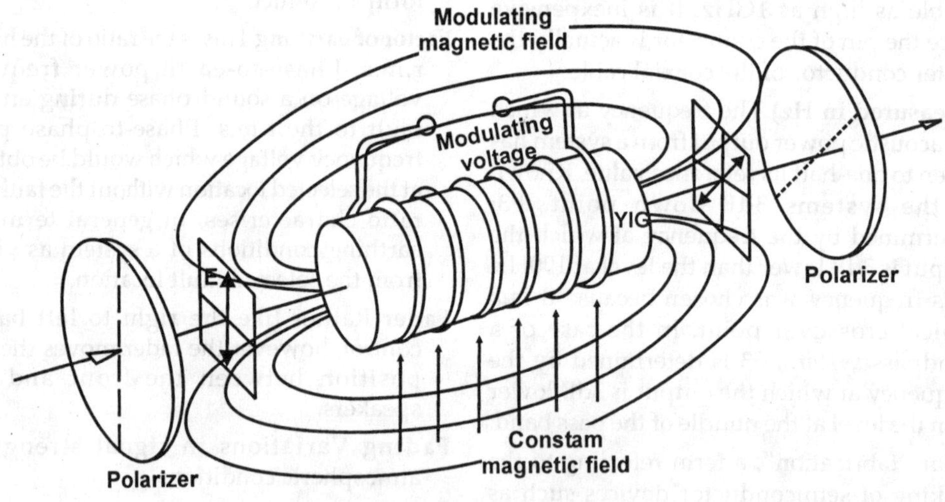

Fig. Faraday rotation

Fast neutrons Neutrons resulting from nuclear fission that have lost little of their energy by collision and therefore travel at high speeds. It is usual to define neutrons with energies in excessof 0.1 mev as fast.

Fast time constant circuit Differentiator circuit in the first video amplifier that allows only the leading edges of target returns, no matter how small or large, to be used.

Fatigue of metals Deterioration of metals owing to repeated stresses above a certain critical value, accompanied by changes in the crystalline structure of the metal.

Fault A circuit condition in which current flows through an abnormal or unintended path.

Fault current A current resulting from a fault.

Fault tolerance The ability of a system or component to continue normal operation despite the presence of hardware or software faults.

Favorites A way of storing web addresses so you can find them easily (Favourites can be found on Micosoft Internet Explorer - they are called "bookmarks" on Netscape).

Fb The tuned frequency of a Vented Enclosure, measured in Hertz. A combination of the resonance frequency of the air in a Port and the resilient pressure of the air in a Ported System. Below this point the Frequency Response of a Tuned Port system rapidly Falls Off, and the system can become unstable if asked to reproduce lower frequencies at high Amplitude.

Feed forward control A formal definition of this control mode is:- 'The transmission of a supplementary signal along a secondary path, parallel to the main forward path, from an earlier to a later stage'. This means that it works to eliminate errors occurring in the first place by forecasting the likely disturbance. However, if not ALL disturbances are forecast correctly then poor control will result. Feedforward control, while conceptually more appealing, significantly escalates the technical and engineering requirements of the overall design. The very sophisticated calculations must reflect an awareness and understanding of the EXACT effects that the disturbances will have on the controlled variable. With such understanding, the feed-forward controllers are able then to compensate for the disturbances. Feed-forward control is reserved for only a very few of the control loops within a plant. Thus, pure feedforward control is rarely encountered and the more common situation is for mainly feedback control with some feedforward control loops included.

Feedback The term is generally applied to electronic amplifiers to which a portion of the output energy is used to reduce or increase the amplification, by reacting on an earlier stage according to the relative phase of the return.

Feedback amplifier An amplifier with an external signal path from its output back to its input.

Feedback control The simplest way to automate the control of a process is through feedback control. Sensors, or measuring devices, are installed to measure the actual values of the controlled variables. These actual values are the transmitted to feedback control hardware which makes a comparison between the setpoint or the desired value of the controlled variable and the measured value of these same variables. Based upon this 'error signal' between the measured and the desired value of the controlled variable, the feedback controller calculates signals that reflect the required value of the manipulated variable. These signals are then transmitted automatically to the final control elements which act to alter the manipulated variable in such a manner as to reduce the error signal. Feedback control acts to eliminate errors. This is in contrast to feed-forward control which operates to stop the error occurring in the first instance.

Feeder All circuit conductors between the service equipment, the source of a separately derived system, or other power supply source and the final branch-circuit overcurrent device.

Feedhorn A horn radiator used to feed a reflector.

Feedthru Device that physically interfaces one side of a separating surface to the other (e.g., an electrical feedthru passes electrical signals from one side of a wall to the other).

Femto (f) Decimal sub-multiple prefix corresponding to 10-15.

Fep Fluorinated Ethylene Propylene, A synthetic type of insulation.

Fermi A unit of length used in nuclear physics. 1 fermi = 10-15 meter.

Fermi dirac statistics The branch of statistical mechanics used with systems of identical particles which have the property that their wave function changes sign if any two particles are interchanged.

Fermi energy The energy level in a solid at which the probability of finding an electron is 1/2. For a metal, the energy corresponding to the highest filled electron state in the valence band at 0 K.

Fermi level The energy level in the semiconductor device at which the probability of finding an electron is 50%. In other words, it's like the water level in a glass. States below the Fermi level are full and those above it are empty.

Ferrimagnetism The type of magnetism occurring in materials in which the magnetic moments of adjacent atoms are anti-parallel, but of unequal strength, or in which the number of magnetic moments oriented in one direction outnumber those in the reverse direction. Typical ferrimagnetic materials are the ferrites.

Ferrite A powdered, compressed and sintered magnetic material having high resistivity. The high resistance makes eddy current losses low at high frequencies.

Ferrite bead Ferrite composition in the form of a bead. Running a wire through the bead increases the inductance of the wire.

Ferrite core inductor An inductor wound on a ferrite core.

Ferrite rod aerial A coil of wire wound on a ferrite material to increase the inductance of the coil. It's signal capturing capability.

Ferrite switch A ferrite device that blocks the flow of energy through a waveguide by rotating the electric field 90 degrees. The rotated energy is then reflected or absorbed.

Ferroelectric material A dielectric material such as Rochelle salt and barium titanate with a domain structure containing dipoles (asymmetric distributions of electrical charge) which spontaneously align. Their domain structure makes them analogueous to ferromagnetic materials. They exhibit hysteresis and usually the piezoelectric effect.

Ferroelectrics Dielectric materials which exhibit properties such as hysteresis which are usually properties of ferromagnetic materials.

Ferromagnetic material A highly magnetic material, such as iron, cobalt, nickel, or their alloys.

Ferromagnetism Ferromagnetism is due to unbalanced electron spin in the inner electron orbits of the elements concerned giving the atoms a resultant magnetic moment. Ferromagnetic materials have very large magnetic permeabilities which vary with the strength of the applied field.

Ferroresonance Resonance resulting when the iron core of an inductive component of an LC circuit is saturated, increasing the inductive reactance with respect to the capacitance reactance.

Ferrous Composed of and or containing iron. A ferrous metal exhibits magnetic characteristics as opposed to non-ferrous material.

Ferrules The cylindrical metallic ends of a cartridge fuse.

Fertile material Isotopes which can be transformed into fissile material by the absorption of neutrons.

FFT (Fast-Fourier Transform) A method by which a system is described using an impulse response. Both frequency and time data can be extracted, with room reflections removed, providing an extremely accurate analysis. Mathematical manipulation of the data is employed to view system parameters from a variety of perspectives.

Fiber optic Relates to transmission of information as modulated light in tiny transparent fibers instead of copper wires.

Fibrous braid An outer covering used to protect a conductor's insulating material. Commonly made from cotton, linen, silk, rayon, or fiberglass.

Fick's first law The diffusion flux is proportional to the concentration gradient. This relationship is employed for steady-state diffusion situations.

Fick's second law The time rate of change of concentration is proportional to the second derivative of concentration. This relationship is employed in non-steady-state diffusion situations.

Fidelity 1. The faithful reproduction of a signal. 2. The accuracy with which a system reproduces a signal at its output that faithfully maintains the essential characteristics of the input signal.

Field The region in which an electrically charged body (electric field), or a magnetised body (magnetic field) exerts its influence.

Field balancing equipment An assembly of measuring instruments for performing balancing operations on assembled machinery which is not mounted in a balancing machine.

Field coil A coil of wire used for magnetising an electromagnet.

Field effect transistor FET A three terminal semiconductor device that depends on the action of an electric field to control its conductivity. In a "FET" the current is from source to drain because a conducting channel is formed by a voltage field between the gate and the source.

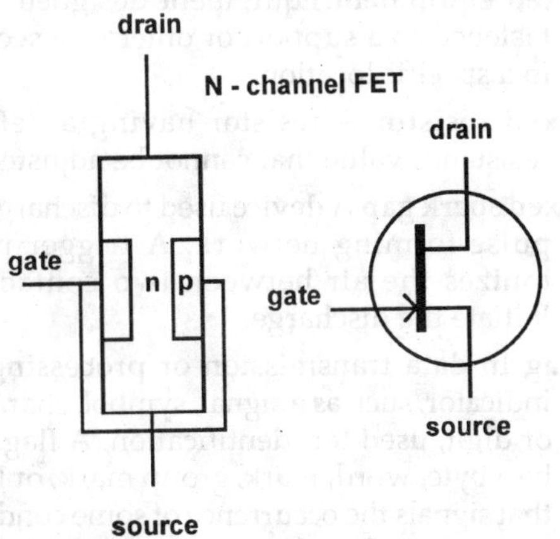

Fig. Field effect transistor FET

Field emission The emission of electrons from an unheated surface as a result of a strong electric field existing at that surface.

Field excitation The creation of a steady magnetic field within the field windings by the application of a dc voltage either from the generator itself or from an external source.

Field interface device (FID) In a BEMS, this serves as a point of consolidation for sensors and controllers.

Field of force A term used to describe the total force exerted by an action-at-a-distance phenomenon such as gravity upon matter, electric charges acting upon electric charges, and magnetic forces acting upon other magnets or magnetic materials.

Field of view A volume in space defined by an angular cone extending from the focal plane of an instrument.

Field weakening The introduction of resistance in series with the shunt wound field of a DC motor to reduce the voltage and current which weakens the strength of the magnetic field and thereby increases the motor speed.

FIFO First-in/first-out memory buffer. The first data into the buffer is the first data out of the buffer.

Filament Thin thread of carbon or tungsten which produces heat or light with the passage of current.

File A collection of related data or records. A file is used to store data, or program, on disk for later use.

Filler An inert foreign substance added to a polymer to improve or modify its properties.

Filling solution A solution of defined composition to make contact between an internal element and a membrane or sample. The solution sealed inside a ph glass bulb is called an internal filling solution. This solution normally contains a buffered chloride solution to provide a stable potential and a designated zero potential point. The solution which surrounds the reference electrode internal and periodically requires replenishing is called the reference filling solution. It provides contact between the reference electrode internal and sample through a junction.

Film ICs Conductive or non-conductive material deposited on a glass or ceramic substrate. Used for passive circuit components, resistors, and capacitors.

Filter A selective network of resistors, capacitors, and inductors that offers comparatively little opposition to certain frequencies, while blocking or attenuating other frequencies. In electronics, a device that transmits only part of the incident energy and may thereby change the spectral distribution of energy: (a) high-pass filters transmit energy above a certain frequency; (b) low-pass filters transmit energy below a certain frequency; (c) bandpass filters transmit energy of a certain bandwidth; (d) band-stop filters transmit energy outside a specific frequency band.

Final circuit The final circuits in an electrical wiring system. A circuit connected directly to current using equipment, or to a socket outlet or socket outlets or other outlet points for the connection of such equipment.

Final control element A device such as a valve or damper that acts to change the value of the manipulated variable. It is positioned by an actuator.

Final power amplifier (FPA) The final stage of amplification in a transmitter.

Financial attributes Attributes that measure the financial health of the company. Key financial attributes include capital requirements, earnings per share of common equity, capitalization ratios, and interest coverage ratios.

Finger Software that allows the user to enter the address of an Internet site to find information about that system's users or a particular user. Some finger addresses return other topic- specific information.

Firewall A system or combination of systems that enforces a boundary between two or more networks. Gateway that limits access between networks in accordance with local security policy. The typical firewall is an inexpensive micro-based Unix box kept clean of critical data, with many modems and public network ports on it, but just one carefully watched connection back to the rest of the cluster.

Firing A high temperature heat treatment that increases the density and strength of a ceramic piece.

Firm energy Power or power-producing capacity covered by a commitment to be available at all times during the period.

Firmware Programmed microprocessor-based controllers that cannot be re-programmed or altered.

Fishbowl To contain, isolate and monitor an unauthorized user within a system in order to gain information about the user.

Fissile material Isotopes which are capable of undergoing nuclear fission. Sometimes the term is restricted to apply only to isotopes which are capable of undergoing fission upon impact with a slow neutron.

Fission products Both stable and unstable isotopes produced as a result of nuclear fission.

Fixed bias A constant value of bias voltage.

Fixed cost The annual (or sometimes monthly) costs associated with the ownership of property such as depreciation, taxes, insurance, and the cost of capital.

Fixed equipment Equipment designed to be fastened to a support or otherwise secured in a specific location.

Fixed resistor A resistor having a definite resistance value that cannot be adjusted.

Fixed spark gap A device used to discharge the pulse-forming network. A trigger pulse ionizes the air between two contacts to initiate the discharge.

Flag In data transmission or processing, an indicator, such as a signal, symbol, character, or digit, used for identification. A flag may be a byte, word, mark, group mark, or letter that signals the occurrence of some condition or event, such as the end of a word, block, or message.

Flame Abusive message posted to a newsgroup as part of an on going argument or discussion.

Flange Mounting endshield with special rabbets and bolt holes for mounting such equipment as pumps and gear boxes to the motor or for overhanging the motor on the driven machine.

Flash memory. It is a non volatile memory technique with fast access times; rewriteable many times and uses a block erase technique as opposed to EEPROM, which erases one bit at a time.

Flash point The lowest temperature at which a substance gives off sufficient inflammable vapour to produce a momentary flash when a small flame is applied.

Flashover Arcing that is caused by the high voltage breakdown of insulation between two conductors, resulting in high current.

Flat line A transmission line that has no standing waves. This line requires no special timing devices to transfer maximum power.

Flat pack An IC package.

Flat rate A fixed charge for goods and services that does not vary with changes in the amount used, volume consumed, or units purchased.

Flat Response An output signal in which fundamental frequencies and harmonics are in the same proportion as those of the input signal being amplified. A flat frequency response would exhibit relatively equal response to all fixed-point frequencies within a given spectrum.

Flatness A term expressed in dB to specify the consistent amplitude of a signal spanning a frequency range. Typical expressions: Flatness is to be +/-1dB across (frequency range), OR Flatness shall have <1.5dB peak to peak.

Fleming valve An earlier name for a diode, or a two-electrode vacuum tube used as a detector.

Fleming's rules If the forefinger, second finger, and thumb of the right hand are extended at right angles to each other, the forefinger indicates the direction of flux, the second finger the direction of the emf and the thumb the direction of the motion in an electric generator. If the left hand is used, the fingers indicate the conditions for an electric motor.

Fletcher-Munson Curves Fletcher and Munson were pioneering researchers who provided the basis of High Fidelity in the '30s. They accurately measured and published a set of plots showing the human's ear's sensitivity to loudness verses frequency. They conclusively demonstrated that human hearing acuity is essentially dependent upon loudness. The curves show the ear most sensitive to sounds in the 3 kHz to 4 kHz area. This means sounds above and below 3-4 kHz must be louder in order to be heard just as loud. For this reason, the Fletcher-Munson curves are referred to as "equal loudness contours." They represent a range of sensitivity from "barely heard," (0 dB SPL) all the way to "painfully loud" (120 dB SPL), usually plotted in 10 dB increments.

Flexible cable A cable whose structure and materials make it suitable to be flexed while in service.

Flexible coaxial line A line made with an inner conductor that consists of flexible wire insulated from the outer conductor by a solid, continuous insulating material.

Flexible cord A flexible cable in which the cross sectional area of each conductor does not exceed 4 mm2.

Flexible load shape The ability to modify the load shape on short notice. When resources are insufficient to meet load requirements, load shifting or peak clipping may be appropriate.

Flexible retail pool a model for the restructured electric industry that features an Independent System Operator (ISO) operating in parallel with a commercial Power Exchange, which allows end-use consumers to buy from a spot market or "pool" or to contract directly with a particular supplier.

Flexible wiring system A wiring system designed to provide mechanical flexibility in use without degradation of the electrical components.

Flip chip A monolithic IC packaging technique that eliminates the need for bonding wires.

Flip flop A device having two stable states and two input terminals (or types of input signals), each of which corresponds with one of the two states. The circuit remains in either state until caused to change to the other state by application of a voltage pulse. A similar bistable device with an input that allows it to act as a single-stage binary counter.

Floating The condition where a common mode voltage exists between an earth ground and the instrument or circuit of interest. (Low of circuit is not at earth potential.)

Floating action This term refers to a controlled device which can stop at any point in it's stroke and can be reversed without completing its stroke. The controller must have a 'dead-spot' or neutral zone in which it sends no signal but allows the device to 'float' in a partly open position. For good operation, this system requires a rapid response in the controlled variable, otherwise it will stop at an intermediate position.

Floating output Ungrounded output of a power supply where either output terminal may be referenced to another specified voltage.

Floppy disk A small, flexible disk carrying a magnetic medium in which digital data is stored for later retrieval and use.

Flow Travel of liquids or gases in response to a force (i.e. Pressure or gravity).

Flow coefficient (C) Parameter used in conjunction with the "flow exponent" in a flow equation.

Flow exponent (n) Parameter which characterises the type of flow through a building (or component) and is used in conjunction with 'flow coefficient' in a 'flow equation'. (When n=1 flow is laminar, and when n=0.5 flow is assumed turbulent). For most openings, n takes a value between these two extremes.

Flow hood A device that measures airflow quantity

Flow network A network of zones or cells of differing pressure connected by a series of flow paths.

A passage/duct for smoke and fumes from a boiler/fire etc.

Flow rate Actual speed or velocity of fluid movement

Flow soldering Flow or wave soldering technique in large scale electronic assembly to solder all the connections on a printed circuit board by moving the board over a wave of molten solder.

Flowmeter A device used for measuring the flow or quantity of a moving fluid.

Fluctuation A surge or sag in voltage amplitude, often caused by load switching or fault clearing.

Flue gas The air exiting from a chimney after combustion and venting from the burner.

Fluidic controller A controller that exploits the DYNAMIC properties of a fluid (e.g., the 'Coanda Effect'), as distinct from hydraulic controllers and pneumatic controllers, which utilise the STATIC properties of a fluid.

Fluorescence property of many substances, of absorbing light of one wavelength (or colour) and emitting light of another wavelength (or colour).

Fluorescent lamp Discharge lamp of the low-pressure mercury type in which most of the light is emitted by a layer of fluorescent material excited by the ultraviolet radiation from the discharge.

Flush (mounting) Mounting a speaker in such a way that the speaker and its Grill do not protrude above the surrounding surface. Usually, this means mounting it at the back of the baffle board (the board the speaker is mounted on).

Flux The magnetic field which is established around an energized conductor or permanent magnet. The field is represented by flux lines creating a flux pattern between opposite poles. The density of the flux lines is a measure of the strength of the magnetic field.

Flux density The number of magnetic lines of force passing through a given area.

Flux linkage The linking of the magnetic flux with the conductors of a coil. The value obtained by multiplying the number of turns in the coil by the flux passing through the coil.

Fluxmeter An instrument for the measurement of magnetic flux.

Fly back converter A switching power supply circuit which normally uses a single transistor. During the first part of the cycle the transistor is on and energy is stored in a transformer primary; during the second part of the cycle this energy is transferred to the transformer secondary and the load.

Flywheel effect The ability of a resonant circuit to operate continuously because of stored energy or energy pulses.

FM Approved An instrument that meets a specific set of specifications established by Factory Mutual Research Corporation.

FM frequency modulation Where voltage levels change the frequency of a carrier wave.

Focus A control that converges beams to produce a sharp display.

Focusing anode An electrode of a CRT that is used to focus the electrons into a tight beam.

Fold back current limiting A power supply output protection circuit whereby the output current decreases with increasing overload, reaching a minimum at short circuit. This minimizes internal power dissipation under overload conditions.

Folded dipole An ordinary half-wave antenna (dipole) that has one or more additional conductors connected across the ends parallel to each other.

Foot (ft) Imperial unit of length. 1 foot = 12 inches = 304.8 mm exactly

Foot candle Unit of illumination at a point one foot distance from a one candela source. (in the imperial system of units) 1 foot candle = 1 lumen per square foot.

Foot lambert Unit of luminance. It is the luminance of a uniform diffuser emitting a foot candle.

Foot pound (ft. Lb.) Unit of energy. 1 ft.lb. = 1.356 J

Forbidden band The energy band in an atom lying between the conduction band and the valence band. Electrons are never found in the forbidden band but may travel back and forth through it. The forbidden band determines whether a solid material will act as a conductor, a semi-conductor, or an insulator.

Force External agency capable of altering the state of rest or motion of a body. (Unit: newton or N)

Forced convection Heat transmission by mechanically induced movement of a fluid.

Forced outage An outage that results from emergency conditions that requires a component to be taken out of service automatically or as soon as switching operations can be performed. The forced outage can be caused by improper operation of equipment or by human error. If it is possible to defer the outage, the outage becomes a scheduled outage.

Forced vibration Vibration of a system caused by an imposed force. Steady-state vibration is an unchanging condition of periodic or random motion.

Forcing function Externally applied time-dependent function, e.g., a step change in the setpoint or disturbance.

Fork bomb Also known as Logic Bomb - Code that can be written in one line of code on any Unix system; used to recursively spawn copies of itself, "explodes" eventually eating all the process table entries and effectively locks up the system.

Form factor A figure of merit which indicates how much rectified current departs from pure (non-pulsating) DC. A large departure from unity form factor (pure DC, expressed as 1.0) increases the heating effect of the motor and reduces brush life. Mathematically, form factor is the ratio of the root-mean square (rms) value of the current to the average (av) current or Irms/ Iav.

Form wound A type of coil in which each winding is individually formed and placed into the stator slot. A cross sectional view of the winding would be rectangular. Usually form winding is used on high voltage, 2300 volts and above, and large motors (449T and above). Form winding allows for better insulation on high voltage than does random (mush) winding.

Former The cylindrical portion of a speaker's voice coil section. A wire is wound around this cylinder to form a coil such that when current interacts with the magnetic field it produces a pumping motion that alternatively compresses and rarifies air, and creates the velocity for such air masses to reach our ears as sound.

Fortin barometer A mercury in glass barometer, which used in conjunction with correction tables enables accurate measurement of atmospheric pressure to be made.

FORTRAN Formula Translation language. A widely used high-level programming language well suited to problems that can be expressed in terms of algebraic formulas. It is generally used in scientific applications.

Forward agc The type of AGC that causes an amplifier to be driven towards saturation.

Forward bias An external voltage that is applied to a PN junction in the conducting direction so that the junction offers only minimum resistance to the flow of current. Conduction is accomplished by majority current carriers (holes in P-type material; electrons in N-type material).

Forward current Current in a circuit of a semiconductor device due to conduction by majority carriers across the pn junction.

Forward path The path that connects the reference value to the controlled variable.

Forward resistance The smaller resistance value observed when you are checking the resistance of a semiconductor.

Fossil fuel Any naturally occurring organic fuel, such as petroleum, coal, and natural gas. Remains of organisms embedded in the surface of the Earth, with high carbon and/ or hydrogen content, used as fuels.

Fossil fuel plant A plant using coal, oil, gas and other fossil fuel as its source of energy.

Foster Seeley discriminator A circuit that uses a double-tuned RF transformer to convert frequency variations in the received FM signal to amplitude variations. Also known as a phase-shift discriminator.

Four element array An antenna array with three parasitic elements and one driven element.

Fourier analysis The expansion of a mathematical function or an experimentally obtained waveform in the form of a trigonometric series.

Fourier number Dimensionless number, equal to Fo=(at)/lo2 , where a is the thermal diffusivity, lo is specific dimension and t is time.

Fourier series Resolution of a periodic function into its direct component, its fundamental sinusoidal component and an infinite series of harmonic sinusoidal components.

Fourier transform An integral transformation from the time domain to the frequency domain.

FPM Flow velocity in feet per minute.

FPS (Feet per second) A measure of flow velocity.

FPS Flow velocity in feet per second.

Fps system The foot pound second system of units is an imperial set of units derived from the fundamental units of the foot, the pound mass and the second.

Frame relay An interface protocol for statistically multiplexed packet-switched data communications in which (a) variable-sized packets (frames) are used that completely enclose the user packets they transport, and (b) transmission rates are usually between 56 kb/s and 1.544 Mb/s (the T-1 rate).

Framing The process of synchronizing a facsimile receiver to a transmitter. This allows proper picture reproduction.

Free convection Heat transmission by movement of a fluid caused by density differences.

Free electron An electron which is not attached to an atom, molecule or ion, but is free to move under the influence of an electric field.

Free energy (G) A thermodynamic quantity that is a function of the enthalpy (H), the Kelvin temperature (T) and the entropy (S) of a system; G=H-TS. At equilibrium, the free energy is at a minimum. Under certain conditions the change in free energy for a process is equal to the maximum useful work.

Free float control A building/plant system in which there is no active control strategy.

Free runing multivibrator A multivibrator that produces a continuous output waveform without any signal input. A square wave generator used to produce a clock signal.

Free sound field A sound field without acoustic boundaries or where the boundaries are so distant as to cause negligible reflections over the frequency range of audible sound. If the boundaries exist but completely absorb the sound then a virtual free field is created, thus anechoic chambers are used to accurately measure loudspeakers for their unique properties.

Free space loss The loss of energy of radio waves caused by the spreading of the wavefront as it travels from the transmitter.

Free-Air Configuration This description usually indicates a speaker that, in the opinion of the manufacturer, is suitable for mounting in only a minimal enclosure, such as a baffle board that separates the back wave from the front.

Freeware Software distributed via the Internet that can be downloaded and utilised for free.

Freezing point The fixed temperature point at which a material changes from a liquid to a solid state. This is the same as the melting point for pure materials. For example, the freezing point of water is 32¼F or 0¼C.

Frequency (f) 1. The number of complete cycles per second existing in any form of wave motion, such as the number of cycles per second of an alternating current.

2. The rate at which the vector that generates a sine wave rotates.

Frequency compensation network Circuit modification used to improve or broaden the linearity of its frequency response.

Frequency counter A circuit that can measure and display the frequency of a signal.

Frequency cutoff The frequency at which the filter circuit changes from an action of rejecting the unwanted frequencies to an action of passing the desired frequencies. Conversely, the point at which the filter circuit changes from an action in which it passes the desired frequencies to an action in which it rejects the undesired frequencies.

Frequency determining network A circuit that provides the desired response (maximum or minimum impedance) at a specific frequency.

Frequency deviation The amount the frequency varies from the carrier frequency.

Frequency diversity Transmitting (and receiving) of radio waves on two different frequencies simultaneously.

Frequency division multiplex (FDM) Transmission of two or more signals over a common path by using a different frequency band for each signal.

Frequency domain analysis A method of representing a waveform by plotting its amplitude against frequency.

Frequency meter Meter used to measure frequency of periodic waves.

Frequency modulated output A transducer output which is obtained in the form of a deviation from a center frequency, where the deviation is proportional to the applied stimulus.

Frequency modulation (fm) Angle modulation in which the modulating signal causes the carrier frequency to vary. The amplitude of the modulating signal determines how far the frequency changes, and the frequency of the modulating signal determines how fast the frequency changes.

Frequency multipliers Special RF power amplifiers that multiply the input frequency.

Frequency of vibration The number of cycles occurring in a given unit of time. RPM - revolutions per minute. CPM- cycles per minute.

Frequency output An output in the form of frequency which varies as a function of the applied input.

Frequency range The measure of a circuit's ability to pass a full amplitude signal over a range of signal frequencies. Normally measured between the point or points where the signal amplitude falls to -3dB below the passband frequency. Normally defines the "bandwidth" of a device or system.

Frequency response curve A curve showing the output of an amplifier (or any other device) in terms of voltage or current plotted against frequency with a fixed-amplitude input signal.

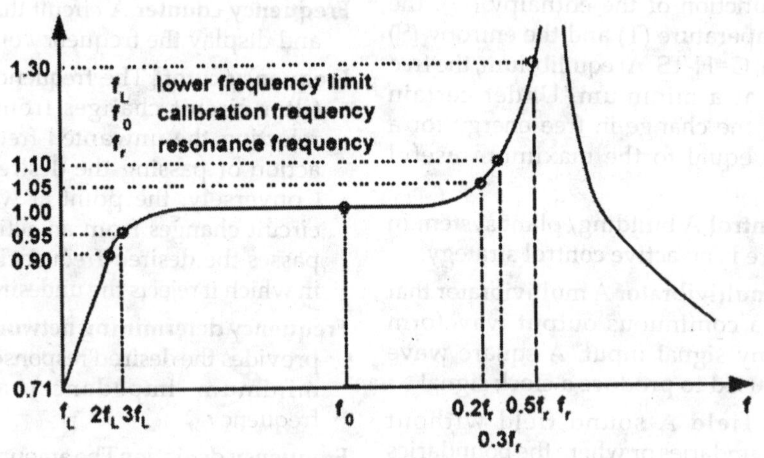

Fig. Frequency response curve

Frequency Response The range of frequencies that a speaker will reproduce (lowest frequency to the highest). While the optimal normal is 20 - 20,000 Hz (Hertz), the range of human hearing for individuals is often much more restricted. A good full-range speaker system however, will reproduce as much of this range as possible in order to cover all variations. Individual Drivers are limited to reproducing only that part of the spectrum for which they were made, so their response will be limited, but still a necessary point to consider when designing a complete sound system.

Frequency scanning Varying the output frequency to achieve electronic scanning.

Frequency shift keying (fsk) Frequency modulation somewhat similar to continuous-wave (cw) keying in AM transmitters. The carrier is shifted between two differing frequencies by opening and closing a key.

Frequency spectrum The frequency spectrum of a signal consists of the plots of the amplitude and phases of the harmonics against frequency.

Frequency stability Refers to the ability of an oscillator to accurately maintain its operating frequency.

Frequency synthesis A process that uses hetrodyning and frequency selection to produce a signal.

Frequency synthesizer 1. A frequency source of high accuracy.

2. A bank of oscillators in which the outputs can be mixed in various combinations to produce a wide range of frequencies.

Internet.

Fuel Any substance which is used for producing heat energy, either by means of the release of its chemical energy by combustion or its nuclear energy by nuclear fission.

Fuel cell A electrochemical energy conversion device for producing electricity converting hydrogen and oxygen into electricity. These cells convert chemical energy directly into electrical energy.

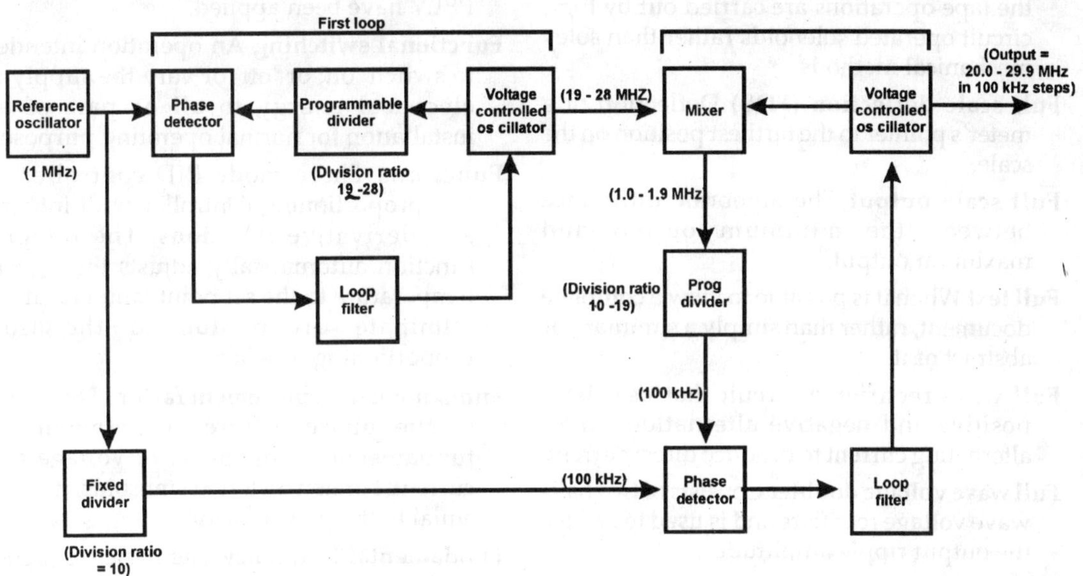

Fig. Frequency synthesizer

Friction Resistance to motion due to contacting surfaces.

Front to back ratio The ratio of the energy radiated in the principal direction compared to the energy radiated in the opposite direction.

Fs Fs or Free Air Resonance is the frequency at which a speaker naturally resonates, like a tuning fork. Sometimes known as ringing response.

FTP File Transfer Protocol. An application program that uses TCP/IP protocol to allow you to move files from a distant computer to a local computer using a network like the

Fuel element An element of nuclear fuel for use in a nuclear reactor, usually uranium encased in a case.

Fuel escalation The annual rate of increase of the cost of fuel, including inflation and real escalation, resulting from resource depletion, increased demand, etc.

Fuel expenses Costs associated with the generation of electricity.

Full bridge A Wheatstone bridge configuration utilizing four active elements or strain gages.

Full duplex circuit A circuit that permits simultaneous transmission in both directions.

Full fanout Distributing the same signal to multiple destinations.

Full load current The largest current that a motor or other device is designed to carry under specific conditions. Also current at rated conditions.

Full load torque That torque of a motor necessary to produce its rated horsepower at full-load speed, sometimes referred to as running torque.

Full Logic Deck A cassette mechanism where the tape operations are carried out by logic circuit operated solenoids rather than soley mechanical methods.

Full scale deflection (FDS) Deflection of a meter's pointer to the farthest position on the scale.

Full scale output The algebraic difference between the minimum output and maximum output.

Full text When it is possible to view a complete document, rather than simply a summary or abstract of it.

Full wave rectifier A circuit that uses both positive and negative alternations in an alternating current to produce direct current.

Full wave voltage doubler Consists of two half-wave voltage rectifiers and is used to reduce the output ripple amplitude.

Function A specific purpose of an entity; its characteristic action.

Function generator Signal generator that can produce sine, square, triangle and sawtooth output waveforms.

Function sequence systems Many digital control systems require control events to occur in a sequence for which an output state produces a change in the input state. That is, where each event is not so much time dependent as dependent on the completion of the previous event. This process of change producing change continues until some overall objective has been met is termed 'function sequence system control'; cf. Time-sequence control.

Functional earth conductor Conductor to be connected to a functional earth terminal.

Functional earth terminal Terminal directly connected to a point of a measuring supply or control circuit or tc a screening part which is intended to be earthed for functional purposes.

Functional earthing Connection to Earth necessary for proper functioning of electrical equipment.

Functional extra low voltage (FELV) An extra low voltage system in which not all of the protective measures required for SELV or PELV have been applied.

Functional switching An operation intended to switch 'on' or 'off' or vary the supply of electrical energy to all or part of an installation for normal operating purposes.

Functions Three mode PID controller. A timeproportioning controller with integral and derivative functions. The integral function automatically adjusts the system temperature to the set point temperature to eliminate droop due to the time proportioning function.

Fundamental displacement factor FDF Cosine of the phase difference between the fundamental components of voltage and current. For non distorted sinusoids, it is also equal to the power factor.

Fundamental frequency The basic frequency or first harmonic frequency.

Fundamental units The units in which physical quantities are measured which are independent from each other.

Fuse A device designed to provide protection for a given circuit or device by physically opening the circuit. Fuses are rated by their amperage and are designed to blow or open when the current being drawn through it exceeds its design rating. They can be fast or slow acting, depending on type.

Fuse carrier The movable part of a fuse designed to carry a fuse link.

Fuse element A part of a fuse, which is designed to melt and thus open a circuit

Fuse linka part of a fuse, which comprises a fuse element and a cartridge (or other container) and is capable of being attached to the fuse contacts

Fusing current This is the minimum current that will cause the fuse element to heat up melt or blow

Fusing factor This is the ratio of the fusing current to current rating

Fuzzy logic Fuzzy logic is basically a multivalued logic that allows intermediate values to be defined between conventional evaluations like yes/no, true/false, black/white, etc. Notions like 'rather warm' or 'pretty cold' can be formulated mathematically and processed by computers. In this way an attempt is made to apply a more human-like way of thinking in the programming of computers, controllers, etc.

G

G The force of acceleration due to gravity equal to 32.1739 ft/sec2 or 386 in./sec2.

Gage factor A measure of the ratio of the relative change of resistance to the relative change in length of a piezoresistive strain gage.

Gage length The distance between two points where the measurement of strain occurs.

Gage pressure Absolute pressure minus local atmospheric pressure.

Gage pressure transducer A transducer which measures pressure in relation to the ambient pressure.

Gain 1. The ratio between the amount of energy propagated from an antenna that is directional compared to the energy from the same antenna that would be propagated if the antenna were not directional.

2. Any increase in the strength of a signal.

Gain bandwidth product The number that results when the gain of a circuit is multiplied by the bandwidth of that circuit. For an operational amplifier, the gain-bandwidth product for one configuration will always equal the gain-bandwidth product for any other configuration of the same amplifier.

Gain margin The factor by which the gain must be increased in order to produce instability.

Galvanic corrosion The preferential corrosion of the more chemically active of two metals electrically coupled and exposed to an electrolyte.

Galvanised iron Iron coated with a layer of zinc to prevent corrosion, usually by hot dipping into the molten metal.

Galvanometer An instrument for detecting, comparing, or measuring small electric currents, but not usually calibrated. Usually depends on the magnetic effect produced by an electric current.

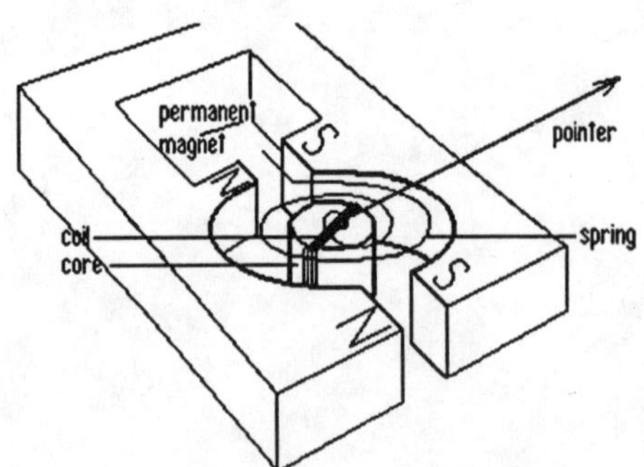

Fig. Galvanometer

Gamma rays High frequency electromagnetic radiation from radio active particles.

Gamma The emitter-to-base current ratio in a common-collector configuration.

Ganged Mechanical coupling of two or more capacitors, switches, potentiometers, or any other adjustable components so that adjusting one control will operate all.

Ganged control A method of control that allows the system to control multiple levels of switching with a single command. Such is the case in a video system where different levels are assigned to the different colours in a video signal (R = Red, G = Green and B = Blue). A single command will control all three levels simultaneously.

Ganged tuning The process used to tune two or more circuits with a single control.

Gas Any aeriform or completely elastic fluid which is not a solid or a liquid. Gasses are produced by heating a liquid beyond its boiling point.

Gas chromatography A process by which gases can be separated from one another. This technique is used to separate tracer gases from each other and from the constituents of air, thus allowing individual quantitative analysis to be performed.

Gas sorption devices Devices used to reduce levels of airborne gaseous compounds by passing the air through materials that extract the gases. The performance of solid sorbents is dependent on the airflow rate, concentration of the pollutants, presence of other gases or vapors, and other factors.

Gas turbine Consists typically of an axial-flow air compressor, one or more combustion chambers (where liquid or gaseous fuel is burned). The hot gases are passed to the turbine and expanded to drive the generator and run the compressor.

Gas turbine plant A plant in which the prime mover is a gas turbine.

Gate As applied to logic circuitry, one of several different types of electronic devices that will provide a particular output when specified input conditions are satisfied. Also, a circuit in which a signal switches another signal on or off.

Gated agc Circuit that permits automatic gain control to function only during short time intervals.

Gated beam detector An FM demodulator that uses a special gated-beam tube to limit, detect, and amplify the received FM signal. Also known as a quadrature detector.

Gateway A site which provides links to Web sites, organised by subject area. These links have been assessed against specified quality criteria and also provide short descriptions of sites.

Gating The process of selecting those portions of a wave that exist during one or more selected time intervals or that have magnitudes between selected limits. Also, the application of a specific waveform to perform electronic switching.

Gauss G An old unit for measuring magnetic flux density (or magnetic induction). 1 G = 10-4 T

Gear ratio Ratio of the motor input speed to the gearhead output speed.

Gearhead The portion of a gearmotor which contains the actual gearing which converts the basic motor speed to the rated output speed.

Gearhead efficiency Actual torque output divided by theoretical torque output (i.e., torque input times the gear ratio).

Gearmotor A gearhead and motor combination to reduce the speed of the motor to obtain the desired RPM's.

Geiger counter Device used to detect nuclear particles.

Gene A given segment of the DNA molecule that contains the code for a specific protein.

General purpose electronic test equipment (GPETE) Test equipment that has the capability, without modification, to generate, modify, or measure a range of electronic functions required to test several equipments or systems of basically different designs.

General purpose motor A general-purpose motor is any motor having a "B" design, listed and offered in standard ratings with standard operating characteristics and mechanical construction for use under usual service conditions without restriction to a particular application or type of application.

Generating station, generating plant The location of prime movers, electric generators, and auxiliary equipment used for converting mechanical, chemical, and nuclear energy into electric energy.

Generating unit Any combination of physically connected generators, reactors, boilers, combustion turbines, or other prime movers operated together to produce electric power.

Generation The process of producing electrical energy by transforming non electrical forms of energy.

Generation dispatch and control Aggregation and dispatching (sending off to some location) generation from various generating facilities, providing backup and reliability of services.

Generator A machine that produces AC electricity from a rotating coil interaction within a magnetic field. In automotive applications, a rectifier is used to convert this output to DC. Also, an electronic device used for converting DC voltage into AC of a given frequency and wave shape. An amplifier is often a form of generator.

Geothermal planta plant in which the prime mover is a steam turbine driven by steam that derives its energy from heat found in rocks or fluids beneath the surface of the earth. The energy is extracted from naturally occurring geothermal fields or by drilling and/or pumping.

Germanium Element 22, used mostly in early semiconductor devices.

Gimbal A mechanical frame, with two perpendicular intersecting axes of rotation, used to support and furnish a gyro wheel with the necessary freedom to tilt in any direction.

Glare Condition of vision in which there is discomfort or a reduction in the ability to see significant objects, or both, due to an unsuitable distribution or range of luminance or to extreme contrasts in space or time.

Glare index A glare index illustrates the probability of the occupants being satisfied with a particular view direction. In modelling and simulation this allows the environmental engineer to predict visual discomfort and thereby ensure that it is avoided.

Glass An amorphous solid obtained when silica is mixed with other compounds, heated above its melting point, and then cooled rapidly.

Glass transition temperature (Tg) The temperature at which, upon cooling, a noncrystalline ceramic or polymer transforms from a supercooled liquid to a rigid glass.

Global points Allows designated points to share their data with other bus connected devices.

Glow discharge Electric discharge in which the secondary emission from the cathode is much greater than the thermionic emission.

Grain boundary The interface separating two adjoining grains having different crystallographic orientations.

Grain growth The increase in average grain size of a polycrystalline material: for most materials, an elevated temperature heat treatment is necessary.

Grain size The average grain diameter as determined from a random cross section.

Gramme ring armature An inefficient type of armature winding in which many of the turns are shielded from the field by its own iron ring.

Granulation noise A type of audible distortion resulting from quantization error in digital transmission modes.

Graph of network The geometric structure of the interconnection of the network elements which completely characterises the number of independent loop currents or the number of independent node-pair voltages necessary to study the network.

Graphic equalizer A multi-band variable equalizer using slide controls as the amplitude adjustable elements. Named for the positions of the sliders "graphically" illustrating the resulting frequency response of the equalizer. Only found on active (amplified)designs. Center frequency and bandwidth are fixed for each band.

Graticule The CRT grid lines that facilitate the location and measurement of oscilloscope traces.

Green ceramic body A ceramic piece, formed as a particulate aggregate, that has been dried but not fired.

Greencap A type of polyester capacitor that the manufacturer dips in green paint to make it distinctive from all other capacitors.

Greenhouse effect The greenhouse effect allows solar radiation to penetrate but absorbs the infrared radiation returning to space. It thus increases the mean global surface temperature of the earth caused by gases in the atmosphere (including carbon dioxide, methane, nitrous oxide, ozone, and chlorofluorocarbon).

Greenhouse gases Greenhouse gases include carbon dioxide, methane, nitrous oxide, hydrocarbons, and chlorofluorocarbons. These provide the greenhouse effect.

Grid The layout of an electrical transmission and distribution system.

Grid bias A constant fixed potential applied between the grid and the cathode of a vacuum tube to establish an operating point.

Grid current The current that flows in the grid-to-cathode circuit of a vacuum tube.

Grid gap tuning A method of changing the center frequency of a resonant cavity by physically changing the distance between the cavity grids.

Grid interconnection: A link between Electricity system and Embedded Generator's Electricity System, made for the purpose of Exporting or Importing Electrical Energy.

Grid leak bias A self-bias provided by a high resistance connected across the grid capacitor or between the grid and cathode.

Grill A barrier meant to prevent damage to the Driver that it covers, or that creates a more attractive appearance.

Gross generation Amount of electric energy produced by generating units as measured at the generator terminals.

Ground 1. The point in a circuit used as a common reference point for measuring purposes.

2. To cònnect some point of an electrical circuit or some item of electrical equipment to earth or to the conducting medium used in lieu thereof.

Ground clutter Unwanted echoes, from surrounding land masses, that appear on a radar indicator.

Ground controlled approach A radar system used to guide aircraft to safe landings in poor visibility conditions.

Ground fault An undesired path that allows current to flow from a line to ground.

Ground fault interruption A unit or combination of units which provides protection against ground fault currents below the trip levels of the breakers of a circuit.

Ground loop A condition created when two or more paths for electricity are created in a ground line, or when one or more paths are created in a shield. Ground loops can create undesirable noise.

Ground plane The earth or negative rail of a circuit. A large or significant mass that presents the effect of earth (ground) to a signal.

Ground plane antenna A type of antenna that uses a ground plane as a simulated ground to produce low-angle radiation.

Ground potential The electrical potential of the earth. A circuit, terminal or chassis is said to be at ground potential when it is used as a reference point for other potential in the system.

Ground range The distance on the surface of the earth between a radar and its target.

Equal to slant range only if both radar and target are at the same altitude.

Ground reflection loss The loss of RF energy each time a radio wave is reflected from the earth's surface.

Ground screen A series of conductors buried below the surface of the earth and arranged in a radial pattern. Used to reduce losses in the ground.

Ground state The most stable energy state of a nucleus, atom or molecule.

Ground waves Radio waves which travel near the surface of the earth.

Grounded Connected to or in contact with earth or connected to some extended conductive body which serves instead of the earth.

Grounded junction A form of construction of a thermocouple probe where the hot or measuring junction is in electrical contact with the sheath material so that the sheath and thermocouple will have the same electrical potential.

Grounded motor A motor with an electrical connection between the motor frame and ground.

Grounded neutral The common neutral conductor of an electrical system which is intentionally connected to earth to provide a current carrying path for the line to neutral load devices.

Grounding A permanent and continuous conductive path to earth with sufficient capacity to carry any fault current liable to be imposed on it, and of a sufficiently low impedance to limit the voltage rise above ground.

Group A collection of units, assemblies, subassemblies, and parts. It is a subdivision of a set or system but is not capable of performing a complete operational function.

Group velocity The forward progress velocity of a wave front in a waveguide.

Grown junction A method of mixing P-type and N-type impurities into a single crystal while the crystal is being grown.

Guarded motor An open motor in which all openings giving direct access to live or rotating parts (except smooth shafts) are limited in size by the design of the structural parts or by screens, grills, expanded metal, etc., to prevent accidental contact with such parts. Such openings shall not permit the passage of a cylindrical rod 1/2 inch in diameter.

Guarding A technique that reduces leakage errors and decreases response time. Consists of a guard conductor driven by a low-impedance source surrounding the lead of a high-impedance signal. The guard voltage is kept at or near the potential of the signal.

Guidance radar A system which provides information that is used to guide a missile to a target.

Gunn diode A semiconductor diode that utilizes the Gunn effect to produce microwave frequency oscillation or to amplify a microwave frequency signal.

Gyro Abbreviation for gyroscope.

Gyroscope A mechanical device containing a spinning mass mounted so that it can assume any position in space.

H parameters (hybrid parameters) Transistor specifications that describe the component operating limits under specific circumstances.

H TYPE T-Junction A waveguide junction in which the junction arm is parallel to the magnetic lines of force in the main waveguide.

Hacker A person who enjoys exploring the details of computers and how to stretch their capabilities. A malicious or inquisitive meddler who tries to discover information by poking around. A person who enjoys learning the details of programming systems and how to stretch their capabilities, as opposed to most users who prefer to learn on the minimum necessary.

Hacking run A hack session extended long outside normal working times, especially one longer than 12 hours.

Hacking Unauthorized use, or attempts to circumvent or bypass the security mechanisms of an information system or network.

Half bridge Two active elements or strain gages.

Half duplex One way at a time data communication; both devices can transmit and receive data, but only one at a time.

Half duplex transmission Transfer of data in two directions but by alternate, one way at a time, independent transmission; cf. Duplex transmission.

Half life The time taken for the activity of a radioactive isotope to decay to half of its original value. In other words the time taken for half the atoms present to disintegrate.

Half power point A point on a waveform or radar beam that corresponds to half the power of the maximum power point.

Half wave dipole antenna An antenna, consisting of two rods (1/4 wavelength each) in a single line, that radiates electromagnetic energy.

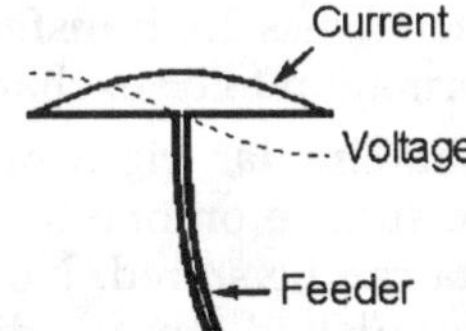

Fig. Half wave dipole antenna

Half wave rectifier A rectifier with only one diode in series with the load. The output is a half-wave rectified voltage with the other half wave being at zero voltage.

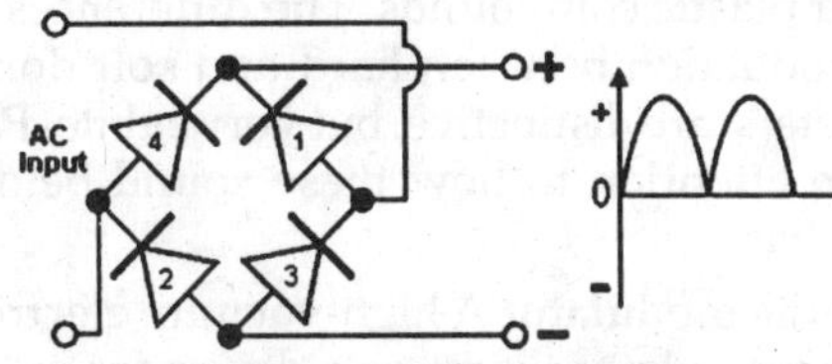

Fig. Half wave rectifier

Half wave symmetry A function has half-wave symmetry when one half of its waveform is exactly the negation of the previous half of the waveform.

Half wave voltage doubler Two half-wave voltage rectifiers connected to double the input voltage.

Halide Binary compound of one of the halogen elements (fluorine, chlorine, bromine or iodine).

Hall effect The phenomenon whereby a force is applied to a moving electron or hole by a magnetic field that is applied perpendicular to the direction of motion. The force direction is perpendicular to both the magnetic field and the particle motion directions.

Hall sensors Feedback device used for commutation.

Halogen lamp Gas-filled lamp containing a tungsten filament and a small proportion of halogens.

Hand over The operation where one earth terminal yields control to another as a satellite moves out of its area of coverage.

Handshake A term used to describe how computers or peripherals can communicate without conflict.A handshake can be hardwired, where status is indicated by on/off electrical levels or software based where control characteristics are transferred along with the information to be exchanged.

Hard disk A flat, circular, rigid plate with a magnetizable surface on one or both sides of which data can be stored. Note: A hard disk is distinguished from a diskette by virtue of the fact that it is rigid.

Hard Dome (Tweeter) A characteristic of some Dome tweeters in which the dome is made of some light, hard metal such as neodymium, titanium, or some of the more rigid plastic compounds. The differences in reproduction between hard and soft dome tweeters, are distinctive, but very subtle. Pay close attention to how these sound before purchase.

Hard tube modulator A high-vacuum electron tube modulator that uses a driver for pulse forming.

Hardcopy Output in a permanent form (usually a printout) rather than in temporary form, as on disk or display terminal.

Hardness The measure of a material's resistance to deformation by surface indentation or by abrasion. There are various scales in use to express hardness. The Mohs scale is qualitative and somewhat arbitrary and ranges from 1 on the soft end for talc to 10 for diamond. Quantitative scales are the Rockwell (HR), Brinell (indicated by HB), Knoop (HK) and Vickers (HV). Knoop and Vickers are referred to as microhardness testing methods on the basis of load and indenter size.

Hardware The tangible electronic (and mechanical) devices which constitute a computer system. This includes both the computer itself, the monitor (with the screen), the keyboard and any peripherals.

Harmonic distortion The presence of harmonics that change the AC voltage waveform from a simple sinusoidal to complex waveform. Harmonic distortion can be generated by a load and fed back to the AC line, causing power problems to other equipment on the same circuit.

Harmonics Also called overtones, these are vibrations at frequencies that are multiples of the fundamental. Harmonics extend without limit beyond the audible range. They are characterized as even-order and odd-order harmonics. A second-order harmonic is two times the frequency of the fundamental; a third order is three times the fundamental; a fourth order is four times the fundamental; and so forth. Each even-order harmonic: second, fourth, sixth, etc.-is one octave or multiples of one octave higher than the fundamental; these even-order overtones are therefore musically related to the fundamental. Odd-order harmonics, on the other hand: third, fifth, seventh, and up-create a series of notes that are not related to any octave overtones and therefore may have an unpleasant sound. Audio systems that emphasize odd-order harmonics tend to have a harsh, hard quality, execept first order loudspeaker crossovers, because they have an absolute minimum of phase shifts (the only type of phase linear speakers).

Harmonized standard A standard which has been drawn up by common agreement between national standards bodies notified to the European Commission by all member states and published under national procedures.

Harness The universal name for a bundle or loom of wires that compose the wiring for a system.

Hartley oscillator An oscillator that uses a tapped inductor in the feedback network.

Harvester Software for collecting and indexing resources and associated metadata according to specified rules.

Hatchback (box or enclosure) An enclosure of such size and shape as to enable its efficient use in hatchback-style vehicles. Normally, these enclosures are somewhat deeper than standard angled enclosures and have either handles on the end panels or no handles at all.

Hazard The potential for harmful effects.

Hazardous duty Empire Magnetics hazardous duty (PT) motors and related products can be designed into a system that meets NFPA class I, II, and/or III requirements.

Head loss The loss of pressure in a flow system measured using a length parameter (i.e., inches of water, inches of mercury).

Headroom A term related to the dynamic range of amplifiers, used to express in dB, the level between the typical operating level and the maximum output level (onset of clipping). For example, a nominal +5 dBu system that clips at +25 dBu has 20 dB of headroom. Because the term depicts a pure ratio, there are no units or reference-level associated with headroom, only relative "dB." Therefore headroom expressed in dB accurately refers to both voltage and power. Which means the example above has both 20 dB of voltage headroom, and 20 dB of power headroom.

Heat Thermal energy. Heat is expressed in units of calories or BTU's.

Heat balance A statement of the heat input.to, and heat loss from, an appliance, plant or structure, intended to account for all sources of heat and equivalent energy.

Heat capacity (Cv at constant volume and Cp at constant pressure) The quantity of heat required to produce a unit temperature rise per mole of material.

Heat Dissipation The function of transfering heat away from a component into the air to prevent damage to the output section of an amplifier or the voice coil of a speaker.

Heat exchanger (air to air) A device designed to transfer heat from two physically separated fluid streams. In buildings, it as generally used to transfer heat from exhaust warm air to incoming cooler outdoor air.

Heat flux The amount of heat passing through any surface per unit time.

Heat flux per unit area/specific heat flow Heat flux related to the unit square surface.

Heat pump (air to air) A device operating on a refrigeration cycle in which both evaporator and condenser are refrigerant/air heat exchangers. As a heating season heat recovery device, the evaporator transfers heat from the exhaust warm air to the refrigerant and the condenser transfers heat from the refrigerant to the incoming air. Arrangements are often made to allow the refrigerant flow to be reversed making the condenser the evaporator and vise versa - thus energy may be recovered in the cooling season.

Heat rate A measure of generating station thermal efficiency and generally expressed as BTU per net kwh. The heat rate is computed by dividing the total BTU content of the fuel burned (or of heat released from a nuclear reactor) by the resulting net kwh generated.

Heat recovery effectiveness Often referred to as heat recovery efficiency. The proportion of heat recovered from otherwise waste heat passing through a heat recovery system. Normally expressed as a percentage.

A device operating on a refrigeration cycle in which the evaporator is a water/ refrigerant heat exchanger and the condenser, a refrigerant/ air heat exchanger. The circuit normally includes an arrangement which allows the refrigerant flow to be reversed thus allowing heat to be transferred in either direction. In one system, a number of small air/water heat pumps installed in various zones around a building are used to transfer heat into or from a common water circuit. Thus heat unwanted in one zone may be transferred to another where it is needed.

Heat sink 1. Thermodynamic. A body which can absorb thermal energy.

2. Practical. A finned piece of metal used to dissipate the heat of solid state components mounted on it.

Heat treating A process for treating metals where heating to a specific temperature and cooling at a specific rate changes the properties of the metal.

Heating The transfer of energy to a space or to the air by the existence of a temperature gradient between the source and the space or air. This process may take different forms,ie, conduction, convection or radiation.

Heating load Diversified total heating loads, by fuel, represent critical plant sizes and hence capital costs. The breakdown of the total, by zone, highlights areas of concern. A comparison of the as-built and reference IPVs for the building can be made.

Heavy oil The fuel oils remaining after the lighter oils have been distilled off during the refining process. Except for start-up and flame stabilization, virtually all petroleum used in steam plants is heavy oil.

Hecto (h) Decimal multiple prefix corresponding to a hundred or 102. This is not a preferred suffix.

Height finding radar A radar that provides target altitude, range, and bearing data.

Helix 1. A spirally wound transmission line used in a traveling-wave tube to delay the forward progress of the input traveling wave.

2. A large coil of wire. It acts as a coil and is used with variable inductors for impedance matching of high-power transmitters.

Helix house A building at a transmitter site that contains antenna loading, coupling, and tuning circuits.

Heme An iron complex.

Hemoglobin A biomolecule composed of four myoglobine-like units (proteins plus heme) that can bind and transport four oxygen molecules in the blood.

HEMT High electron mobility transistor.

Henry (H) The electromagnetic unit of inductance or mutual inductance. The inductance of a circuit is 1 henry when a current variation of 1 ampere per second induces 1 volt. In electronics, smaller units are used, such as the millihenry (mH), which is one-thousandth of a henry (H), and the microhenry (μH) which is one-millionth of a henry.

Henry's law The amount of gas dissolved in a solution is directly proportional to the pressure of the gas above the solution.

HEPA An acronym for high efficiency particulate arrestance (filters).

HERTZ (Hz) Unit of frequency. One hertz is equal to one cycle per second.

Hertz antenna A half-wave antenna that is installed some distance above ground and positioned either vertically or horizontally.

Heterodyne detection The use of an a.f. voltage to distinguish between available signals. The incoming cw signal is mixed with locally generated oscillations to give an a.f. output.

Heterodyning 1. The process of mixing two frequencies across a nonlinear impedance.

2. The process of mixing the incoming signal with the local oscillator frequency. This produces the two fundamentals and the sum and difference frequencies.

Heuristic A method of solving mathematical problems for which no algorithm exists. Involves the narrowing down of the field of search for a solution by inductive reasoning from past experience of similar problems.

Hexadecimal Refers to a base sixteen number system using the characters 0 through 9 and A through F to represent the values. Machine language programs are often written in hexadecimal notation.

Hexadecimal system Pertaining to the number system with a radix of sixteen. It uses the ten digits of the decimal system and the first six letters of the English alphabet.

Hi Pot test A test that applies a high voltage to a conductor to assure the integrity of the surrounding insulation.

High end speaker A speaker that utilizes high-performance drivers, a well-built enclosure incorporating quality materials and finish, and well-designed crossover circuitry with narrow tolerance components. High-end speakers reproduce sound with high accuracy.

High Fidelity A method of sound reproduction that emphasizes the highest possible adherence to the exact character of the original sound. This is a principal that must be paramount in every stage of the recording, transportation, and re-creation of the signal to be fully operational. It is usually the listener's equipment wherein this objective is most seriously compromised.

High frequency The band of frequencies from 3 megahertz to 30 megahertz.

High intensity discharge HID lamp A type of lamp that may consist of mercury vapor, metal halide, high pressure sodium, or low pressure sodium.

High Level Input An audio input configured to accept speaker level signals.

High level modulation Modulation produced in the plate circuit of the last radio stage of the system.

High Pass Filter An electronic filter of a type commonly incorporated in Crossover circuits that permits the passage of high frequencies while suppressing lower ones. The place in the frequency spectrum where this occurs is called the crossover point and is different for each set of Drivers being considered. The most basic form of such filter is a non-polarized capacitor. Typical values for such a unit would be in the range of 1 to 100 microfarads.

High Power Output Speaker level outputs driven by an amplifier, typically at least 15 watts RMS per channel.

High tension(HT) Lethal voltage in the kilovolt range and above.

High voltage direct current transmission, hvdc transmission power transmission carried out at high voltage direct current.

High Voltage Switching Power Supply An amplifier's internal power supply that converts the vehicle's 12 volts to higher voltage for improved dynamic range and higher amp output power.

High voltage test A test which consists of the application of a specified voltage higher than the rated voltage between windings and frame, or between two or more windings, for the purpose of determining the adequacy of insulating materials and spacing against breakdown under normal conditions. (It is not the test of the conductor insulation of any one winding.

Hi-Pot Test A dielectric test performed by applying a high voltage for a specified time to two isolated points in a device to determine adequacy of insulating materials.

HIPPI [High-Performance Parallel Interface] The HIPPI specification defines the mechanical, electrical and signalling protocol for a high performance simplex interface [data transmits in one direction only]. An additional HPPI interface is required to transmit data in a full-duplex mode.

Histogram A type of graphical representation, used in statistics, in which the frequency distribution is expressed by rectangles.

Hits Different definitions depending on context:

Hits per scan The number of times an RF beam strikes a target per antenna revolution.

Hold Meter HOLD is an external input which is used to stop the A/D process and freeze the display. BCD HOLD is an external input used to freeze the BCD output while allowing the A/D process to continue operation.

Hold Up time The time under worst case conditions during which a power supply's output voltage remains within specified limits following the loss or removal of input power.

Holding torque Maximum torque the motor can provide to hold itself in a fixed position; also called static torque.

Hole A fictitious mobile particle that behaves as though it is a positively charged particle; holes are produced in the valence band when electrons from the valence band are promoted to the conduction band or an acceptor level of a p-type dopant.

Hole flow In the valence band, a process of conduction in which electrons move into holes, thereby creating other holes that appear to move toward a negative potential. (The movement of holes is opposite the movement of electrons.)

Hologram Three-dimensional picture created with a laser.

Holography The science dealing with three-dimensional optical recording.

Homogeneous and heterogeneous assays A homogeneous assay does not require a separation step to remove free antigen from bound antigen and relies upon the fact that the function of the label is modified upon binding, leading to a change in signal intensity. Because of high background signal a heterogeneous approach incorporating a separation step of bound and unbound makes the detection limit lower, approaching the values obtained by RIA. The homogeneous assay is less technically demanding.

Hooke's law Defines the basis for the measurement of mechanical stresses via the strain measurement. The gradient of Hooke's line is defined by the ratio of which is equivalent to the Modulus of Elasticity E (Young's Modulus).

Horizontal axis On a graph, the straight line axis that is plotted from left to right.

Horizontal deflection plates A pair of parallel electrodes that moves the electron beam from side to side in a CRT.

Horizontal pattern The part of a radiation pattern that is radiated in all directions along the horizontal plane.

Horizontal plane An imaginary plane that is tangent (or parallel) to the earth's surface at a given location.

Horizontally polarized wave Electromagnetic wave that has the electric field in the horizontal plane.

Horizontally polarized Waves radiated with their E field component parallel to the earth's surface.

Horn A type of speaker system now principally used for high-frequency reproduction, but which is capable of full range sound in its largest format. As with those huge alpine horns, even the smallest transducer or diaphragm can produce the lowest sound if the horn is long enough and large enough at its business end.

Horn radiator A tapered, tubular or rectangular microwave antenna that is widest at the open end.

Horn Tweeter A driver consisting of a relatively small emitter surmounted by a curvilinear or exponential horn. This is an effective system for radiating high frequencies in a variety of situations. The size and shape of the horn will usually dictate the pattern and use of the driver. Long horns with narrow apertures, tend to have the narrowest radiation pattern, and are very useful in large listening rooms, especially where highly directional effects, such as surround sound requires, are mandatory. Shallower versions have more general applications, especially in car stereo applications where a wider field of coverage is desired, along with a robust driver that can withstand severe environments. In such situations, a Piezo driven emitter (driver) is highly desirable for its ability to handle high-energy inputs on a variable basis. Horns can be driven by a number of different driver types: Dynamic ,(magnet and coil) Piezo, Electrostatic, Ribbon, and even Gas Plasma have been used effectively for this purpose.

Horsepower The measure of rate of work. One horsepower is equivalent to lifting 33,000 pounds to a height of one foot in one minute. The horsepower of a motor is expressed as a function of torque and rpm. For motors the following approximate formula may be used: where HP = horsepower, T = torque (in. lb.ft.), and RPM = revolutions per minute..

Horseshoe magnet A permanent magnet or electromagnet bent into the shape of a horseshoe or having a U-shape to bring the two poles near each other.

Host A single computer or workstation; it can be connected to a network.

Host based Information, such as audit data from a single host which may be used to detect intrusions.

Hostile input An unselected input carrying a signal which causes unwanted interference and coupling in a desired output.

Hot carrier A carrier, which may be either a hole or an electron, that has relatively high energy with respect to the carriers normally found in majority-carrier devices.

Hot carrier diode A semiconductor diode in which hot carriers are emitted from a semiconductor layer into the metal base. Also called hot-electron diode. An example is the Schottky barrier diode.

Hot junction The junction of two dissimilar metals in a thermocouple circuit that is used to measure an unknown temperature. Also known as measurement junction.

Hot wire instrument An electrical instrument which depends upon the expansion (or change of resistance) of a wire which is heated by the passage of an electric current.

Hot wire meter movement A meter movement that uses the expansion of a heated wire to move the pointer of a meter; measures dc or ac.

House service meter Energy meter at a consumer's premises, measuring power in kwh.

Hub A wiring concentrator used in local area networks.

Humidification The process of transferring a mass of water to the atmospheric air.

Humidistat A device used in control systems for switching plant to maintain a relative humidity at some setpoint. The output signal is usually sent via a relay device to the final control element.

Humidity The measure of moisture in the atmosphere. In building simulation it is often refered to as the relative humidity within a space.

Humphreys model Humphreys model for thermal comfort. Humphreys equation is a fit to considerable data for climate-controlled and non-climate controlled buildings.

Hunting Prolonged self-sustained oscillation of undesirable amplitude.

Hv: high voltage exceeding 1000 V between conductors and 600 V between conductors and earth.

HVAC An acronym for Heating, Ventilating, and Air-Conditioning.

Hybrid circuit Circuit that combines two technologies (passive and active or discrete and integrated components) onto one microelectronic circuit. Passive components are usuall made by thin film techniques, while active components are made with semiconductor techniques.

Hybrid ICs Two or more integrated circuit types, or one or more integrated circuit types and discrete components on a single substrate.

Hybrid junction A waveguide junction that combines two or more basic T-junctions.

Hybrid ring A hybrid-waveguide junction that combines a series of E-type T-junctions in a ring configuration. When properly terminated, energy is transferred from any one branch into any two of the remaining three branches.

Hybrid step motor Class of step motors that utilizes a permanent magnet to polarize soft iron pole pieces around the rotor.

Hydrocarbons (HC) Chemical compounds made up entirely of carbon and hydrogen. When combusted they release carbon dioxide, nitrogen oxide, sulphur dioxide and water vapour. The carbon dioxide emissions released to the atmosphere are the primary cause of the "greenhouse effect" leading to global warming of the Earth's atmosphere.

Hydroelectric plant A plant in which the turbine/generators are driven by the kinetic energy of water. One common type of hydropower plant involves using a dam to store water in a reservoir and when released spins a turbine, creating electricity.

Hydrogen ion activity (ah+) Activity of the hydrogen ion in solution. Related to hydrogen ion concentration (CH+) by the activity coefficient for hydrogen (f H+).

Hydrometer An instrument used to measure specific gravity. In batteries hydrometers are used to indicate the state of charge by the specific gravity of the electrolyte.

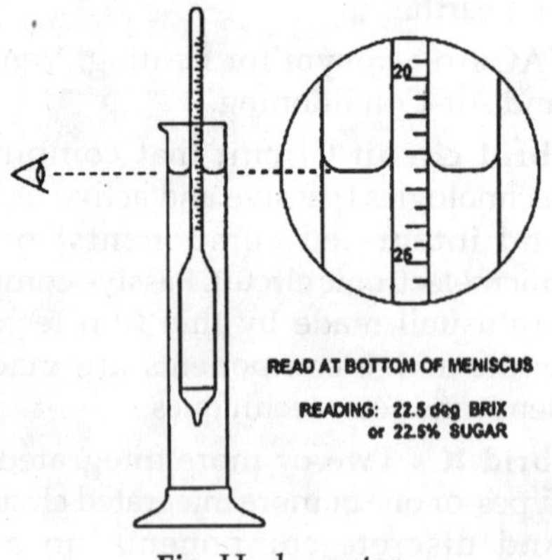

Fig. Hydrometer

Hyper Transport A Point-to-Point interface with at least two unidirectional links. The HyperTransport as a data rate of 800M Bytes/ps using 8 bit pairs and a 400MHz clock.

Hyperbaric Resonator Impulse An audible event characterized by a sudden and often unexpected pressure wave emanating from a vent port with a driver whose energetic and propulsive energy transitions, are generated by a process involving the chemical modification of legumes. This phenomenon frequently produces olfactory irritation in associated groups in consequence. This event is also know as a Farsical Audio Report Transient.

Hypercardioid A narrower heart-shaped pick-up pattern than that of cardioid microphones.

Hysteresis Loop A hysteresis loop shows the relationship between the induced magnetic flux density B and the magnetizing force H. It is often referred to as the B-H loop

Hysteresis (Electrode Memory) When an electrode system is returned to a solution, equilibrium is usually not immediate. This phenomenon is often observed in electrodes that have been exposed to the other influences such as temperature, light, or polarization.

Hysteresis loss The resistance offered by materials to becoming magnetized (magnetic orientation of molecular structure) results in energy being expended and corresponding loss. Hysteresis loss in a magnetic circuit is the energy expended to magnetize and demagnetize the core.

I

IC Abbreviation for "integrated circuit".

IC Synchros Obsolete synchros with reverse rotation and limited torque capabilities.

IC voltage regulator Three terminal device used to hold the output voltage of a power supply constant over a wide range of load variations.

Ice point The temperature at which pure water changes from a liquid to a solid (freezes). 32¼F (0¼C).

Icon A graphic functional symbol display. A graphic representation of a function or functions to be performed by the computer.

IDEA (International data encryption algorithm) A private key encryption-decryption algorithm that uses a key that is twice the length of a DES key.

Ideal control Control that involves no time lags and with all control elements behaving linearly.

Ideal current source A source which maintains the source current at a predefined value independent of the load conditions.

Ideal dependent source An active element in which the source voltage or current is controlled precisely by another voltage or current.

Ideal driver A speaker driver that delivers excellent performance in terms of polar response, power handling, efficiency, frequency range and reliability.

Ideal op amp An operational amplifier with infinite open-loop gain, infinite input impedance and zero output impedance.

Ideal source An ideal independent source is an active element that provides a specified voltage or current that is completely independent of the remaining circuit elements.

Ideal voltage source A source which maintains the source voltage at a predefined value independent of the load conditions. In other words the terminal voltage is maintained equal to the internal emf.

Idempotent law In Boolean algebra, combining a quantity with itself either by logical addition or logical multiplication will result in a logical sum or product that is the equivalent of the quantity (for example, A + A = A; A

A = A).

Identification In most instances, the following information will help identify a motor: 1. Frame designation (actual frame size in which the motor is built).

2. Horsepower, speed, design and enclosure.

3. Voltage, frequency and number of phases of power supply.

4. Class of insulation and time rating.

5. Application.

Identity law In Boolean algebra, the law which states that any expression is equal to itself.

Illuminance The quantity of light at one point on a surface. It is the quotient of the luminous flux incident on an element of the surface containing the point, and the area of that element. (Unit lux, lx)

Illumination Emission of optical radiation by the process of thermal radiation.

Image frequency An undesired frequency capable of producing the desired frequency through heterodyning.

Imaginary Numbers with negative squares. S-1 is the base of such numbers.

Imaginary axis The y-axis in the complex plane.

Imaginary operator j An multiplier or operator with a magnitude of unity and an anticlockwise rotation of 90o. It also has the value S-1 in the complex domain.

Imaging Imaging describes the extent to which an audio system reproduces the directional cues that enable the listener to locate the instruments and vocalists as they were positioned during recording and mixing. Good imaging creates a listening experience that seems natural and lifelike. Since directional cues in sound come mainly in the higher frequencies, the key to attaining the best possible imaging is to have equal and unobstructed path lengths between the tweeters and the listener's ears. That's one of the reasons why matched component speakers, with their versatile tweeter placement, sound as good as they do.

Impact standards An industrial standard for impact stress. Cosel power supplies are designed on the Japanese JISC 0041 standards.

Impedance (Z) Measured in ohms it is the total opposition to the flow of current offered by a circuit. Impedance consists of the vector sum of resistance and reactance.

Impedance and electrical phase Electrical resistance of a driver and a complete loudspeaker in an alternating current environment; sound is delivered to the speaker in the form of an alternating current. The impedance of a driver and the loudspeaker is measured in Ohms and usually varies over its frequency range. When the impedance varies the electrical phase does it too. Electrical phase shifts requires the power amplifier is able to handle large variations in its the output. Therefore it is very important that the impedance shifts are soft and it is best when impedance of a speaker is linear.

Impedance coupling Coupling of two signal amplifier circuits through the use of an impedance such as a inductor.

Impedance electrical The electrical impedance defines the relation between current and voltage in a circuit. In a simple case, the impedance may be purely resistive such that the current and voltage are directly related by a constant of proportionality (the "resistance"). When the circuit incorporates reacive components (inductance, capacitance), the impedance is a complex quantity.

Impedance matching Matching the output impedance of a source to the input impedance of a load to attain maximum power transfer.

Imperfect mixing The combination of two or more substances such that the parts of one are unevenly distributed among the parts of another.

Implosion The inward bursting of a CRT because of high vacuum. The opposite of explosion.

Impregnation The process of filling the pores of paper and similar material in order to improve its insulation properties.

Impulse A disturbance of the voltage waveform that is less than about one millisecond. Voltages can rise to hundreds or even thousands of volt in a very short period of time. An impulse may be additive or subtractive.

Impulse function A mathematical function with zero magnitude other than at zero time, where it has an infinite magnitude. The magnitude of an impulse function is defined as its time integral.

Impulse generator In most impulse generators, certain capacitors are charged in parallel through high series resistances, and then discharged through a combination of resistors and capacitors, giving rise to the required surge waveform (usually double exponential) across the test device.

Impulse response Behaviour of a circuit when the excitation is the unit impulse function. The excitation function may be a voltage or a current.

In Circuit meter A meter permanently installed in a circuit; used to monitorcircuit operation.

In phase When two or more waves of the same frequency have their positive and negative peaks occuring at the same time.

In the early years of telephony and cinema, the first equalizers were fixed units designed to compensate for losses in the transmission and recording of audio signals. Hence, the term equalizer described electronic circuits that corrected for these losses and attempted to make the output equal to the input. Equalizers permit the modification the frequency response spectrum of the signal passing through them; that is, they modify the amplitude versus frequency characteristics.

Incandescence State of a material when heated to the point where it emits light. (red hot or white hot).

Incandescent (electric) lamp Lamp in which light is produced by means of an element heated to incandescence by the passage of an electric current.

Incandescent light A gas filled (argon) bulb containing a metallic filament (tungsten) that produces light when a sufficient voltage is applied; an ordinary light bulb.

Inch A measure of length in the imperial system. It is now defined as follows. 1 inch = 25.400 mm

Incidence matrix A connection matrix having elements 1, -1 or 0 dependent on whether a particular connection is present and having the same sign as the reference, has the opposite sign to the reference or not connected at all.

Incident wave 1. The wave that strikes the surface of a medium.

2. The wave that travels from the sending end to the receiving end of a transmission line.

Incoherent Refers to radiation on a broad band of frequencies.

Incremental control A form of modulating control where the control device is sent an increase - hold - decrease signal from two binary outputs working as a pair. with this type of control it is possible to provide a form of proportional+integral (PI) control without position feedback from the actuator of the controlled device.

Incremental Refers to an output that is measured in known increments from a known reference point; e.g., a step motor provides incremental motion based on an initial position and an incremental encoder provides incremental feedback information based on a known zero point.

Independent power producer IPP Private entrepreneurs who develop, own or operate electric power plants fueled by alternative energy sources such as biomass, cogeneration, small hydro, waste-energy and wind facilities. Organisations with generating capacity that are not associated with traditional electricity utilities.

Independent system operator ISO An entity responsible for the reliable operation of the grid and provision of open transmission access to all market participants on a non-discriminatory basis.

Index of refraction(n)+A2678 The degree of bending of an RF wave when passing from one medium to another.

Index The number indicating the power to which the quantity is raised.

Indexer Electronics that convert high level computer or PLC commands to step and direction pulses.

Indicating instrument A piece of equipment in which the output is given as the deflection of a needle or the reading of a counter.

Indicator Equipment in radar that provides a visual presentation of target position information.

Indicators & warnings (I & W) I & W refers to how an event or series of events can provide enough information to classify it as an incident.

Indirect contact Contact of persons or livestock with exposed conductive parts which have become live under fault conditions.

Indirect control This means automatic operation of a control device located in the energy flow stream via an intermediary system.

Indirect lighting Lighting by means of luminaires with a light distribution such that not more than 10 per cent of the emitted luminous flux reaches the working plane direct, assuming that this plane is unbounded.

Indirectly heated cathode Same as the directly heated cathode with one exception: The hot filament raises the temperature of the sleeve around the filament; the sleeve then becomes the electron emitter.

Induced charge An electrostatic charge produced on an object by the electric field that surrounds a nearby object.

Induced current Current caused by the relative motion between a conductor and a magnetic field.

Induced electromotive force The electromotive force induced in a conductor because of the relative motion between the conductor and a magnetic field.

Induced failure The failure of a component resulting from an induced condition that causes a malfunction of the device.

Induced voltage Voltage generated in a conductor when subjected to a moving magnetic field.

Inductance [Henry, H] That property of an electric circuit which tends to oppose change in current in the circuit. One henry (H) is the inductance of a closed circuit in which an electromotive force of 1 volt is produced when the electric current in the circuit varies uniformly at the rate of 1 ampere per second.

Inductance bridge An ac bridge circuit used to measure an unknown value of inductance.

Induction field The electromagnetic field that is produced about an antenna when current and voltage are present on the same antenna.

Induction heating A form of heating in which electrically conducting material is heated as a result of the electric currents induced in it by an alternating magnetic field.

Induction losses The losses that occur when the electromagnetic field around a conductor cuts through nearby metallic objects and induces a current into that object.

Induction machine Induction machines run at a speed slightly different to synchronous speed as the difference speed, known as slip, is required to generate torque. Induction motors run at sub-synchronous speed where as induction generators run at super-synchronous speed.

Induction meter The induction meter depends on the torque produced by the reaction between a flux (whose value depends on the value of the current in one coil) and the eddy currents which are induced in a non-magnetic disc (usually aluminium) by another flux (produced by current in a second coil). Since the action depends on induction, they can be used to measure alternating quantities only. The meter would have a deflection proportional to the product of the two currents.

Induction motor An induction motor is an alternating current motor in which the primary winding on one member (usually the stator) is connected to the power source and a secondary winding or a squirrel-cage secondary winding on the other member (usually the rotor) carries the induced current. There is no physical electrical connection to the secondary winding, its current is induced.

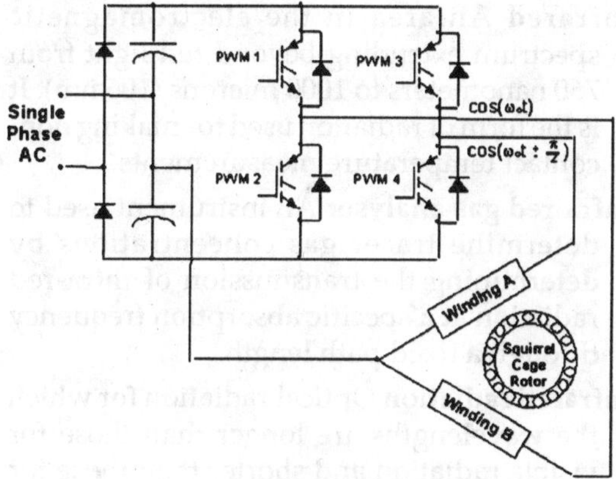

Fig. Induction motor

Inductive circuit Circuit having greater inductive reactance than capacitive reactance.

Inductive coupling The transfer of energy from one circuit to another by virtue of the mutual inductance between the circuits. Coupling of two coils by means of magnetic lines of force. In transformers, coupling applied through magnetic lines of force between the primary and secondary windings.

Inductive load Generated by a wire coil such as a relay, solenoid, motor or contactor. In an Inductive Load when the thermostat opens, a high voltage will be generated until all the current that is stored in the coil is discharged across the contacts. This can result in severe contact degradation.

Inductive reactance The opposition to the flow of an alternating current caused by the inductance of a circuit, expressed in ohms. Identified by the symbol X L.

Inductor A coil or component with the properties of inductance.

Industrial building A building in which the main purpose is to provide space for manufacturing and assembly processes. These are characterised by high levels of activity both mechanical and human, and often by the generation of internal pollution and heat.

Industrial conditioning Industrial conditioning refers to a process which requires a controlled atmosphere. A typical specification would provide for an internal environment of 21 oC +or- 0.5 oC and 50% relative humidity +or- 2.5% at all external conditions. Industrial conditioning has clearly defined limits as opposed to comfort conditioning which is based on statistical surveys of occupants feelings.

Inert A material or gas that does not react chemically or biologically; for example, military thermostats are backfilled with an inert gas to prevent oxidation.

Inertia The physical tendency of a body in motion to remain in motion and a body at rest to remain at rest unless acted upon by an outside force (Newton's First Law of Motion).

Inertia match Refers to the ratio of the reflected load inertia to the rotor inertia; a general rule is that this ratio should be no more than 10:1 for an inertia match.

Infiltration rate The rate at which outside air infiltrates a room or building. Equivalent to the fresh air change rate, usually expressed in air changes per hour (ach) or litres per second (l/s).

Infinite 1. Extending indefinitely, endless.

2. Boundless, having no limits.

3. An incalculable number.

Infinite Baffle An infinite baffle speaker design is defined as an enclosure that contains a greater volume of air than the Vas requirement of the driver. An infinite baffle system can easily be applied to an automobile. This is accomplished by mounting the speakers on a board and using the trunk of the vehicle as the other walls of the enclosure. It is important that the enclosure be tightly sealed such that no air moves from the front to the back of the cone. Look for speakers where the Qts is greater than .6, and a Vas figure lower than the volume available, when selecting a woofer for an infinite baffle system.

Infinitesimal A very small quantity tending to zero, without actually being zero.

Infinity A very large quantity greater than any assignable quantity and tending to the inverse of zero.

Information assurance (IA) Information Operations that protect and defend information and information systems by ensuring their availability, integrity, authentication, confidentiality, and non-repudiation. This includes providing for restoration of information systems by incorporating protection, detection, and reaction capabilities.

Information operations (IO) Actions taken to affect adversary information and information systems while defending one's own information and information systems.

Information provider The publisher or owner of a resource.

Information retrieval by entering one or more keywords into a search engine.

Information security The result of any system of policies and/or procedures for identifying, controlling, and protecting from unauthorized disclosure, information whose protection is authorized by executive order or statute.

Information superiority The capability to collect, process, and disseminate an uninterrupted flow of information while exploiting or denying an adversary's ability to do the same.

Information warfare (IW) 1. Actions taken to achieve information superiority by affecting adversary information, information based processes, and information systems, while defending our own information, information based processes, and information systems. Any action to deny, exploit, corrupt, or destroy the enemy's information and its functions, protect themselves against those actions; and exploiting their own military information functions.

2. Information Operations conducted during time of crisis or conflict to achieve or promote specific objectives over a specific adversary or adversaries.

Infralow frequency The band of frequencies from 300 Hz to 3,000 Hz.

Infrared An area in the electromagnetic spectrum extending beyond red light from 760 nanometers to 1000 microns (106 nm). It is the form of radiation used for making non-contact temperature measurements.

Infra-red gas analyser An instrument used to determine tracer gas concentrations by determining the transmission of infra-red radiation at a specific absorption frequency through a fixed path length.

Infrared radiation Optical radiation for which the wavelengths are longer than those for visible radiation and shorter than those for radio waves. It corresponds to invisible heat radiation.

Infrasonic Sounds below 15 Hz. (Subsonic)

Inherent regulation (also 'self-regulation') Many simple processes are characterised by the fact that a new steady- state is reached automatically, i.e., without interference by manual or automatic control.

Inhibit To stop an action or block data from passing.

Inhibitor A chemical substance that, when added in relatively low concentrations, slows down a chemical reaction.

Initial unbalance Initial unbalance is that unbalance of any kind that exists in the rotor before balancing.

Inorganic Not belonging to the large class of carbon compounds which are termed organic.

Input Process variable information that is supplied to the instrument.

Input bias current The current that flows at the input due to internal circuitry and bias voltage. Also, the current that must be supplied to the high input measuring terminal (with zero input signal and offset voltage) to reduce the output indication to zero.

Input bus A circuit path on the input side of a switching array which connects to the inputs of one or more crosspoint switches. Each input connector leads to an input bus.

Input end The end of a two-wire transmission line that is connected to a source.

Input impedance The resistance of a panel meter as seen from the source. In the case of a voltmeter, this resistance has to be taken into account when the source impedance is high; in the case of an ammeter, when the source impedance is low.

Input isolation On a switching card, the isolation from signal high to low (or guard) for a two-pole circuit. Specified as resistance and capacitance.

Input line filter A low-pass or band-reject filter at the input of a power which reduces noise fed to the supply or created by the supply. This filter may be external to the power supply.

Input offset current The difference between the two currents that must be supplied to the input measuring terminals of a differential instrument to reduce the output indication to zero (with zero input voltage and zero offset voltage).

Input offset voltage The voltage that must be applied directly between the input measuring terminals, with bias current supplied by a resistance path, to reduce the output indication to zero.

Input output isolation Circuit techniques that provide DC isolation between input and output circuitry of a power supply.

Input Overload Distortion Distortion caused by too great an input signal being sent to an amplifier or preamplifier. It is not affected by volume control settings and often occurs when mics are positioned too close to the sound source. This distortion may be controllable through the use of an attenuator or pad.

Input resistance (Impedance) The input resistance of a ph meter is the resistance between the glass electrode terminal and the reference electrode terminal. The potential of a ph-measuring electrode chain is always subject to a voltage division between the total electrode resistance and the input resistance.

Input scaling The ability to scale input readings (readings in percent of full scale) to the engineering units of the process variable.

Input Sensitivity SPL (sound pressure level) A speaker will produce given one watt of power as measured from one meter away given a typical input frequency (usually 1kHz unless otherwise noted on the speaker). Typical sensitivities for car audio speakers are around 90dB/Wm. Some subwoofers and piezo horns claim over 100dB/Wm. However, some manufacturers do not use true 1W tests, especially on low impedance subwoofers. Rather, they use a constant voltage test which produces more impressive sensitivity ratings.

Input type The signal type that is connected to an input, such as thermocouple, RTD, linear or process.

Input voltage range The range of input values for which a power supply or device operates within specified limits.

Input Voltage The power voltage provided to an amplifier. While most cars can be expect to reliably produce 12 volts, amplifiers are sometimes measured at higher voltages; up to 15 volts. In this way, higher power can be developed, albeit at the expense of longevity. It is a practice that allows higher power figures to be claimed.

Input/output (I/O) The process of transferring data to and from a computer-controlled system using its communication channels, operator interface devices, data acquisition devices, or control interfaces.

Inrush current Measurement of the initial current flow in Capacitive or Lamp applications. Inrush currents can be as much as 15 times the rated current flow of the application.

Inrush current limiting The characteristics of a circuit that limits inrush current when a power supply or the circuit is energized.

Insertion Loss The loss of voltage (or power), as measured in dB, resulting from placing a resistor (or some other power absorbing network or component) between a voltage or power source (amp) and its load impedance (speaker.) It is the ratio of the voltage (or power) absorbed in the load without the resistor (or network) to that when the network is inserted. For example, if the voltage across a load is 2 volts without a network and 1 volt with the network, then the insertion loss is stated as 6 dB.

Insertion point (in lithography context) Adaptation of a new lithography technique is referred to as the insertion point of that technique.

Insolation The magnitude of solar energy which is incident on a particular building element (W/m2). Both the direct and diffuse components must be considered.

Installation An electrical installation is a combination of electrical equipment installed to fulfil a specific purpose and having coordinated characteristics.

Instantaneous amplitude The amplitude at any given point along a sine wave at a specific instant in time.

Instantaneous automatic gain control (IAGC) A circuit that can vary the gain of the radar receiver with each input pulse to maintain a nearly constant output peak amplitude.

Instantaneous frequency The rate of change of phase angle (in rad/s) or additionally divided by 2p (in Hz).

Instantaneous value The value of an alternating current or voltage at any specified instant in a cycle.

Institutional building A building with mixed occupational activities where special requirements arising from those activities may be needed; such buildings include hospitals, prisons etc.

Instructed person A person adequately advised or supervised by skilled persons to enable him/her to avoid dangers which electricity may create.

Instrument transformer A transformer specifically designed to be used with instruments. Their design ensures high accuracy for the quantity to be measured.

Insulated When a non conducting material is used to isolate conducting materials from one another.

Insulating material Material that will prevent the flow of current due to its chemical composition.

Insulation A material used to prevent the leakage of electricity from a conductor and to provide mechanical spacing or support as protection against accidental contact with the conductor.

Insulation class Since there are various ambient temperature conditions a motor might see and different temperature ranges within which motors run and insulation is sensitive to temperature; motor insulation is classified by the temperature ranges at which it can operate for a sustained period of time.

Insulation co-ordination Insulation co-ordination now comprises the selection of the electric strength of the various equipment in relation to the voltages which can appear on the system for which the equipment is intended. The overall aim is to reduce to an economically and operationally acceptable level the cost and disturbance caused by insulation failure and resulting system outages.

Insulation failure Fault between the phase conductor and non-current carrying metallic parts of an electrical equipment, as a result of which high voltages may appear on the frames of equipment and may be dangerous to a person coming in contact with it.

Insulation level It defines the level of insulation with regard to power frequency and with regard to surges. For equipment rated at less than 300 kv, it is a statement of the Lightning impulse withstand voltage and the short duration power frequency withstand voltage. For equipment rated at greater than 300 kv, it is a statement of the

Switching impulse withstand voltage and the power frequency withstand voltage.

Insulation materials materials with low heat conductivity

Insulation or insulator A non-conductive protective covering for electrically active parts and wire that prevents short circuits and other unwanted interference.

Insulation resistance -The ohmic resistance of insulation. It degrades quickly as humidity increases. Lower insulation resistance provides a path for leakage current to ground. This is very critical when making measurements on semiconductor components where picoamp measurements are being made.

Insulation systems Five specialized elements are used, which together constitute the motor's Insulation system. The following are typical in an AC motor: 1. TURN-TO-TURN INSULATION between separate wires in each coil. (Usually enamel on random wound coils of smaller motors - tape on "form wound" coils of larger motors.)

2. Phase-to-Phase Insulation between adjacent coils in different phase groups. (A separate sheet material on smaller motors - not required on form wound coils because the tape also performs this function.)

3. phase-to-ground insulation between windings as a whole and the "ground" or metal part of the motor. (A sheet material, such as the liner used in stator slots, provides both di-electric and mechanical protection.)

4. Slot Wedge to hold conductors firmly in the slot.

5. Impregnation to bind all the other components together and fill in the air spaces. (A total impregnation, applied in a fluid form and hardened, provides protection against contaminants.

Insulation/economic thickness of insulation thickness, giving the largest energy savings for the lowest investment costs

Insulator 1, Material of such low conductivity that the flow of current through it can usually be neglected.

2. A device having high electrical resistance; used for supporting or separating conductors so as to prevent undesired flow of current from the conductors to other objects.

Integral action The action of a control element whose output signal changes at a rate which is proportional to its input signal size. The integral term may be considered as the 'memory' of the controller, looking at past errors. It acts to remove offset. It can be used in conjunction with derivative control action (I+D) or, more commonly, with proportional action (P+I) or proportional and derivative action (P+I+D).

Integral action time (IAT) In a control system having P+I control action, the time interval in which the part of the output signal due to integral action increases by an amount equal to the part of the output signal due to the proportional action, when the deviation is unchanging.

Integral gain Defined as the ratio of proportional gain to integral action time.

Integral wind up It is possible for a final control element to be in a 'fully open' position but the controlled variable to be some way from the setpoint. For example, such an occurrence often occurs during the morning heat-up period in a heating system. The error will persist for some time and the integral action term will become enormously large. This error is called the 'integral wind-up error'.

Integrated When two or more components are combined into a circuit and then incorporated into a single package.

Integrated circuit (IC) 1. A circuit in which many elements are fabricated and interconnected by a single process (into a single chip), as opposed to a "nonintegrated" circuit in which the transistors, diodes, resistors, and other components are fabricated separately and then assembled.

2. Elements inseparably associated and formed on or within a single substrate.

Integrating conversion An analogue to digital conversion process where the output results in a digital representation of the integral of the input signal over a specified time interval.

Integrating meter A meter whose output is proportional to the integrated value of a quantity over time. They are usually with rotating discs where the revolutions correspond to the time of integration.

Integrator An op amp whose output is proportional to the integral of the input signal.

Integrity Assuring information will not be accidentally or maliciously altered or destroyed.

Intelligent step control This is a microprocessor based step controller or sequencer, which can be programmed or selected for different operational modes such as binary switching, and equalised run time. A second advantage which this type of step controller has over the motorised version is its ability to provide easily adjustable time delays.

Intensity (Of sound) The measurement of the amplitude of sound energy. Generally synonymous with loudness.

Intentional ventilation Ventilation provided through the use of purpose provided openings, such as through windows or airbricks.

Interaction space The region in an electron tube where the electrons interact with an alternating electromagnetic field.

interchangeability The maximum difference in indicated value between two sensors chosen at random, connected to the same interface and exposed to identical conditions.

Interchangeability error A measurement error that can occur if two or more probes are used to make the same measurement. It is caused by a slight variation in characteristics of different probes.

Interconnection voltage The nominal voltage at which the grid interconnection is made.

Interelectrode capacitance The capacitance between the electrodes of an electron tube.

Interface A device or protocol which facilitates the linking of any two devices or systems; or when used as a verb ('to interface'), the process of linking.

Interference Any disturbance that produces an undesirable response or degrades a signal.

Interlock A device connected in such a way that the motion of one part is held back by another part.

Intermediate frequency (IF) A lower frequency to which an RF echo is converted for ease of amplification.

Intermediate frequency amplifier In a superheterodyne radio it amplifies a fixed frequency lower than the received radio frequency and higher than the audio frequency.

Intermediate level maintenance (SM&R Code I) Direct support and technical assistance to user organizations. Tenders and shore-based repair facilities.

Intermediate load (electric systems) A load in the range from base load to peak load.

Intermediate power amplifier The amplifier between the oscillator and final power amplifier.

Intermittent A fault occuring at random intervals of time. Intermittent problems are often difficult to locate because of the random nature. They often don't occur when the technician is present.

Intermittent duty A requirement of service that demands operation for alternate intervals of

1. load and no load; or
2. load and rest; or
3. load, no load and rest; such alternate intervals being definitely specified.

Intermittent heating It can be demonstrated both theoretically and by actual tests, that a great deal of heat can be conserved by cutting off the heat supply in any type of building during periods of non-occupancy.

Intermodulation Distortion A species of Distortion that results when one set of frequencies is superimposed on, or is modified by, another to produce a third frequency not present in the original signal. Quantifies the distortion products of nonlinearities in the unit under test that causes complex waves to produce beat frequencies, i.e., sum and difference products not harmonically related to the fundamentals.

Internal combustion engine IC engine An engine in which energy supplied by a burning fuel is directly transformed into mechanical energy by the controlled combustion of the fuel in an enclosed cylinder behind a piston. Usually used in petrol and diesel engines.

Internal discharge A discharge occurring within a material.

Internal fan pressurization The building's own mechanical ventilation system can be used to provide the required pressure differential from within. The supply fans are operated while all return and exhaust fans are turned off, and all return dampers are closed, so that air can only leave through the doors, windows and other leakage sites.

Internal pressure The pressure inside a building envelope or space. Usually expressed with respect to outside or atmospheric pressure.

Internal pressure distribution The pattern of static pressure variation at various points inside a building due to variations in air density and air flow into and out of the building.

Internal reference electrode (Element) The reference electrode placed internally in a glass electrode.

Internal resistance Every source has some resistance in series with the output current. When current is drawn from the source some power is lost due to the voltage drop across the internal resistance. Usually called output impedance or output resistance.

Internal trigger A software-generated event that starts an operation.

Interoperability A broad term, encompassing many of the issues where distributed databases and resources on the Internet might be made to work together, offering the user the ability to access, cross-search and cross-browse them from a single interface.

Interpolation A process of filling in intermediate values or terms between known values or terms.

Interpoles Small auxiliary poles, placed between main field poles, whose magnetic field opposes the armature field and cancels armature reaction. Interpoles accomplish the same thing as compensating windings.

Interpreter A system program that converts and executes each instruction of a high-level language program into machine code as it runs, before going onto the next instruction.

Interrupt A signal to the CPU indicating that the board detected the occurrence of a specified condition or event.

Interrupt level A specific priority that ensures that high priority interrupts get serviced before low priority interrupts.

Interruptible loads Loads that can be interrupted in the event of capacity or energy deficiencies on the supplying system.

Interruptible power Power whose delivery can be curtailed by the supplier, usually under some sort of agreement by the parties involved.

Interruptible rate Tariff rate for the provision of power at a lower rate to large industrial and commercial consumers who agree to reduce their electricity use in times of peak demand.

Intersection law In Boolean algebra, the law which states that if one input to an AND gate is already true, then the output will depend upon the state of the other inputs only.

Interstitial diffusion a diffusion mechanism that causes atomic motion from interstitial site to interstitial site.

Interzonal air flow The process of air exchange between internal zones of a building.

Intrinsic Characterizes pure undoped semiconductor; electrical conductivity depends only on temperature and the band gap energy.

Intrinsic material A semiconductor material with electrical properties essentially characteristic of ideal pure crystal. Essentially silicon or germanium crystal with no measurable impurities.

Intrinsic semiconductors A semiconductor material that is essentially pure.

Intrinsic stand off ratio A unijunction transistor (UJT) rating used to determine the firing potential of the device.

Intrinsically safe An instrument which will not produce any spark or thermal effects under normal or abnormal conditions that will ignite a specified gas mixture.

Intrusion detection Pertaining to techniques which attempt to detect intrusion into a computer or network by observation of actions, security logs, or audit data. Detection of break-ins or attempts either manually or via software expert systems that operate on logs or other information available on the network.

Invert To change a physical or logical state to its opposite state.

Inverter A circuit with one input and one output. Its function is to invert or reverse the input. When the input is high, the output is low, and vice versa. The inverter is sometimes called a NOT circuit, since it produces the reverse of the input.

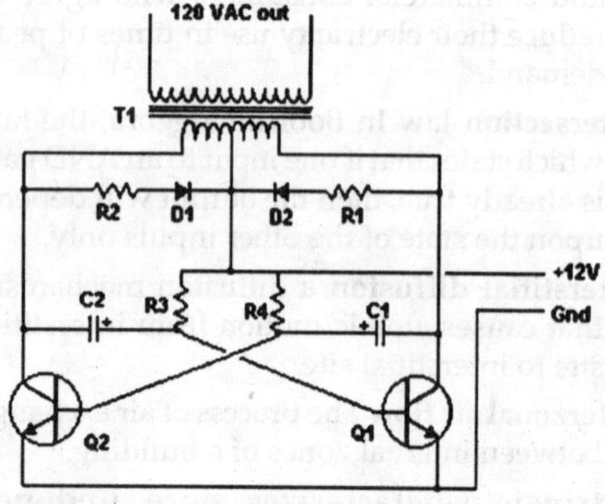

Fig. Inverter

Inverting amplifier An amplifier that has a 180° phase shift from input to output.

Inverting input In an operational amplifier (op amp) the input that is marked with a minus sign. A signal applied at the inverting input will be given 180° phase shift between input and outptu.

Ionization 1. The process of producing ions.

2. The electrically charged particles produced by high-energy radiation, such as light or ultraviolet rays, or by the collision of particles during thermal agitation.

Ionization point The potential required to ionize the gas of a gas-filled tube. Sometimes called firing potential.

Ionize To make an atom or molecule of an element lose an electron, as by X-ray bombardment, and thus be converted into a positive ion. The free electron may attach itself to a neutral atom or molecule to form a negative ion.

Ionophore A macro-organic molecule capable of specifically solubilizing an inorganic ion of suitable size in organic mediums.

Ionosphere The most important region of the atmosphere extending from 31 miles to 250 miles above sea level. Contains four cloud-like layers that affect radio waves.

Ionospheric storms Disturbances in the earth's magnetic field that make communications practical only at lower frequencies.

IP address Every computer on the Internet has a unique numerical IP address assigned to it, such as 123.456.78.9.

IP Splicing/Hijacking An action whereby an active, established, session is intercepted and co-opted by the unauthorized user. IP splicing attacks may occur after an authentication has been made, permitting the attacker to assume the role of an already authorized user. Primary protections against IP splicing rely on encryption at the session or network layer.

IP Spoofing An attack whereby a system attempts to illicitly impersonate another system by using IP network address.

IPTS-48 International Practical Temperature Scale of 1948. Fixed points in thermometry as specified by the Ninth General Conference of Weights and Measures which was held in 1948.

IPTS-68 International Practical Temperature Scale of 1968. Fixed points in thermometry set by the 1968 General Conference of Weights and Measures.

IRIS A metal plate with an opening through which electromagnetic waves may pass. Used as an impedance-matching device in waveguides.

Irradiation Exposure to radiation of any kind.

Irreversible process A process that is not fully reversible.

ISA bus Industry Standard Architecture. The 16-bit wide bus architecture used in most MS-DOS and Windows computers. Sometimes called the AT bus.

ISDN Abbreviation for integrated services digital network. An integrated digital network in which the same time-division switches and digital transmission paths are used to establish connections for different services.

Islanded operation The situation that arises when a part of the electrical system is disconnected from the main grid and is energised by one or more generators connected to it.

Islanding The process whereby a power system is separated into two or more parts, with generators supplying loads connected to some of the separated systems.

Isokeraunic level Contours of constant keraunic level or thunder days.

Isolated junction A form of thermocouple probe construction in which the measuring junction is fully enclosed in a protective sheath and electrically isolated from it. Commonly called an ungrounded junction.

Isolated outputs Output signals where a common reference is not connected to either input terminal.

Isolation The degree to which a device can separate the electrical environment of its input from its output, while allowing the desired transmission to pass across the separation. A function intended to cut off for reasons of safety the supply from all, or a discrete section, of the installation by separating the installation or section from every source of electrical energy.

Isolation transformer A transformer with physically separate primary and secondary windings. An isolation transformer does not transfer unwanted noise and transients from the input circuit to the output windings.

Isolation voltage The maximum AC or DC specified voltage that may be continuously applied between isolated circuits.

Isolator A mechanical switching device which, in the open position, complies with the requirements specified for isolation. An isolator is otherwise known as a disconnector.

Isometric diagram A diagram showing the outline of a ship, aircraft, or equipment and the location of equipment and cable runs.

Isomorphous Having the same structure. In the phase diagram sense, isomorphicity means having the same crystal structure or complete solid solubility for all compositions.

Isothermal At a constant temperature. In an isothermal process heat is, if necessary, supplied or removed from the system at just the right rate to maintain constant temperature.

Isothermal surface Surface with even temperature in all its points.

Isotropic Having identical values of a property in all crystallographic directions.

Isotropic radiation The radiation of energy equally in all directions.

IT system, a system having no direct connection between live parts and Earth, the exposed conductive parts of the electrical installation being earthed.

J

Jack A receptical for a plug used to interconnect electronic devices.

Jacket The outer sheath which protects a cable

Jackscrew A device used for leveling the positioning of a motor. These devices are adjustable screws fitting on the base or motor frame. Also a device for removing endshields from a motor assembly.

Jitter Jitter is the difference [deviation] between the expected occurrence of a signal edge and the time the edge actually occurs [phase variation]. Jitter may also be expressed as the movement of a signal edge from its ideal position in time [the expected position]. Abrupt and unwanted variations of one or more signal characteristics, such as the interval between successive pulses, the amplitude of successive cycles, or the frequency or phase of successive cycles.

Johnson noise Thermal Noise.

Joule (J) SI unit of work or energy. One joule is defined to be the work done by a force of one newton acting to move an object through a distance of one meter in the direction in which the force is applied.

Joule's equivalentmechanical equivalent of heat. 4.185 J/cal

Journal A journal is that part of a rotor that is in contact with or supported by a bearing in which it revolves.

Joystick A peripheral device used with personal computers to translate physical movement in two axis into electrical signals used by the computer.

Junction Contact or connection between two or more wires or cables. The area where the p-type material and n-type material meet in a semiconductor.

Junction box A box with a cover that serves the purpose of joining different runs of wire or cable and provides space for the connection and branching of the enclosed conductors.

Junction diode A two-terminal device containing a single crystal of semiconducting material that ranges from P-type at one terminal to N-type at the other.

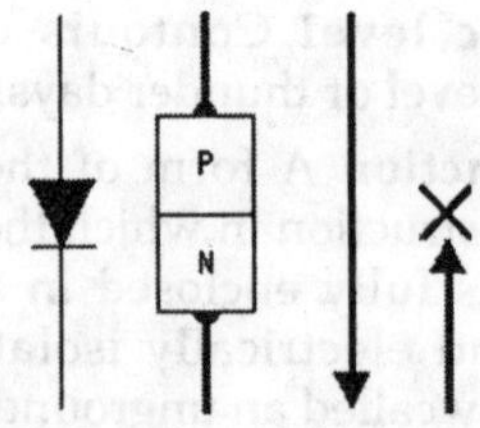

Fig. Junction diode

Junction transistor A bipolar transistor constructed from interacting PN junctions. The term is used to distinguish junction transistors from other types, such as field-effect and point-contact.

K When referring to memory capacity, two to the tenth power (1024 in decimal notation).

K Type connector A small type of threaded coaxial signal connector typically used in higher frequency applications. This connector is typically usable as high as 40GHz. It may be mated by an SMA connector with much lower performance.

Kapton A voice coil design in which multiple layers are used on a Kapton coated coil form to create the rotor element in the speaker driver's linear motor. This allows greater power handling, and cooler operation.

Karnaugh map An arrangement of cells representing the combinations of variables in a Boolean expression and used for a systematic simplification of the expression.

KC Kilocycle {obsolete, now kHz}.

Keep alive voltage DC voltage applied to a tr gap electrode to produce a glow discharge that allows the tube to ionize faster when the transmitter fires.

Kelvin (K) An absolute temperature scale. Zero Kelvin is absolute zero. No degree symbol (¼) is used with the Kelvin scale. (0¼C = 273.15K, 100¼C = 373.15K).

Kelvin contacts A means for testing or making measurements in electronic devices and circuits, particularly when low values are being measured. Two sets of leads are used at each test point, similar with respect to thickness, material and length; one set carries the test signal and the other connects with the measuring instrument. The effect of resistance in the leads is thus eliminated.

Kelvin double bridge It is a double bridge arrangement, which is an extension of the Wheatstone bridge, for the precise measurement of low resistance. The errors due to contact and lead resistances are eliminated by the additional bridge incorporated.

Keraunic level Number of days in the year in which thunder is heard.

Kernel A module of a program that forms a logical entity or performs a unit function.

Key A symbol or sequence of symbols (or electrical or mechanical correlates of symbols) applied to text in order to encrypt or decrypt.

Key escrow The system of giving a piece of a key to each of a certain number of trustees such that the key can be recovered with the collaboration of all the trustees.

Keyed oscillator transmitter. A transmitter in which one stage is used to produce the RF pulse.

Keyer 1. A device that changes dc pulses to mark and space modulation for teletypewriter transmissions.

2. A synchronizer.

Keying relays Relays used in radio transmitters where the ordinary hand key cannot accommodate the plate current without excessive arcing.

Keypad A panel usually made of metal or plastic with numbered push-button switches (like a touch-tone telephone) designed to provide access to certain types of control functions. Security, CD changers, and cellular systems represent typical examples

Keystroke monitoring A specialized form of audit trail software, or a specially designed device, that records every key struck by a user and every character of the response that the AIS returns to the user.

Keyword A word or words which can be searched for in documents or menus.

KHz (kilohertz) A frequency of one thousand (1,000) cycles per second.

Kilo A prefix meaning one thousand, 103.

Kilobyte (kB) 210 (= 1024, or about one thousand) bytes of information.

Kilogram (kg) The kilogram is the SI unit of mass. It is a fundamental unit. It is equal to the mass of the international prototype of the kilogram preserved in France.

Kilohertz (kHz) One thousand cycles per second.

Kilovolt ampere The practical unit of apparent power, which is 1,000 volt-ampere.

Kilowatt hour A unit of energy when one kilowatt of power is expended for one hour. Example A radiator bar is usually rated at 1,000 watts and this switched on for one hour consumes one kilowatt-hour of electricity.

Kilowatt hour meter A meter used by electric companies to measure the amount of electric power used by a customer.

Kinetic energy Energy associated with mass in motion, i.e., 1/2 rv2 where r is the density of the moving mass and V is its velocity.

Kinetic molecular theory A model that assumes that an ideal gas is composed of tiny particles (molecules) in constant motion.

Kirchhoff's laws 1. The algebraic sum of the current flowing toward any point in a circuit and the current flowing away from it is zero.

2. The algebraic sum of the products of the current and resistance in each of the conductors in any closed path in a network is equal to the algebraic sum of the electromotive forces in the path.

Kirchhoff"s current law The sum of the currents flowing into a point in a circuit is equal to the sum of the currents flowing out of that same point.

Kirchhoff"s voltage law The algebraic sum of the voltage drops in a closed path circuit is equal to the algebraic sum of the source voltages applied.

Kirchoff's Current Law (KCL) A law stating that the total current entering a point or junction in a circuit must equal the sum of the current leaving that point or junction.

Klydonograph An instrument for the detection and recording of the occurrence of lightning in transmission lines.

Klystron power amplifier A multicavity microwave electron tube that uses velocity modulation.

Knee of the curve The point of maximum curvature of a magnetization curve. (Shaped like the knee of a leg that is bent.)

Knee voltage The voltage at which a curve joins two relatively straight portions of a characteristic curve. For a PN junction diode, the point in the forward operating region of the characteristic curve where conduction starts to increase rapidly. For a zener diode, the term is often used in reference to the zener voltage rating.

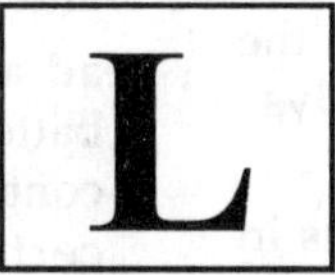

L C Oscillator Oscillator in which the frequency determining components are an inductor and capacitor.

Label or marker A problem endemic in immunoassays is the absence of a chemical signal created by the antibody-antigen binding, in contrast with an enzyme-substrate binding reaction which produces a chemical reaction product. As a result of this absence, the use of a label or marker is usually required to detect the bound antibody-antigen complex. Several markers have been established for use in immunoassays. Examples of such markers are: - Particles (e.g. latex, gold particles, erythrocytes) - Metal and dye sols (e.g. Au, Palanil®; Luminous Red G) - Chemiluminescent and bioluminescent compounds (e.g. Luciferase/luciferins, Luminol and derivatives, Acridinium esters, Peroxidase) - Electrochemical active species (ions, redox species, ionophores) Ï%- Fluorophores (e.g. Dansyl chloride DANS, Rare earth metal chelates, Umbelliferones) Ï%- ChromophoresÏ%- Enzymes (e.g. Alkaline phosphatase, â-D-Galactosidase, Peroxidase), substrates, cofactorsÏ%- LiposomesÏ%- Iodine-125, tritium, 14C, 75Se, 57Co.

Lag A delay in the effect of a changed condition at one point in the system, or some other condition to which it is related. Also, the delay in response to the sensing element of a control loop due to the time required for the sensing element to sense a change in the sensed variable.

Lagging load An inductive load with current lagging voltage.

Lambert An old unit of luminance. The luminance of a uniform diffuser of light which emits one lumen per square centimetre. 1 lambert = 10-4 lux

Laminar flow Flow in which fluid moves smoothly. In this flow form cross stream momentum transfer takes place by viscous action alone and mixing between flow strata does not occur.

Laminated core A core built up from thin sheets of metal insulated from each other and used in transformers.

Laminations The steel portion of the rotor and stator cores made up of a series of thin laminations (sheets) which are stacked and fastened together by cleats, rivets or welds. Laminations are used instead of a solid piece in order to reduce eddy-current losses.

Lamp Device that produces light.

LAN (Local area network) A computer communications system limited to no more than a few miles and using high-speed connections (2 to 100 megabits per second). A short-haul communications system that connects ADP devices in a building or group of buildings within a few square kilometers, including workstations, front-end processors, controllers, switches, and gateways.

Lap winding An armature winding in which opposite ends of each coil are connected to adjoining segments of the commutator so that the windings overlap.

Laplace transform An integral transformation of a function from the time domain to the complex frequency domain. Used to analyse transients in circuits.

Large motors Usually refers to AC motors in 5,000 series frames and above and to 500 series frames and larger in DC.

Large opening Hole or gap in a building envelope which is generally purpose made, for example, a door window or vent.

Large scale integration (LSI) The combining of about 1,000 to 10,000 circuits on a single chip. Typical examples of LSI circuits are memory chips and microprocessor.

Laser Device that produces a very narrow intense beam of light. The name is an axcronym for "light amplification by stimulated emission of radiation.

Laser diode A solid-state semiconductor device that is capable of emitting coherent light.

Laser trimming A method for adjusting the value of thin- or thick-film resistors by using a computer-controlled laser system.

Latch A non-clocked flip-flop.

Latching In relay or switching technology, this refers to the ability to keep the contact status in place even if power is removed from the equipment.

Latent heat Quantity of heat required to effect a change of state of a unit mass of a substance from solid to liquid (latent heat of fusion) or from liquid to vapour (latent heat of vaporisation) without change of temperature.

Latent heat transfer Heat added or removed during a change of state of a substance ie solid, to a liquid to a gas or vice versa, the temperature remaining constant.

Lattice A regular network of fixed points arranged in a geometrical pattern of straight lines.

Law of magnetism Like poles repel; unlike poles attract.

Leaching Washing out of a soluble constituent of a material.

Lead The angle advance between voltage signal and the corresponding current signal.

Lead acid cell A cell in an ordinary storage battery in which electrodes are grids of lead containing an active material consisting of certain lead oxides that change in composition during charging and discharging. The electrodes or plates are immersed in an electrolyte of diluted sulfuric acid.

Lead inductance The inductance of the lead wires connecting the internal components of an electron tube.

Lead sheath A continuous jacket of lead molded around a single conductor or multiple conductor cable. Generally used to ensure conductors are protected from water or extensive moisture.

Leading load A capacitive load with current leading voltage.

Leakage The loss of all or parts of a useful agent, as of the electric current that flows through an insulator or the magnetic flux that passes outside useful flux circuits.

Leakage area The actual open area of a hole or gap.

Leakage current. Error current that can degrade sensitive measurements. Leakage current is any unwanted current that flows when test voltage is applied. The ideal leakage current is zero. Leakage currents can originate in instruments, cables, or the device being tested. Even high resistance paths between low current conductors and nearby voltage sources can generate significant leakage currents.

Leakage factor Ratio of the total flux to the useful flux is defined as the leakage factor. It has a value of around 1.2 for electrical machines.

Leakage flux Magnetic flux lines produced by the primary winding that do not link the turns of the secondary winding.

Leakage inductance A small inductance associated with the flux of a transformer winding which are not magnetically coupled to the other windings of the transformer.

Leakage path A route by which air enters or leaves a building or flows through a component.

Leakage rate The maximum rate at which a fluid is permitted or determined to leak through a seal. The type of fluid, the differential Limits of Error A tolerance band for the thermal electric response of thermocouple wire expressed in degrees or percentage defined by ANSI specification MC-96.1 (1975).

Leakage resistance. The resistance calculated from the dependence of leakage current on applied voltage. leakage resistance = V / leakage current

Leapfrog attack Use of user id and password information obtained illicitly from one host to compromise another host. The act of TELNETing through one or more hosts in order to preclude a trace (a standard cracker procedure).

Leased line A dedicated circuit, typically supplied by the telephone company, that permanently connects two or more user locations. These lines are used to transmit data.

Least cost alternatives The lowest cost option for providing for incremental demands. In least cost planning to serve electric demands, the least cost alternatives are often construed broadly to include demand-side management as well as various generation and purchased power options.

Least significant BIT (LSB). The lowest order bit in a digital quantity.

Least significant digit (LSD) The LSD is the digit whose position within a given number expression has the least weighting power.

Least-squares line The straight line for which the sum of the squares of the residuals (deviations) is minimized.

LED Light Emitting Diode. A diode that emits light when current is passed through it. It has two leads cathode (k) and anode.

Left hand rule for generators A rule or procedure used to determine the direction of current flow in a generator.

Left hand rule If fingers of the left hand are placed around a wire so that the thumb points in the direction of electron flow, the fingers will be pointing in the direction of the magnetic field being produced by the conductor.

LEG One connection in an electric circuit.

Lens Any device which causes a beam of rays to converge or diverge on passing through it.

Lenz's law The current induced in a circuit due to a change in the magnetic field is so directed as to oppose the flux, or to exert a mechanical force to oppose the motion.

Level detector An op-amp circuit that compares two inputs and provides a DC output indicating the polarity relationship between the inputs. A comparitor.

Lewis acid An electron-pair acceptor.

Lewis base An electron-pair donor.

Library catalogue A searchable database of items held in a library's collection.

Lie detector Piece of electronic equipment also called a polygraph used to determine whether a person is telling the truth by looking for dramatic changes in blood pressure, body temperature, breathing rate, heart rate and skin moisture in response to questions.

Life (lifetime) The length of time the sensor can be used before its performance changes.

Life cycles The endurance rating of a device expressed in number of operations with the stated electrical load applied.

Life test A test in which a device is subjected to actual or accelerated use to obtain an estimate of life expectancy.

Lifetime The time from the creation of an electron hole pair until recombination occurs.

Light Any radiation capable of causing a visual sensation direct i.e. Visible electromagnetic radiation in the wavelength range 400 to 700 nano meter.

Light emitting diode (Led) A semiconductor p-n junction device that is optimized to release light of approximately the band gap energy when electrons fall from the conduction band to the valence band.

Light oil Lighter fuel oils distilled off during the refining process. Virtually all petroleum used in internal combustion and gas turbine engines use light oil.

Light rays Light waves emitting from a source in straight lines.

Light year A measure of astronomical distances. 1 light year = 9.461 x 1015 m

Lighthouse tube An electron tube shaped like a lighthouse that is designed to handle large amounts of power at uhf frequencies.

Lightning Any form of visible electric discharge between thunder clouds or between a thunder cloud and the earth. Ninety percent of all lightning never touches the ground - it occurs inside the thunder cloud or jumps from cloud to cloud.

Lightning arrester A device used to pass large impulses to ground. It is vital that this device be placed upstream from the equipment to be protected.

Lightning rod A grounded metallic rod set up on a structure (like a building) to protect it from lightning.

Limit control A control system in which, for reasons of comfort or efficiency, the controlled variable is prevented from exceeding or dropping below a certain value.

Limit of detection The smallest measurable input. This differs from resolution, which defines the smallest measurable change in input. For a temperature measurement, this would provide an indication of the lowest temperature a sensor could generate an output in response to.

Limiter A device that prevents (limits) a waveform from exceeding a specified value.

Limits of error A tolerance band for the thermal electric response of thermocouple wire expressed in degrees or percentage defined by ANSI specification MC-96.1 (1975).

Lin log amplifier An amplifier in which the response is linear for weak signals and logarithmic for large signals.

Line A designation of one or more power-carrying conductors for power distribution. The brown (or red) wire is the line conductor, the blue (or black) wire is the neutral, and the green-yellow (or green) wire is the ground. The voltage difference between the line conductor and the neutral is the supply voltage.

Line conditioner A line conditioner contains multiple protection devices in one package to provide, for example, electrical noise isolation and voltage regulation.

Line of force A line in an electric or magnetic field that shows the direction of the force.

Line of sight Straight line from a radar antenna to a target.

Line pulsing modulator Circuit that stores energy and forms pulses in the same circuit element, usually the pulse-forming network (pfn).

Line regulation The ability of a voltage regulator to maintain a constant voltage when the regulator input voltage varies.

Line regulator Power conversion equipment that regulates and/or changes the voltage of incoming power.

Line to line A term used to describe a given condition between conductors of a multiphase feeder.

Line to neutral A term used to describe a given condition between a phase conductor and a neutral conductor.

Line voltage The voltage between two lines (or phases) of a three phase system is defined as the line-to-line voltage or more commonly as the line voltage. For a balanced three phase system, the line voltage is Ö3 times the phase-to-neutral voltage.

Linear 1. Referring to mechanical movement, the ability of the voice coil to move in and out in the air gap without moving side-to-side. Non-linear movement can damage the voice coil.

2. Referring to woofer response, the ability to maintain power or movement without loss of drive force.

3. Referring to enclosure port operation, the relationship bewteen the amount of air moving through the port vs. the amount of air moved by the cone. Non-linear response in a port can cause audible distortion.

Linear actuator Refers to an assembly that transfers the rotary motion of a motor to linear motion.

Linear circuit One whose output is linearly related to the input. A circuit which obey's Ohm's law.

Linear control system A control system in which the transfer function between the controlled condition and the command signal is independent of the amplitude of the command signal. The characteristics of a linear system may be described by a linear differential equation of the first degree, which remains true over the range being considered.

Linear current booster (LCB) An electronic circuit that matches PV output directly to a motor. Most commonly used in direct water pumping.

Linear impedance An impedance in which a change in current through a device changes in direct proportion to the voltage applied to the device.

Linear input A process input that represents a straight-line function.

Linear motor A form of motor (normally induction motor) in which the stator and rotor are linear instead of cylindrical and parallel instead of coaxial.

Linear phase response Any system which accurately preserves phase relationships between frequencies.

Linear regulator An active device connected in series or shunt with the load of a power supply in such a way that the feedback to the control device changes its voltage drop as required to maintain a constant DC output voltage or current.

Linear scale A scale in which the divisions are uniformly spaced.

Linear supply An electronic power supply employing linear regulation techniques.

Linear taper (aka B-taper) Always 50% resistance at the 50% travel point

Linearity The closeness of a calibration curve to a specified straight line. Linearity is expressed as the maximum deviation of any calibration point on a specified straight line during any one calibration cycle.

Linearization, square root The extraction of a linear signal from a nonlinear signal corresponding to the measured flow from a flow transmitter. Also called square root extraction.

Link A hypertext connection between two documents, image maps, graphics and so on.

Liquid A state of matter between a solid and a gap. Liquids assume the shape of the vessel containing them, other than the top surface which will assume a horizontal position when free to air. They are only slightly compressible.

Liquid cooling system Source of cooling for high-heat producing equipments, such as microwave components, radar repeaters, and transmitters.

Liquid junction potential The potential difference existing between a liquid-liquid boundary. The sign and size of this potential depends on the composition of the liquids and the type of junction used.

Lissajous pattern A combined, simultaneous display of the amplitude and phase relationships of two input signals on a CRT.

Live part A conductor or conductive part intended to be energised in normal use, including a neutral conductor but, by convention, not a PEN conductor.

Live Term used to describe a circuit or piece of equipment that is on and has current flow within it.

Load Any passive electrical device connected to a power source may be called by the general term of "load". It is the amount of electric power delivered or required at any specific point or points on a system. The requirement originates at the energy consuming equipment of the consumers.

Load center A limited geographical area where large amounts of power are used by consumers.

Load current Current drawn from a source by a load.

Load diversity The load condition that exists when the peak demands of a variety of electric consumers occur at different times.

Load duration curve LDC A curve that displays load values on the horizontal axis in descending order of magnitude against percent of time (on the vertical axis) the load values are exceeded.

Load factor The ratio of the average load supplied (kW) during a designated period to the peak or maximum load (kW) occurring during that period.

Load forecast Estimate of electrical demand or energy consumption at some future time.

Load impedance The impedance presented to the output terminals of a transducer by the associated external circuitry.

Load isolator A passive attenuator in which the loss in one direction is much greater than that in the opposite direction. One example is a ferrite isolator for waveguides that allows energy to travel in only one direction.

Load line Locus of instantaneous operating points used to find the exact operating values of voltage and current.

Load management Techniques used by utilities to manage daily and/or seasonal fluctuations in customer demand. Managing the level and shape of demand for electrical energy so that demand conforms to present supply situations and long-run objectives and constraints.

Load profile Information on a consumer's usage over a period of time, sometimes shown as a graph.

Load regulation (Dynamic) The change in output voltage expressed as a percent for a given step change in load current. Initial and final current values and the rates of change must be specified. The rate of change shall be expressed as current/unit of time, e.g., 20 amperes per microsecond. The dynamic regulation is expressed as a +/- percent for a worst case peak-to-peak deviation for DC supplies, and worst case rms deviation for AC supplies.

Load resistance Resistance of a load.

Load shape A curve showing power (kW) supplied (on the horizontal axis) plotted against time of occurrence (on the vertical axis), and illustrating the varying magnitude of the load during the period covered.

Load shifting A load shape objective that involves moving loads from peak periods to off-peak periods.

Loading effect Connection of a meter (an ammeter in series or a voltmeter in shunt) to a circuit to make a measurement alters the original circuit in that it draws some energy from the circuit. The error caused is known as the loading effect.

Lobe An area of greater signal strength in the transmission pattern of an antenna.

Local action A continuation of current flow within a battery cell when there is no external load. Caused by impurities in the electrode.

Local lighting Lighting for a specific visual task, additional to and controlled separately from the general lighting.

Locked rotor current Steady state current taken from the line with the rotor at standstill (at rated voltage and frequency). This is the current seen when starting the motor and load.

Locked rotor torque The minimum torque that a motor will develop at rest for all angular positions of the rotor (with rated voltage applied at rated frequency).

Lockout A mechanical device which may be set to prevent the operation of a push-button or other device.

Locus The line that can be drawn through adjacent positions satisfying a given criteria.

Log A chronological record of a series of events. Each event is identified by time, day,value, and physical reference.

Log taper Often used an an audio taper since its 50% rotation point has 10% resistance

Logarithmic receiver Receiver that uses a linear logarithmic amplifier (lin-log) instead of a normal linear amplifier.

Logarithmic scale A method of displaying data (in powers of ten) to yield maximum range while keeping resolution at the low end of the scale.

Logging capacity The number of data elements that can be stored by an outstation when undertaking a log.

Logic Science of dealing with the principle and applications of gates, relays and switches.

Logic bomb Also known as a Fork Bomb - A resident computer program which, when executed, checks for a particular condition or particular state of the system which, when satisfied, triggers the perpetration of an unauthorized act.

Logic circuit The primary control information processor in digital equipment; made up of electronic gates and so named because their operation is described by simple equations of a specialized logic algebra.

Logic diagram In computers and data processing equipment, a diagram representing the logical elements and their interconnections without necessarily expressing construction or engineering details.

Logic element The smallest building blocks that can be represented by operators in an appropriate system of symbolic logic. Typical logic elements are the AND-gate and the flip-flop, which can be represented as operators in a suitable symbolic logic. Also a device that performs the logic function.

Logic instruction Any instruction that executes a logic operation that is defined in symbolic logic, such as AND, OR, NAND, or NOR.

Logic level State of a voltage variable. States high and low correspond to the two usable voltage levels of a digital device.

Logic operation A nonarithmetical operation in a computer, such as comparing, selecting, making references, matching, sorting, and merging, where the logical yes or no quantities are involved.

Logic switch A diode matrix or other switching arrangement that is capable of directing an input signal to one of several outputs.

Logic symbol A symbol used to represent a logic element graphically. Also a symbol used to represent a logic operator.

Logon To sign on to a computer system.

LOM protection: loss of mains protection

Long term stability The maintenance of repeatability over a long period of time e.g. one year.

Long wire antenna An antenna that is a wavelength or more long at its operating frequency.

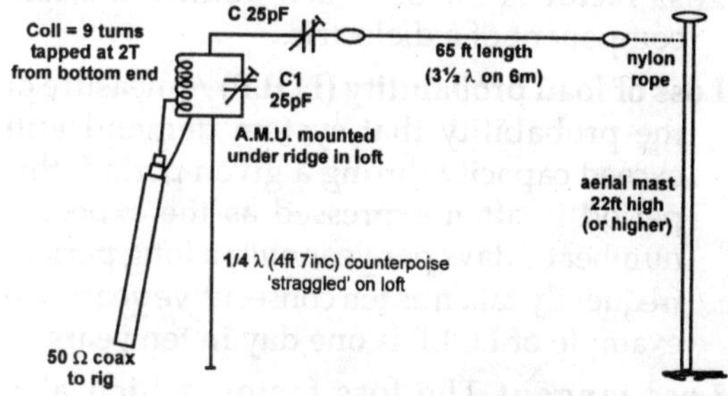

Fig. Long wire antenna

Longitudinal waves Those waves in which the disturbance (back and forth motion) takes place in the direction of propagation. Sometimes called compression waves.

Loop A curved conductor that connects the ends of a coaxial cable or other transmission line and projects into a waveguide or resonant cavity for the purpose of injecting or extracting energy.

Loop current A component of current common to the complete loop.

Loop or mesh A closed path of elements in a circuit.

Loop resistance The total resistance of a thermocouple circuit caused by the resistance of the thermocouple wire. Usually used in reference to analogue pyrometers which have typical loop resistance requirements of 10 ohms.

Looping input An input which passes a video signal in and out of a device without terminating the cable or affecting the signal quality. Looping inputs use two connectors normally wired together with no active components between them. If the looping feature is not used, a 75 ohm terminator should be placed on the second connector, or a provision for switching in a 75 ohm termination internally.

Loose coupling Inefficient coupling of energy from one circuit to another that is desirable in some applications. Also called weak coupling.

Loss angle Angle of deviation of the current from the ideal current for a dielectric for a sinusoidal input.

Loss factor A factor which defines the loss component of a dielectric.

Loss of load probability (LOLP) A measure of the probability that system demand will exceed capacity during a given period; this period is often expressed as the expected number of days per year over a long period, frequently taken as ten consecutive years. An example of LOLP is one day in ten years.

Loss tangent The loss factor, which also corresponds to the tangent of the loss angle.

Losses A motor converts electrical energy into a mechanical energy and in so doing, encounters losses. These losses are all the energy that is put into a motor and not transformed to usable power but are converted into heat causing the temperature of the windings and other motor parts to rise.

Lossy Compression A type of data compression which permanently discards data that humans supposedly "cannot hear" to create much smaller audio, video and image file sizes. When the file is decompressed by the recipient, this compression method replaces the data for the sections it removed with calculated values to restore the file. The decompressed file is similar but not identical to the original file.

Loudspeaker A device used for converting electrical signals into sound waves.

Low ambient controllers/head pressure controllers When equipment is required to provide cooling in cold weather the external unit will over condense i.e. will reject too much heat. To overcome this problem it is necessary to fit a low ambient controller also known as a head pressure controller. This device slows down the condenser fan or uses other methods to prevent over condensing. Some equipment has this fitted as standard others do not.

Low energy house A house designed or retro-fitted with energy savings in mind, to provide overall lower energy consumption.

Low Frequency Refers to radio frequencies within the 30-300 kHz band. In audio it usually refers to frequencies in the 20-160 Hz band.

Low level modulation Modulation produced in an earlier stage than the final.

Low line The lowest specified input operating voltage.

Low noise earth An earth connection in which the level of conducted or induced interference from external sources does not produce an unacceptable incidence of malfunction in the data processing or similar equipment to which it is connected. The susceptibility in terms of amplitude/frequency characteristics varies depending on the type of equipment.

Low pass filter A filter that passes a majority of the low frequencies on to the next circuit and rejects, or attenuates, the higher frequencies. Also called a high-frequency discriminator.

Low pressure mercury (vapor) lamp Mercury vapor lamp, with or without a coating of phosphor, in which during operation the partial pressure of the vapor does not exceed 100 pa.

Low pressure sodium (vapor) lamp Sodium vapor lamp in which the partial pressure of the vapor during operation does not exceed 5 pa.

Lower frequency cutoff The lowest frequency a circuit will pass.

Lower sideband All difference frequencies below that of the carrier.

Lowest usable frequency The minimum operating frequency that can be used for communications between two points.

L-Pad A low resistance (commonly 8 ohms) potentiometer used primarily to control the input delivered to a speaker. A remote site volume control.

LSD (Least-Significant Digit) The rightmost active (non-dummy) digit of the display.

LS-TTL Compatible For digital input circuits, a logic 1 is obtained for inputs of 2.0 to 5.5 V which can source 20 μA, and a logic 0 is obtained for inputs of 0 to 0.8 V which can sink 400 μA. For digital output signals, a logic 1 is represented by 2.4 to 5.5 V with a current source capability of at least 400 μa; and a logic 0 is represented by 0 to 0.6 V with a current sink capability of at least 16 MA. "LS" stands for low-power Schottky.

LS-TTL Unit Load A load with LS-TTL voltage levels, which will draw 20 μA for a logic 1 and -400 μA for a logic 0.

Lubrication In order to reduce wear and avoid overheating certain motor components require lubricating (application of an oil or grease). The bearings are the major motor component requiring lubrication (as per manufacturer's instructions). Excess greasing can however damage the windings and internal switches, etc.

Lumen (lm) SI unit for measuring the flux of light. One lumen is equal to the luminous flux emitted in unit solid angle (steradian) by uniform point source having a luminous intensity of 1 candela.

Luminaire Apparatus that distributes, filters or transforms the light given by a lamp or lamps and which includes all the items necessary for fixing and protecting these lamps and for connecting them to the supply circuit.

Luminaire efficiency The ratio of total lumen output of a luminaire to the lumen output of the lamps, expressed as a percentage.

Luminescence defined as the mission of light from a substance in an electronically excited state. Depending on whether the excited state is singlet or triplet, the emission is called fluorescence (less than one second decay) or phosphorescence (longer than 1 second decay). Depending on the source, molecules get the needed extra energy from different types of luminescence are distinguished: radioluminescence, photoluminescence (in the same category are fluorescence and phosphorescence), chemiluminescence and bioluminescence, electrochemiluminescence, sonochemiluminescence and thermoluminescence.

Luminous efficacy Quotient of the luminous flux emitted by a source and the power consumed. (Unit lumen per watt, lm/W) Visible light output of a luminary measured relative to power input.

Luminous flux The quantity derived from radiant flux by evaluating the radiation according to its action upon the standard photometric observer. (Unit lumen, lm).

Luminous intensity distribution Distribution of the luminous intensities of a lamp or luminaire in all spatial directions.

Lumped constants The properties of inductance, capacitance, and resistance in a transmission line.

Lumped impedance tuning The insertion of an inductor or capacitor in series with an antenna to electrically lengthen or shorten the antenna.

Lux (lx) SI unit for measuring the illumination of a surface. One lux is defined as an illumination of one lumen per square meter.

Lv low voltage Not exceeding 1000 V between conductors and 600 V between conductors and earth.

M Mega; one million. When referring to memory capacity, two to the twentieth power (1,048,576 in decimal notation).

Machine language Instructions that are written in binary form that a computer can execute directly. Also called object code and object language.

Magic-T Junction. A combination of H-type and E-type T-junctions.

Magnet Body that can be used to attract or repel magnetic materials.

Magnet Boot A rubber or plastic cover for the magnet housing for protection or appearance, mostly the latter.

Magnet wire Wire coated with an enamel insulation and used in coils, relays, transformers, motor windings, and so forth.

Magnet/Magnet Structure A combination of magnetic material and connected field concentrators that creates the magnetic field within which the voice coil interacts to produce sound. Magnetic materials have changed greatly over the years to produce much higher concentrations of magnetic fields (rated in gauss) with lighter and smaller volumes of material.

In marketing speakers, a great deal of hype is often applied to the question of magnet weight. But many of these claims should be treated with skepticism. With greater and greater concentrations of gauss fields being developed from ever lighter and smaller mass metallurgical materials, the only good measure of adequate power handling is the manufacturer's RMS Wattage rating. Hefty magnets may look impressive, but while capable,they are no longer an essential index to a speaker's power capacity.

Magnetic amplifier An electromagnetic device that uses one or more saturable reactors to obtain a large power gain. This device is used in servosystems requiring large amounts of power to move heavy loads.

Magnetic circuit breaker Circuit breaker that is tripped or activated by use of an electromagnet.

Magnetic coils Applications involving Magnetic Coils such as solenoids, relays and contactors are inductive loads which do not usually produce high Inrush Currents. However, when this type of load is turned off, the magnetic field collapses resulting in arcing across open contacts. This can cause serious contact deterioration.

Magnetic core Material that exists in the center of the mahnetic coil to either physically support the windings (non-magnetic material) or to concentrate the magnetic flux (magnetic material).

Magnetic field 1. The region in which the magnetic forces created by a permanent magnet or by a current-carrying conductor or coil can be detected.

2. The field that is produced when current flows through a conductor or antenna.

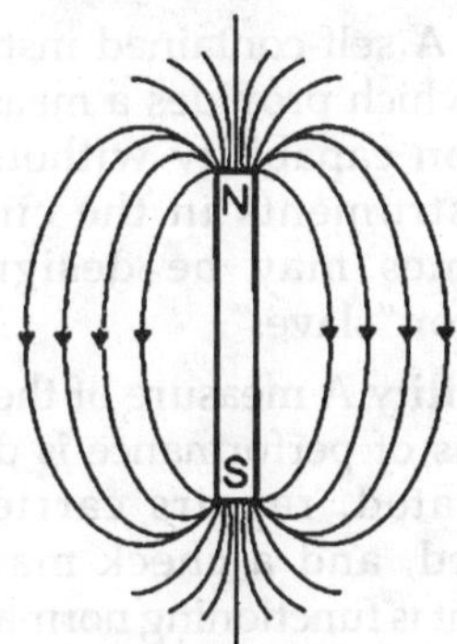

Fig. Magnetic field

Magnetic field strength Magnetic field produced by a current, independent of the presence of magnetic material. The units of H are ampere-turns per meter, or just amperes per meter.

Magnetic flux A measure of quantity of magnetism, taking account of the strength and the extent of a magnetic field.

Magnetic flux density or magnetic induction The magnetic field produced in a substance by an external magnetic field. The units of B are tesla (T). One tesla is the magnetic flux density given by a magnetic flux of 1 weber per square meter. One weber is a magnetic flux that, linking a circuit of 1 turn, would produce in it an electromotive force of 1 volt if it were reduced to zero at a uniform rate in 1 second. Both B and H are field vectors. One henry (H) is the inductance of a closed circuit in which an electromotive force of 1 volt is produced when the electric current in the circuit varies uniformly at the rate of 1 ampere per second. The magnetic field strength and flux density are related according to: $B = \mu H$, where μ is the permeability.

Magnetic induction Generating a voltage in a circuit by the creation of relative motion between a magnetic field and the circuit. The relative motion can be the result of physical movement or the rise and fall of a magnetic field created by a changing current.

Magnetic leakage The passage of magnetic flux outside the path along which it can do useful work.

Magnetic lines of force Imaginary lines used for convenience to designate the direction in which magnetic forces are acting as a result of magnetomotive force.

Magnetic mechanism The magnetic mechanism uses a solenoid with an iron piece to operate the circuit breaker in the event of an overcurrent.

Magnetic microphone A microphone in which the sound waves vibrate a moving armature. The armature consists of a coil wound on the armature and located between the pole pieces of a permanent magnet. The armature is mechanically linked to the diaphragm.

Magnetic polarity It is a fundamental principle of a winding that adjacent poles must be wound to give opposite magnetic polarity. This does not mean that the coils actually have to be wound in this direction before being placed into the stator. It does mean that the winding must be connected so that, if the current proceeds through one pole in a clockwise direction, it must proceed through the next pole in a counterclockwise direction. This principle is used to determine the correctness of connection diagrams.

Magnetic poles The section of a magnet where the flux lines are concentrated; also where they enter and leave the magnet.

Magnetic shield A piece of magnetic material used to carry the magnetic flux around an object to prevent the object from being affected by the magnetic field.

Magnetic stack One set of soft iron pole pieces and permanent magnet; step motors typically have one, two, and three magnetic stack versions for each frame size.

Magnetic susceptibility **(+m)** The proportionality constant between the magnetization M and the magnetic field strength H. The magnetic susceptibility is unitless.

Magnetic trip element A circuit breaker trip element that uses the increasing magnetic attraction of a coil with increased current to open the circuit.

Magnetisation curve The relationship between the magnetic flux density and the applied magnetic field (or the magnetic flux and the applied mmf) is called the magnetisation curve.

Magnetism The property possessed by certain materials by which these materials can exert mechanical force on neighboring masses of magnetic materials and can cause currents to be induced in conducting bodies moving relative to the magnetized bodies.

Magnetizing force Also called magnetic field strength. It is the magnetomotive force per unit length at any given point in a magnetic circuit.

Magneto hydrodynamics MHD A method of generating electricity by subjecting the free electrons in a high velocity flame or plasma to a strong magnetic field. The free electron concentration in the flame is increased by the thermal ionisation of added substances of low ionisation potential.

Magnetostrictive material A material that changes dimension in the presence of a magnetic field or generates a magnetic field when mechanically deformed.

Magnetron oscillator An electron tube that provides a high power output. Theory of operation is based on interaction of electrons with the crossed electric and magnetic fields in a resonant cavity.

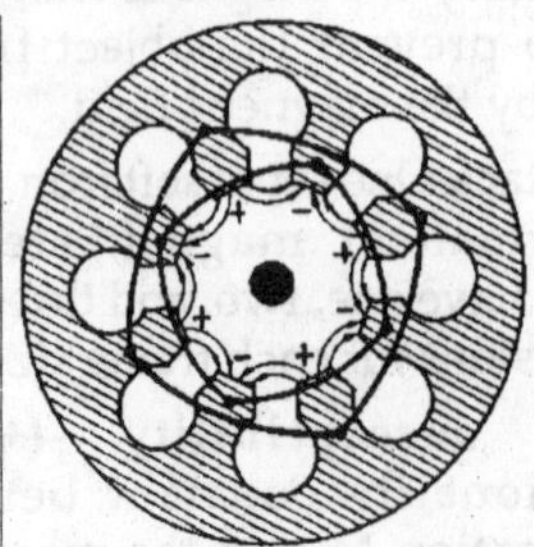

Fig. Magnetron oscillator

Main earthing terminal The terminal or bar provided for the connection of protective conductors, including equipotential bonding conductors, and conductors for functional earthing, if any, to the means of earthing.

Main switch The principal (or main) switch in an electrical installation.

Mainframe A self-contained instrument in a cabinet, which provides a measurement or connection capability without requiring other instruments in the circuit. Some mainframes may be designated as a "master" or "slave."

Maintainability A measure of the speed with which loss of performance is detected, the fault located, repairs carried out and completed, and a check made that the equipment is functioning normally again; cf. availability and reliability.

Maintenance factor Ratio of the average illuminance on the working plane after a specified period of use of a lighting installation to the average illuminance obtained under the same conditions for a new installation.

Major lobe The lobe in which the greatest amount of radiation occurs.

Majority carriers The conduction band electrons in an n-type material and the valence band holes in a p-type material. Produced by pentavalent impurities in n-type material and trivalent impurities in p-type material.

Majority voting A form of discriminatory control. Sensed values are input to the controller, which establishes, for each input signal, a corresponding 'dummmy' output signal. The 'resultant' controller output signal is determined from the majority of the dummy outputs. For example, if there are 3 input sensors to an on-off controller, and 2 result in an ON status and 1 results in an OFF status, the resultant controller output signal will indicate an ON status.

Make before break Connecting a new circuit before disconnecting the present circuit.

Makeup air Outdoor air supplied to replace exhausted air, brought in through a ventilation system without previous circulation in the system.

Malicious code Hardware, software, of firmware that is intentionally included in a system for an unauthorized purpose; e.g. a Trojan horse.

Malleability Capacity of being hammered out into thin sheets.

Mandrel (Balancing Arbor) An accurately machined shaft on which work is mounted for balancing.

Manual reset (Adjustment) The adjustment on a proportioning controller which shifts the proportioning band in relationship to the set point to eliminate droop or offset errors.

Manual transfer switch A switch designed so that it will disconnect the load from one power source and reconnect it to another source while at no time allowing both sources to be connected to the load simultaneously.

Marconi antenna A quarter-wave antenna that is operated with one end grounded and is positioned perpendicular to the earth.

Marginal cost The sum that has to be paid for the next increment of product of service. The marginal cost of electricity is the price to be paid for kilowatt-hour above and beyond those supplied by presently available generating capacity irrespective of sunk costs.

Mark An interval during which a signal is present. Also the presence of an RF signal in cw keying. The key-closed condition (presence of data) in communications systems.

Market based price A price set by the mutual decisions of many buyers and sellers in a competitive market.

Marking The state where a circuit is closed and current flows in teletypewriter operation.

Martensite A metastable iron phase supersaturated in carbon that is the product of a diffusionless (athermal) transformation from austenite.

Mash An acronym for Multi-stage noise shaping. This is the name given to the digital advances that were introduced after the CD specification had been established.

Mass flow rate Volumetric flowrate times density, i.e. Pounds per hour or kilograms per minute.

Mass storage A device like a disk or magtape that can store large amounts of data readily accessible to the central processing unit.

Master oscillator In a transmitter, the oscillator that establishes the carrier frequency of the output.

Master oscillator power amplifier (MOPA) A transmitter in which the oscillator is isolated from the antenna by a power amplifier.

Master station A station that can select and transmit a message to a sleve station.

Master/Master mainframe A mainframe that has control of other mainframes (slaves) in a serial chain. A master/slave combination has one bus address and appears as one mainframe with increased capacity.

Master/Slave operation Interconnection of two or more regulated supplies in which one (the master) controls the other (the slaves).

Matched impedance Condition that occurs when the output impedance of a souce is equal to the input impedance of a load.

Matching Connection of two components or circuits so that maximum power is transferred between the two.

Matching transformer A device used to convert impedance between two levels. A common use is between a 75 ohm impedance and a 300 ohm impedance.

Matrix An arrangement of signal circuits in which input buses are represented by parallel vertical lines and output buses as overlapping horizontal lines (or visa versa), forming a grid-like array. Crosspoint switches at each crossing point connect inputs to outputs. Also referred to as a switching array, or crosspoint switch.

Maximum allowable input The maximum DC plus peak AC value (voltage or current) that can be applied between the high and low input measuring terminals without damaging the instrument.

Maximum demand The largest of all demands of the load (usually expressed in kVA or MVA) that has occurred within a specified period of time.

Maximum elongation The strain value where a deviation of more than ±5% occurs with respect to the mean characteristic (diagram of resistance change vs strain).

Maximum load The highest specified output power rating of a supply specified under worst case conditions.

Maximum load impedance The largest load that the output device can operate. Usually specified in ohms.

Maximum operating temperature The highest ambient temperature at which the power supply will continuously operate safely and within specifications.

Maximum power rating The maximum operating power at which a device can operate safely or with expected normal operating life.

Maximum power transfer A theorem that states that maximum power will be transferred from source to load when input impedance of the load equals the output impedance of the source.

Maximum usable frequency Maximum frequency that can be used for communications between two locations for a given time of day and a given angle of incidence.

Maxwell Unit of magnetic flux. One maxwell equals one magnetic line of force.

Mbps Megabits per second (millions of bits per second). A measure of digital data transmission rate.

MCM MultiChip Module; the interconnection of two or more semiconductor chips in a semiconductor-type package.

MDF MDF is one of the best materials to loudspeaker cabinets. MDF is for example better than massive wood, aluminium, marble or other hard materials, because of its unique sound absorbing qualities.

Mean Numerical average of data values.

Mean active repair time (MART) This applies to repairable items, and means the average time an item may be expected to be out of service for maintenace and repair, given that the require tools/parts are to hand.

Mean radiant temperature Mean radiant temperature is not equal - but may be calculated from - the reading of a globe thermometer.

Mean time between failures (MTBF) This applies to repairable items, and means the that if an item fails, say, five times over a period of use totalling 1000 hours, the mean time between failures would be 1000/5, or 20 hours. The MTBF is a measure of the likelihood that equipment will break down in a given period.

Mean time to repair (MTTR) This applies to repairable items, and means the average time an item may be expected to be out of service for maintenance and repair, including any order/delivery times.

Means and swings method The CIBSE method of determining the peak inside environmental temperature.

Measurand A physical quantity, condition, or property that is to be measured.

Measured value The physical quantity that is measured by the transducer, i.e., the input to the transducer.

Measuring junction The thermocouple junction referred to as the hot junction that is used to measure an unknown temperature.

Mechanical extract ventilation system A ventilation system in which air is extracted from a space so creating an internal negative pressure. Supply air is drawn through adventitious or purpose provided openings.

Mechanical hysteresis The difference of the indication with increasing and decreasing strain loading, at identical strain values of the specimen.

Mechanical rotation frequency The speed in revolutions per minute of armatures in electric motors and engine-driven generators; blade speed in turbines.

Mechanical scanning The reflector, its feed source, or the entire antenna is moved in a desired pattern.

Mechanical ventilation system A ventilation system in which air is extracted from a space so creating an internal negative pressure. Supply air is drawn through adventitious or purpose provided openings.

Mechatronics The synergistic combination of precision mechanical engineering with electronic control.

Medical electrical equipment Electrical equipment, provided with no more than one connection to a particular supply mains, and intended to diagnose, treat or monitor the patient under medical supervision and which makes physical or electrical contact with the patient and/or transfers energy to or from the patient and/or detects such energy transfer to or from the patient. The equipment includes those accessories as defined by the manufacturer which are necessary to enable the normal use of the equipment.

Medium The vehicle through which a wave travels from one point to the next. Air, water, and wood are examples.

Medium altitude orbit An orbit from 2,000 to 12,000 miles above the earth. The rotation rate of the earth and satellite are quite different, and the satellite moves quickly across the sky.

Medium effect (f m) For solvents other than water the medium effect is the activity coefficient related to the standard state in water at zero concentration. It reflects differences in the electrostatic and chemical interactions of the ions with the molecules of various solvents. Solvation is the most significant interaction.

Medium frequency The band of frequencies from 300 kHz to 3 MHz.

Mega A prefix meaning one million, 106; also MEG.

Megawatt hour One thousand kilowatt-hours or one million-watt hours.

Meggar test A measure of an insulation system's resistance. This is usually measured in megohms and tested by passing a high voltage at low current through the motor windings and measuring the resistance of the various insulation systems.

Megger or megohmeter A high resistance range specially designed ohmmeter for measuring insulation resistance of conductors and other electrical equipment.

Megohm One million ohms. Written as "M".

Megohmmeter A meter that measures very large values of resistance; usually used to check for insulation breakdown in wires.

Melting point The constant temperature at which the solid and liquid phase of a substance are in equilibrium at a given pressure.

Membrane The pH-sensitive glass bulb is the membrane across which the potential difference due to the formation of double layers with ion-exchange properties on the two swollen glass surfaces is developed. The membrane makes contact with and separates the internal element and filling solution from the sample solution.

Memory The word most commonly used to refer to a system's ability to retain specific information, particularly; time, stations,.and other preference settings that are electronically stored and governed.

Menu A list of options from which the user selects an action to be performed by entering a letter or moving the cursor.

Mercury cell Primary cell using a mercuric oxide cathode, a zinc anode and a potassium hydroxide electrolyte.

Mercury vapour lamp Lamp emitting a strong bluish-white light by the passage of an electric current through mercury vapour in a bulb.

Mercury wetted relay A reed relay in which the contacts are wetted by a film of mercury (Hg). Usually has a required operating position to avoid liquid mercury from shorting the contacts; other types are position insensitive. This type of relay is usually higher power and longer life, but at a higher dollar cost. Another benefit of this type of contact is the repeatability of contact resistance and virtually no contact bounce.

Mesh analysis A method of analysis of circuits based on defining mesh currents as the variables.

Metabolism Biochemical reactions by which energy is made available for an organism to use. Includes all chemical transformations that occur in an organism from the time a nutrient substance enters until it has been used and the waste products eliminated.

Metadata Data which provides information about an Internet resource and which is commonly used to aid information retrieval on the Internet.

Metal A material with a partially filled energy band; metals are generally malleable, ductile, good reflectors of electromagnetic radiation, and good conductors of heat and electricity; metals are usually identified by having electrical conductivities that decrease with increasing temperature.

Metal film resistor A resistor in which a film of metal oxide or alloy is deposited on an insulating substrate.

Metal halide lamp Discharge lamp in which the major portion of the light is produced by the radiation from a mixture of a metallic vapor (for example, mercury) and the products of the dissociation of halides (for example, halides of thallium, indium or sodium).

Metal oxide field effect transistor (MOSFET) A field effect transistor in which the insulating layer betwen the gate electrode and the channel is a metal oxide layer.

Metal oxide resistor A metal film resistor in which an oxide of metal (such as tin) is deposited as a film onto the substrate.

Metal oxide semiconductor field-effect transistor See MOSFET.

Metal Tape EQ An equalization circuit that compensates for the unique frequency response characteristics of metal tape.

Metal vapor lamp Discharge lamp such as the 'mercury (vapor) lamp' and the 'sodium (vapor) lamp' in which the light is mainly produced in a metallic vapor.

Metallic armor A protective covering for wires or cables. Made as a woven wire braid, metal tape, or interlocking metal cover. Made from steel, copper, bronze, or aluminum.

Metallic rectifier Also known as a dry-disc rectifier. A metal-to-semiconductor, large area, contact device in which a semiconductor is sandwiched between two metal plates. This asymmetrical construction permits current to flow more readily in one direction than the other.

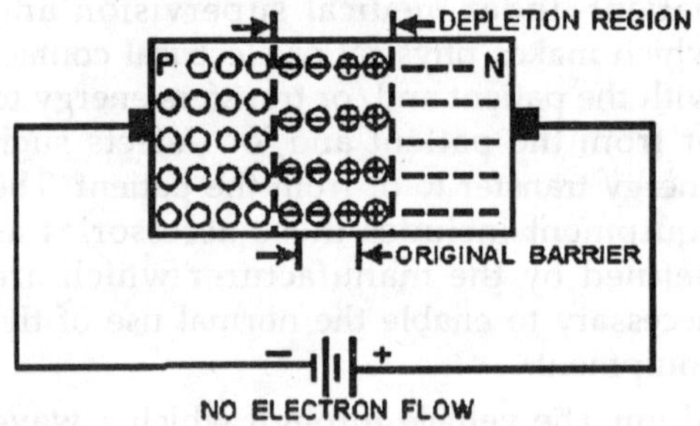

Fig. Metallic rectifier

Metallic, insulator A shorted quarter-wave section of transmission line.

Meter (M) The meter is the SI unit of length. It is a fundamental unit. It is defined as the length of the path travelled by light in vacuum during a time interval of 1/299 792 458 of a second (1983)

Meter FSD current Value of meter current needed to cause the needle to deflect to its maximum position (full scale deflection).

Meter movement The part of the meter that moves to indicate some value.

Meter resistance DC resistance of the meter's armature coil.

Meter shunt A resistor placed in parallel with the meter terminals; used to provide increased range capability.

Metering Monitoring of energy or water consumption or other data over a period of time.

Method of correction A procedure whereby the mass distribution of a rotor is adjusted to reduce unbalance, or vibration due to unbalance, to an acceptable value. Corrections are usually made by adding material to, or removing it from, the rotor.

Metric A random variable x representing a quantitative measure accumulated over a period.

Mezzanine The termed used to describe a daughter card which plugs into another card [motherboard] which is then plugged a backplane.

Micro (M) Decimal sub-multiple prefix corresponding to one-millionth or 10-6.

Microamp One millionth of an ampere, 10-6 amps, μA.

Microbar 1. A unit of atmospheric pressure equal to one millionth of a bar.

2. A place to consume the product of microbreweries.

Microcircuit A small circuit having high equivalent-circuit-element density, which is considered as a single part composed of interconnected elements on or within a single substrate to perform an electronic-circuit function.

Microcircuit module An assembly of microcircuits or a combination of microcircuits and discrete components that perform one or more distinct functions.

Microcomputer A computer which is physically small. It can fit on top of or under a desk; based on LSI circuitry, computers of this type are now available with much of the power currently associated with minicomputer systems.

Microelectronics The solid-state concept of electronics in which compact semiconductor materials are designed to function as an entire circuit or sub-assembly rather than as circuit components.

Microenvironment a particular part of the large environment that is in some way whole by itself. Used to describe a subset of the global environment such as the indoor environment.

Micron A unit describing wavelength. One micron is equivalent to 1 x 10-6 meters.

Microphone An electromechanical transducer that converts sound pressure into an electrical signal.

Microprocessor A multiple semiconductor IC device that can be dedicated or programmed to perform a variety of tasks in many different systems. These exist in virtually all consumer and commercial electronic devices of more than rudimentary functionality.

Microprocessor based control A control circuit that operates on low voltage and uses a microprocessor to perform logic and control functions, such as operating a relay or providing an output signal to position an actuator. Electronic devices are primarily used as sensors. The controller often furnishes flexible DDC and BEMS control functions.

Microprocessor Chip containing the logical elements for performing calculations and carrying out stored instructions.

Microsoft internet explorer The name of a Web browser - a means of accessing information on the World Wide Web - Microsoft is the company that provides it.

Microstepping A method of proportioning current so that the full steps of the step motor are broken into smaller, microsteps; a typical microstepping drive can provide up to 25,000 steps per revolution.

Microstructure In material engineering the structural features of a material such as grain boundaries, grain size and structure, subject to observation under a microscope, selective etching etc. In MEMS microstructure unfortunately is also used to designate a micromachined feature.

Microvolt (μV) One 10-6 of a volt (one millionth in the US). Mil One thousandth of an inch, or 0.001 inches in decimal form.

Microwave Band of very short wavelength radio waves within the UHF, SHF and EHF bands.

Microwave region The portion of the electromagnetic spectrum from 1,000 MHz to 100,000 MHz.

Mid point bias An amplifier biased at the center of its DC load line.

Midband gain Gain of an amplifier operating within its bandwidth.

Midbass Those frequencies roughly between 100 and 300 Hertz. (CPS)

MIDI Acronym for Musical Instrument Digital Interface, a standard adopted by the electronic music industry for controlling devices, such as synthesizers and sound cards, that produce music. At minimum, a MIDI representation of a sound includes values for the note's pitch, length, and volume, but can also include additional characteristics, such as attack and decay time.

MIDI files A computer file format containing musical information and performance data capable of being used in MIDI capable devices.

Midrange A Driver that is usually much smaller than a woofer, but with a surface area greater than the typical tweeter. It reproduces the mid frequency range from approximately 300 to 5000 Hertz. This optimum range can vary considerably from one driver to the next, thus giving the system designer more flexibility in choosing Crossover points for the other drivers.

Mil A unit of small length in the imperial system equal to one-thousandth of an inch. 1 mil = 2.54 x 10-5 m

MIL foot A unit of measurement for conductors (diameter of 1 mil, 1 foot in length.).

Mile Unit of distance in the imperial system. 1 mile = 1.609 km

Miles per hour (mph) Unit of speed in the imperial system. 1 mph = 0.44704 m/s

Military specifications (MIL-SPEC) Technical requirements and standards adopted by the Department of Defense that must be met by vendors selling materials to DOD.

Military standards (MILSTD) Standards of performance for components or equipment that must be met to be acceptable for military systems.

Miller indices A set of 3 integers (4 for hexagonal) that designate crystallographic planes, as determined from reciprocals of fractional axial intercepts.

Miller's theorem A theorem that allows you to represent a feedback capacitor as equivalent input and output shunt capacitors.

Milli (m) Decimal sub-multiple prefix corresponding to one-thousandth or 10-3.

Milliampere (mA) One thousandth of an ampere.

Milliamps A unit of measurement of electric current equal to 1/1000th of an ampere. The milliampere is the most common unit used when measuring quiescent current drain in consumer audio electronics.

Millimeter One thousandth of a meter, symbol mm.

Millivolt Unit of electromotive force. It is the difference in potential required to make a current of 1 millampere flow through a resistance of 1 ohm; one thousandth of a volt, symbol mV.

Mimic display A method of representing an HVAC system in the form of a graphic display. It usually shows the current values or status of plant.

Mimicking Synonymous with Impersonation, Masquerading or Spoofing.

Mineral insulated thermocouple A type of thermocouple cable which has an outer metal sheath and mineral (magnesium oxide) insulation inside separating a pair of thermocouple wires from themselves and from the outer sheath. This cable is usually drawn down to compact the mineral insulation and is available in diameters from375 to010 inches. It is ideally suited for high-temperature and severe-duty applications.

Mineral Substance occurring naturally in the Earth.

Mineral oil General name given to the various mixtures of natural hydrocarbons.

Miniature circuit breaker mcb A device designed to perform the same function as a fuse but resettable. When the circuit breaker activates, or trips out, it disconnects the circuit. The fault condition must be found and rectified before the mcb can be reset. Usually for ratings less than 63 A.

Miniature electronics Modules, packages, pcbs, and so forth, composed exclusively of discrete components.

Minimum discernible signal (MDS) The weakest input signal that produces a usable signal at the output of a receiver. The weaker the input signal, the more sensitive the receiver.

Minimum load current The smallest load current required insuring proper operation of an output-switching device.

Minimum load The lowest specified current to be draw on a constant voltage power supply for the voltage to be in a specified range.

Minimum operating temperature The lowest ambient temperature at which a power supply or other device will continuously operate safely and within specifications.

Minimum starting temperature The lowest ambient temperature at which a power supply or other device will turn on and operate safely.

Minimum ventilation requirement The minimum quantity of outdoor or conditioned air entering a building, which is needed to maintain acceptable indoor air quality.

Minor lobe The lobe in which the radiation intensity is less than that of a major lobe.

Minor scale division On an analogue scale, the smallest indicated division of units on the scale.

Minority carriers Either electrons or holes, whichever is the less dominant carrier in a semiconductor device. In P-type semiconductors, electrons are the minority carriers; in N-type semiconductors, the holes are the minority carriers.

Minority current A very small current that passes through the base-to-collector junction when this junction is reverse biased.

MIPS (Millions of Instructions per Second) A measure of computing power.

Mismatch Term used to describe a difference between the output impedance of a source and the input impedance of a load. A mismatch prevents the maximum transfer of power from source to load.

Misuse detection model The system detects intrusions by looking for activity that corresponds to a known intrusion techniques or system vulnerabilities. Also known as Rules Based detection.

Mixing fan Small fan used to aid the mixing of room air and tracer gas before and/or during ventilation rate measurements.

Mixing flow ventilation mixing flow describes a method of air distribution from an air conditioning or ventilation system. It is the most widely used method of supplying air into an atmosphere being air conditioned. This form of air distribution commonly uses ceiling diffusers or wall grilles at high level. Most air conditioning and air handling units are manufactured for the mixed flow air distribution market.

Mks system The metric system in which the meter, kilogram and the second are the fundamental units.

MN taper (" balance control") Special taper developed for home stereo. Consists of two sections (one for each channel) operating opposite each other. Exactly one-half of each section is a zero resistance surface (i.e., solid-copper or equivalent), the next 50% of travel is linear taper. Therefore for one channel rotating the slider through the first 50% of travel does not change the level at all, while the other channel is reduced from full to zero, and vice-versa, with the middle position (usually featuring a center-detent) always passing full signal to each channel.

Mobile substation A movable substation which is used when a substation is not working or additional power is needed.

Mockingbird A computer program or process which mimics the legitimate behavior of a normal system feature (or other apparently useful function) but performs malicious activities once invoked by the user.

Mode shifting In a magnetron, the inadvertent shifting from one mode to another during a pulse.

Mode skipping Operation in which the magnetron fires randomly, rather than firing on each successive pulse as desired.

Moderator A substance used in nuclear reactors to reduce the speed of fast neutrons produced by nuclear fission.

Modular circuitry A technique where printed circuit boards are stacked and connected together to form a module.

Modular packaging Circuit assemblies or subassemblies packaged to be easily removed for maintenance or repair.

Modular power A physically descriptive term used to describe a power supply or other system made up of a number of separate subsections, such as an input module, power module, or filter module.

Modulated wave A complex wave consisting of a carrier and a modulating wave that is transmitted through space.

Modulating A control action that adjusts by minute increments and decrements.

Modulating wave An information wave representing intelligence.

Modulation factor (M) An indication of relative magnitudes of the RF carrier and the modulating signal.

Modulation Process by which an information signal (audio for example) is used to modify some characteristic of a higher frequency wave known as a carrier (radio for example).

Modulation index The ratio of frequency deviation to the frequency of the modulating signal.

Modulator 1. A device that produces modulation; that is, a device that varies the amplitude, frequency, or phase of an ac signal.

2. A circuit used in servo-systems to convert a dc signal to an ac signal. The output ac signal is a sine wave at the frequency of the ac reference voltage. The amplitude of the output is directly related to the amplitude of the dc input. The circuit's function is opposite to that of a demodulator.

3. In radar, it produces a high-voltage pulse that turns the transmitter on and off.

Modulator switching device Controls the on (discharge) and off (charge) time of the modulator.

Module A circuit or portion of a circuit packaged as a removable unit. A separable unit in a packaging scheme displaying regularity of dimensions.

Modulus of elasticity (E) The ratio of stress to strain when deformation is totally elastic. Also the Young's modulus.

Molality A measure of concentration expressed in mols per kilogram of solvent.

Molding (plastics) Shaping a plastic material by forcing it, under pressure at a high temperature, into a mold cavity.

Mole (mol) The mole is the SI unit of the amount of substance. It is defined as the amount of substance of a system which contains as many elementary entities as there are atoms in 0.012 kilogram of carbon 12; (1971)

Molecule Smallest particle of a compound that still retains its characteristics.

Momentary overvoltage or swell An increase in voltage outside the normal tolerance for a few seconds or less. Voltage swells are often caused by sudden load decreases or turn-off of heavy equipment.

Monitoring The process of collecting, analysing, and reporting data.

Mono (monaural) The operation of an amplifier in one channel for both input and output. Can refer to an amplifier with only one channel of amplification or operating in bridged mode. For low frequency amplification applications, it provides better phase coherence and less distortion than stereo operation.

Monochromatic radiation Radiation characterized by a single frequency. In practice, radiation of a very small range of frequencies that can be described by stating a single frequency.

Monoclonal antibodies Produced by injecting animals to elicit a response from lymphocytes to produce antibodies. Lymphocytes which produce antibodies with strong binding capability can be isolated and used to produce only one kind of antibody (monoclonal) on a permanent basis once the lymphocytes are immortalized. This is accomplished by fusing them (combining them genetically) with cancer cells which have the distinction of living indefinitely in a culture. Monoclonal antibodies can be produced repeatedly and collected for use in immunodetection.

Monolithic circuit A circuit where all elements (resistors, transistors, and so forth) associated with the circuit are fabricated inseparably within a continuous piece of material (called the substrate), usually silicon.

Monolithic IC ICs that are formed completely within a semiconductor substrate. Silicon chips.

Monopoly The only seller with control over market sales.

Monopulse (Simultaneous) lobing A radar receiving method using two or more (usually four) partially overlapping lobes. Sum and difference locate the target with aspect to the axis of the antenna.

Monopulse radar A radar that gets the range, bearing, and elevation position data of a target from a single pulse.

Monostable Circuit with two states. Only one state is stable.

Monostable multivibrator A nultivibrator with one stable output state. When triggered, the circuit output will switch to the unstable state for a predetermined period of time and then return to the stable state. A timer.

Monovalent Ion An ion with a single positive or negative charge (H+, C1-).

Moore's law after Gordon Moore:"The number of transistors per computer chip will double roughly every two years".

Mortality rate The number of operating hours elapsed before a certain percentage of the lamps fail.

MOS Metal Oxide Semiconductor. Technology used in the manufacture of semiconductors.

MOSFET (Metal Oxide Semiconductor Field Effect Transistor) A type of large output transistor used in the final stages of many power amplifiers, and commonly found in most car and home amplifiers today.

These field-effect transistors are controlled by voltage rather than current, like a bipolar transistor. MOSFETs have a significantly higher switching speed than bipolar transistors. They generate almost no loss (little heat generation), which lends the power supply fast response, excellent linearity, and high efficiency.

Mosfet transistors are most often discrete devices, used with smaller driver transistors and other devices, to convert a small signal to a large one. They are highly stable and efficient, compared to the bipolar types that preceded them.

Mosfet A semiconductor device that contains diffused source and drain regions on either side of a P- or N-channel area. Also contains a gate insulated from the channel area by silicon-oxide. Operates in either the depletion or the enhancement mode.

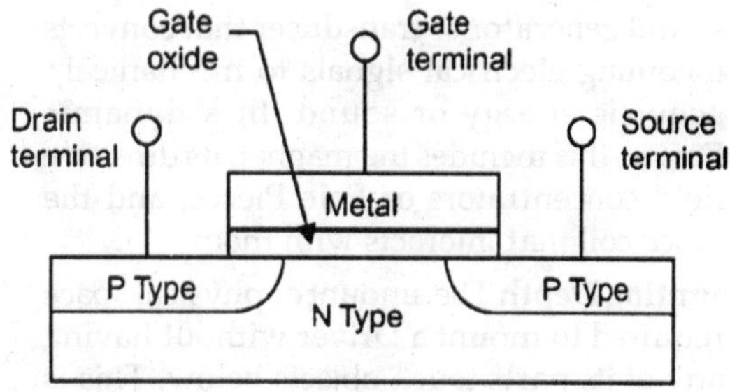

Fig. Mosfet

Most significant digit (MSD) The MSD is the digit whose position within a given number expression has the greatest weighting power.

Motherboard The pc board of a computer that contains the bus lines and edge connectors to accommodate other boards in the system. In a microcomputer, the motherboard contains the microprocessor and connectors for expansion boards.

Motor A device that takes electrical energy and converts it into mechanical energy to turn a shaft.

Motor efficiency Mechanical power output of the motor divided by electrical power input to the motor.

Motor enclosures Reliance Electric offers the industry's widest selection of motor/ generator enclosures to meet specific operating needs.

Motor load Any device driven by a motor. Typical loads are drills, saws, water pumps, rotating antennas, generators, and so forth. The speed and power capabilities of a motor must be matched to the speed and power capabilities of the motor load.

Motor reaction The force created by generator armature current that tends to oppose the normal rotation of the armature.

Motor starters Large resistive devices placed in series with dc motor armatures to prevent the armature from drawing excessive current until armature speed develops counter emf. The resistance is gradually removed from the circuit either automatically or manually as motor speed increases.

Motor Structure In speakers, the complete sound generator or transducer that converts incoming electrical signals to mechanical/ acoustic energy or sound. In a dynamic Driver, this includes the magnet, its directive field concentrators or Pole Pieces, and the voice coil that interacts with them.

Mounting Depth The amount of physical space required to mount a Driver without having any of its parts touch objects below. This is particularly relevant to car speakers where such mounting spaces may be sharply limited as to their ability to accommodate deep speakers with large magnets. Door panels with movable windows are a typical example of where care must be taken in selecting speakers to be mounted.

Mounting error The error resultant from installing the transducer, both electrical and mechanical.

Mounting Ring Often used to describe the circular gasket seal incorporated into the edge of a speaker, traditionally the term "mounting ring" refers to a separate device placed between a Driver and the surface on which it is mounted, for the effect of raising the speaker so that larger and deeper speakers can be accommodated in limited spaces.

Moving coil loudspeaker Loudspeaker that uses a moving "voice coil" placed within a fixed magnetic field. Audio frequency current in the voice coil causes movement which is mechanically transferred to the speaker cone. Also known as a dynamic loudspeaker.

Moving coil meter In this instrument, a moving coil is suspended between the poles of a permanent magnet. When a current is passed through the coil, the coil becomes an electromagnet and tries to align with the permanent magnet. The deflecting becomes proportional to the current.

Moving coil microphone Microphone that uses a moving coil within a fixed magnetic field. Dynamic microphone.

Moving coil pick up Dynamic phonograph pick-up in which the stylus causes a coil to move within a fixed magnetic field.

Moving iron meter An instrument working on the principle of a moving iron placed within an electromagnet getting an induced emf and with the deflection being proportional to the square of the current. The meter is calibrated with the square root of the deflection and hence has basically a non-linear scale.

Moving vane meter movement A meter movement that uses the magnetic repulsion of the like poles created in two iron vanes by current through a coil of wire; most commonly used movement for ac meters.

MP3, mp3 The file extension for MPEG, audio layer 3. Layer 3 is one of three coding schemes (layer 1, layer 2 and layer 3) for the compression of audio signals. Layer 3 uses perceptual audio coding and psychoacoustic compression to remove all superfluous information (that, in the opinion of the developers, the human ear doesn't hear anyway). It also adds an algorithm that increases the frequency resolution 18 times higher than that of layer

2. The result is mp3 encoding shrinks the original sound data from a CD by a factor of 12 without sacrificing sound quality.

MSD (Most-Significant Digit) The leftmost digit of the display.

MTBF (Mean Time Between Failure) The average length of time between system failures, exclusive of infant mortality and rated end-of-life.

MTTR MTTR An abbreviation for Mean Time To Repair, a theoretical period of time need to repair a piece of equipment given certain circumstances.

Muddiness Poor sonic performance in a speaker, resulting either from distortion or from a particular frequency being too predominant, thereby obscuring higher frequencies.

Mueller bridge A high-accuracy bridge configuration used to measure three-wire RTDthermometers.

Multi level Usually indicating that separate levels of switching are included different levels are required. This is typical when switching RS-232 type data lines. These could be up to 25 levels deep. Some types of video signals are divided into different levels, one for each primary colour (Red, Green and Blue). For most applications of this nature, this type of switching array is gang controlled.

Multi speed motors A motor wound in such a way that varying connections at the starter can change the speed to a predetermined speed. The most common multi-speed motor is a two speed although three- and four-speeds are sometimes available. Multi-speed motors can be wound with two sets of windings or one winding. They are also available either constant torque, variable torque or constant horsepower.

Multi tracking The process of recording a multi-part performance on separate tracks at different times which allows the engineer to subsequently combine, balance and process those tracks during mixdown.

Multiconductor More than one conductor, as in a cable.

Multicouplers Couplers that patch receivers or transmitters to antennas. They also filter out harmonics and spurious responses and impedance-match the equipment.

Multielectrode tube An electron tube normally classified according to its number of electrodes (the multi-electrode tube contains more than three electrodes).

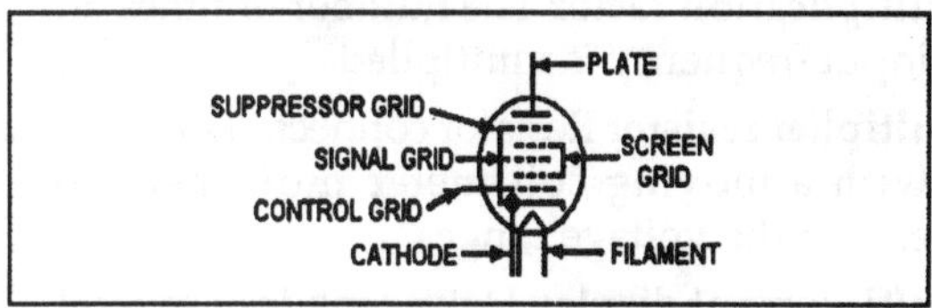

Fig. Multielectrode tube

Multielement array An array that consists of one or more arrays and is classified as to directivity.

Multielement parasitic array An array that contains two or more parasitic elements and a driven element.

Multihost based auditing Audit data from multiple hosts may be used to detect intrusions.

Multiloop servosystem A servo-system that contains more than one servo loop; each loop is designed to perform its own function.

Multimedia The combination of plain text, pictures, sound and even moving video clips.

Multimeter A common term used to describe a VOM. A multimeter usually has the ability to measure volts, ohms, and amperes or milliamperes.

Multipath The multiple paths a radio wave may follow between transmitter and receiver.

Multiple output power supply A power supply with two or more outputs.

Multiple tracer gas technique A Measurement method using two or more tracer gases. This method is often used to evaluate interzonal airflows.

Multiplex A technique which allows different input (or output) signals to use the same lines at different times, controlled by an external signal. Multiplexing is used to save on wiring and I/O ports.

Multiplexer A type of field panel used in BEMS to minimize data transmission costs by using time-sharing transmission techniques.

Multiplexing A method for simultaneous transmission of two or more signals over a common carrier wave.

Multiplication factor The number of times an input frequency is multiplied.

Multiplier resistor Resistor connected in series with a moving coil meter movement to extend the voltage ranges.

Multisegmant display Device made of several light emitting diodes arranged in a numeric or alphanumeric pattern. By lighting selected segments numeric or alphabet characters can be displayed.

Multispeed synchro systems Systems that transmit data at different transmission speeds; for example, dual-speed and tri-speed synchro systems.

Multiunit tube An electron tube containing two or more units within the same envelope. The multi-unit tube is capable of operating as a single-unit tube, or each unit can operate as a separate tube.

Multivibrator A building block in which the output is either High or Low. There are three foms of multivibrator Monostable, bi-stable and astable. The monostable has only one stable state, the bi-stable has two stable states and the astable is free-running (no stable states).

Multivibrator modulator An astable multivibrator used to provide frequency modulation. The modulating af voltage is inserted in series with the base return of the multivibrator transistors to produce the frequency modulation.

Mute A control found on receivers, some mixers, and certain signal processing units that silences (mutes) a signal path, or output. 2. Someone with much to say but lacking the hardware or software to do so.

Mutual flux The total flux in the core of a transformer that is common to both the primary and secondary windings. The flux links both windings.

Mutual inductance A circuit property existing when the relative position of two inductors causes the magnetic lines of force from one to link with the turns of the other. The symbol for mutual inductance is M.

MUX Device for combining several signals or data streams into a single flow.

MxN This term is another way to express a type of switching configuration. It is the same as a matrix or crosspoint switch where a number of inputs (M) can connect to a number of outputs (N). This term is common since when giving the dimensions of a matrix, it is usually is expressed as the number of input (M) by the number of outputs (N). For example, a matrix with 8 inputs and 32 outputs would commonly be expressed as an 8x32 matrix. The MxN term is also used since it doesn't explicitly indicate "in" and "out" since some type of matrices are bi-directional. When one axis has only a single port, this is commonly expressed as a 1xN array.

N Type connector A larger threaded coaxial connector with high power handling and good high frequency characteristics. Typically usable to 12.5GHz, but some manufacturers offer connectors usable to 18GHz.

N type semiconductor A semiconductor compound formed by doping an intrinsic semiconductor with a pentavalent element. An n-type material contains an excess of conduction band electrons.

Nameplate The plate on the outside of the motor describing the motor, HP, voltage, RPM's, efficiency, design, enclosure, etc.

NAND circuit A combination of a NOT function and an AND function in a binary circuit that has two or more inputs and one output. The output is logic 0 only if ALL inputs are logic 1; it is logic 1 if ANY input is logic 0.

NAND Gate Gate whose output is 0 when both inputs are 1, and 1 in all other cases.

Nano One thousand millionth. 1/1,000,000,000, e.g. 1 nanofarad - 1nF. 1nF is equal to 1,000pF is equal to 0.001uF and one million uF equals one Farad.

Nanovoltmeter A sensitive DC voltmeter (typically one decade more sensitive than a digital multimeter) with a low thermal input connection.

Natural response The natural response of a circuit refers to be behaviour of the circuit (in terms of voltage and current), in the absence of external excitation.

Natural ventilation The movement of outdoor air into a space through intentionally provided openings, such as windows and doors, or through non powered ventilators or by infiltration.

Navy service "A" Motors designed to meet requirements of MIL M-17059 or MIL M-17060 for high shock and service and are essential to the combat effectiveness of a ship. These motors are usually made of nodular iron.

Near synchronous orbit An orbit in which the satellite rotates close to but not exactly at the same speed as the earth.

NEC, national electrical code A set of rules and regulations that are put out by the National Fire protection Association. Generally accepted as the building wiring standard in the US.

Negation The process of inverting the value of a function or variable.

Negative alternation That part of a sine wave that is below the reference level.

Negative charge A charge that has more electrons than protons .

Negative clamper A circuit that clamps the upper extremity of the output waveshape to a dc potential of 0 volts.

Negative electrode A terminal or electrode having more electrons than normal. Electrons flow out of the negative terminal of a voltage source.

Negative feedback A feedback signal 180° out of phase with an amplifier input signal. Used to increase amplifier stability, bandwidth and input impedance. Also reduces distortion.

Negative ground A system where the negative terminal of the source is connected to the system's metal chassis.

Negative ion An atom having a greater number of electrons in orbit than there are protons in the nucleus.

Negative LCD A liquid crystal display employing a dark backfield with lit elements. This results in a primarily dark display, which improves cosmetic integration with a dark coloured headunit.

Negative logic The form of logic in which the more positive voltage level represents a logic 0, FALSE, or LOW and the more negative voltage represents a logic 1, TRUE, or HIGH.

Negative pressure A condition that exists when less air is supplied to a space than is exhausted from the space, so the air pressure within that space is less than that in surrounding areas.

Negative resistance A resistance such that when the current through it increases the voltage drop across the resistance decreases.

Negative resistance element A component having an operating region in which an increase in the applied voltage increases the resistance and produces a proportional decrease in current. Examples include tunnel diodes and silicon unijunction transistors.

Negative sequence A balanced set of three phase components which have the same magnitude, opposite sequence to the original unbalanced set, and phase angle diferring from each other by 120o. The frequency is of course the same as the original unbalanced three phase system.

Negative temperature coefficient A characteristic of a semiconductor material, such as silver sulfide, in which resistance to electrical current flow decreases as temperature increases.

Neon bulb Glass envelope filled with neon gas which when ionized by an applied voltage will glow red.

Nernst equation A mathematical description of electrode behavior E is the total potential, in millivolts, developed between the sensing and reference electrodes; Ex varies with the choice of electrodes, temperature, and pressure 2.3RT/nF is the Nernst factor (R and F are constants, n is the charge on the ion, including sign, T is the temperature in degrees Kelvin), and ai is the activity of the ion to which the electrode is responding.

Nernst factor (S, Slope); The term 2.3RT/nF is the Nernst equation, which is equal (at T = 25°C) to 59.16 mV when n = 1 and 29.58 mV when n - 2, and which includes the sign of the charge on the ion in the term n. The Nerst factor varies with temperature.

Net capability Maximum load carrying ability of the equipment, excluding station use.

Net generation Gross generation minus the energy consumed at the generating station for its use.

Net Volume The amount of airspace that is enclosed within a speaker's enclosure. This does not include the airspace taken up by bracing, vents, or the speaker itself.

Netiquette The cultural and social rules on the Internet. Ignoring them may result in being flamed or castigated in public.

Netscape The name of a Web browser - a means of accessing information on the World Wide Web - and the company that provides it.

Network A group of computers that are connected in some fashion. Most school networks are known as LANs, or Local Area Networks, because they are networks linking computers in one small area. The Internet could be referred to as a WAN, or a Wide Area Network, because it connects computers in more than one local area.

Network based Network traffic data along with audit data from the hosts used to detect intrusions.

Network level firewall A firewall in which traffic is examined at the network protocol (IP) packet level.

Network security Protection of networks and their services from unauthorized modification, destruction, or disclosure, and provision of assurance that the network performs its critical functions correctly and there are no harmful side-effects. Network security includes providing for data integrity.

Network security officer Individual formally appointed by a designated approving authority to ensure that the provisions of all applicable directives are implemented throughout the life cycle of an automated information system network.

Network technique Theoretical method for estimating the magnitude of air infiltration, ventilation and interzonal air movement, using a model which considers a building to comprise a number of enclosed spaces, each at its own internal pressure and linked by flow paths.

Network weaving Another name for "Leapfrogging".

Neutral 1. In a normal condition, hence neither positive nor negative. A neutral object has a normal number of electrons (the same number as protons).

2. The teletypewriter operation where current flow represents a mark and no flow represents a space.

Neutral atom An atom in which the number of negative charges (electrons in orbit) is equal to the number of positive charges (protons in the nucleus).

Neutral conductor A conductor connected to the neutral point of a system and contributing to the transmission of electrical energy. The term also means the equivalent conductor of an IT or d.c. System unless otherwise specified in the Regulations and also identifies either the mid wire of a three wire d.c. Circuit or the earthed conductor of a two wire earthed d.c. Circuit.

Neutral point displacement voltage: The voltage between the real or virtual neutral point and the earth.

Neutral pressure level Level at which the air pressure difference, derived from the stack effect between inside and outside a building is zero.

Neutral wire The conductor of a polyphase circuit or a single-phase three wire circuit that is intended to have a ground potential. The potential difference between the neutral and each of the other conductors are approximately equal in magnitude and equally spaced in phase.

Neutralization The process of counteracting or "neutralizing" the effects of interelectrode capacitance.

Neutron Subatomic particle in the nucleus of an atom and having no electrical charge.

Newton (N) SI unit of force. One newton is equal to the force required to accelerate a body with the mass one kilogram by one metre per second per second.

Newton's second law of motion If an unbalanced outside force acts on a body, the resulting acceleration is directly proportional to the magnitude of the force, is in the direction of the force, and is inversely proportional to the mass of the body.

Ni cad Nickel-Cadmium cell. Type of rechargeable battery.

Nibble One half of a byte.

Nickel cadmium cell A secondary cell that uses a nickel oxide positive electrode and a cadmium negative electrode.

Nicrosil/Nisil A nickel chrome/nickel silicone thermal alloy used to measure high temperatures. Inconsistencies in thermoelectric voltages exist in these alloys with respect to the wire gage.

Night purge An energy management function (rarely used in the U.K.) in which 100% outdoor air is introduced to the building prior to starting up the air-conditioning equipment, thereby reducing the cooling load.

Night set-back control An energy management function which acts to reduce the occupancy temperature by a few degrees, with the heating ticking over to maintain it. It is not usually as effective compared to the modern practice of intermittent heating, where the system is switched off overnight.

No load condition The condition that exists when an electrical source or secondary of a transformer is operated without an electrical load.

Nodal analysis A method of analysis of circuits based on defining node voltages as the variables.

Node point of connection between two or more elements or branches in a network.

Nodular iron (Ductile iron) Special cast iron with a crystalline formation which makes it capable of handling high shock.

Noise 1. In reference to sound, an unwanted disturbance caused by spurious waves that originate from man-made or natural sources.

2. In radar, erratic or random deflection or intensity of the indicator sweep that tends to mask small echo signals.

Noise figure An expression of noise generated with in a device specified in dB. This parameter is important in RF application such as a receive antenna switching system and IF signal routing. The lower the noise figure, the better.

Noise filter A filter to attenuate electrical noise which comes from a power supply, an inverter, or some other noise generating device.

Noise gate An expander with a fixed "infinite" downward expansion ratio. Used extensively for controlling unwanted noise, such as preventing "open" microphones and "hot" instrument pick-ups from introducing extraneous sounds into the system. When the incoming audio signal drops below the user set-point (the threshold point) the expander prevents any further output by reducing the gain to "zero." The actual gain reduction is typically on the order of -80 dB, thus once audio falls below the threshold, effectively the output level becomes the residual noise of the gate. Common terminology refers to the gate "opening" and "closing." Another popular application uses noise gates to enhance musical instrument sounds, especially percussion instruments. Judicious setting of a noise gate's attack (turn-on) and release (turn-off) times adds "punch," or "tightens" the percussive sound, making it more pronounced.

Noise limiter Circuit that clips the peaks of the noise spikes in a receiver.

Noise suppression The use of components to reduce electrical interference that is caused by making or breaking electrical contact, or by inductors.

Nominal value The stated or objective value of a quantity of component, which may not be the actual value measured.

Nominal voltage A nominal value assigned to a circuit or system for the purpose of conveniently designating its voltage rating. The actual voltage at which a circuit operates can vary from the nominal within a range that permits satisfactory operation of equipment. A suitable approximate value of voltage used to designate or identify a System.

Non blocking Non-Blocking A term with multiple and conflicting industry usage. 1. May be used to express the ability to connect a single input of a switching array to multiple outputs simultaneously without any input loading or mismatches. This usually results in a constant signal loss because of the use of power dividers (signal splitters) to configure the non-blocking switching array. Non-blocking switching arrays can also be achieved using impedance shifting in place of power dividers. Also referred to as Full Fanout.

2. In multi-stage switching arrays (tri-stage or 3-stage), it refers to the ability to route and

input to an output at all times (no blocking due to unavailable middle stages).

Non coincidental peak load The sum of two or more peak loads on individual systems, not occurring in the same time period.

Non discretionary security The aspect of DOD security policy which restricts access on the basis of security levels. A security level is composed of a read level and a category set restriction. For read-access to an item of information, a user must have a clearance level greater then or equal to the classification of the information and also have a category clearance which includes all of the access categories specified for the information.

Non equilibrium The condition in a semiconductor device or material when there is still a tendency for its macroscopic properties to change with time. In other words, there still remains some settling to occur before equilibrium is achieved. Devices in which current always flows, like diodes and bipolar transistors, do not ever reach equilibrium.

Non firm power Power supplied or available under terms with limited or no assured availability.

Non inductive load A non-inductive load is a load in which the current is in phase with the voltage across the load.

Non linear A characteristic which does not follow a straight line.

Non linear scale A scale in which the divisions are not equally spaced.

Non minimum phase systems In terms of feedback control theory, minimum phase systems are stable systems having zeroes in the left-half side of the complex s-plane, i.e. all denominator terms contribute phase lag whilst numerator terms contribute phase lead. Systems with zeroes in the right half of the s-plane (non-minimum phase systems) are inherently unstable (due to additional phase lag).

Non repudiation Method by which the sender of data is provided with proof of delivery and the recipient is assured of the sender's identity, so that neither can later deny having processed the data.

Nondegenerative parametric amplifier A parametric amplifier that uses a pump signal frequency that is higher than twice the frequency of the input signal.

Non-inverting input The terminal on an operational amplifier that is identified by a plus sign.

Nonlinear device A device in which the output does not rise and fall in direct proportion to the input.

Nonlinear impedance An impedance in which the resulting current through the device is not proportional to the applied voltage.

Non-Lossy Compression A form of data compression which seeks out chunks of data which are identical, replacing them with markers called keys. In this way, the file is reduced in size, and when it is decompressed by the recipient, the keys are replaced with the large chunks of data that were originally there (this is called Run Length Encoding). Using non-lossy compression, the uncompressed file is identical to the original file.

Nonresonant line A transmission fine that has no standing waves of current or voltage.

NOR gate A logic circuit that outputs a 1 only when each one of its inputs is a 0.

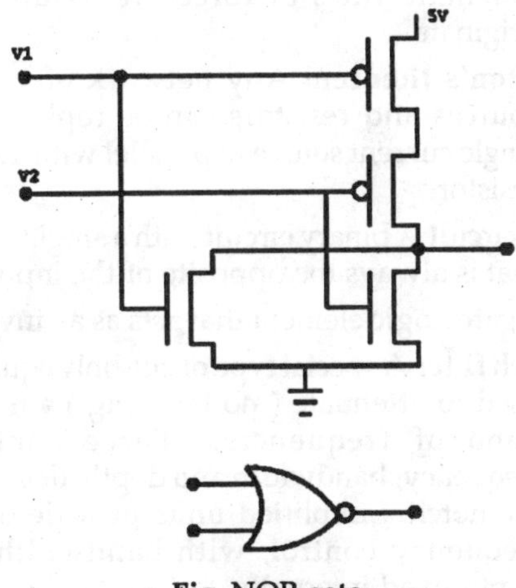

Fig. NOR gate

Normal hydrogen electrode A reversible hydrogen electrode (Pt) in contact with hydrogen gas at 1 atmosphere partial pressure and immersed in a solution containing hydrogen ions at unit activity.

Normal mode noise noise signal which appears between a set of phase conductors.

Normal mode rejection ratio (NMRR). The ability of an instrument to reject interference(usually of line frequency) across its input terminals. Usually expressed in decibels at a frequency.

Normalised leakage area Equivalent leakage area expressed per unit area of building envelope.

Normal closed Designation which states that the contacts of a switch or relay are closed or connected when at rest. When activated, the contacts open or separated.

Normally open Designation which states that the contacts of a switch or relay are normally open or not connected. When activated the contacts close or become connected.

Normally open and normally closed The terms "Normally Open" and "Normally Closed" are applied to a magnetically operated switching device (such as a contactor or relay, or to the contacts thereof) to signify the position taken when the operating magnet is de-energized.

North pole Pole of a magnet out of which magnetic lines of force are assumed to originate.

Norton's theorem Any network of voltage sources and resistors can be replace by a single current source in parallel with a single resistor.

Not circuit A binary circuit with a single output that is always the opposite of the input.

Not gate Logic element that acts as an inverter.

Notch filter A special type of cut-only equalizer used to attenuate (no boosting) a narrow band of frequencies. Three controls: frequency, bandwidth and depth, determine the notch. Simplified units provide only a frequency control, with bandwidth and depth fixed internally.

N-type Characterizes a semiconductor containing predominantly mobile electrons.

Nuclear energy Energy produced in the form of heat during the fission process in a nuclear reactor. When released in sufficient and controlled quantity, this heat energy may be used to produce steam to drive a turbine-generator and thus be converted to electrical energy.

Nuclear fission A nuclear reaction in which a heavy atomic nucleus splits into two parts, and at the same time emitting neutrons and releasing very large amounts of nuclear energy.

Nuclear fuel Fissionable materials that have been enriched to such a composition that, when placed in a nuclear reactor, will support a self-sustaining fission chain reaction, producing heat in a controlled manner for process use.

Nuclear fusion A nuclear reaction between light atomic nuclei as a result of which a heavier nucleus is formed and a large quantity of nuclear energy is released.

Nuclear power power released in exothermic (a reaction which gives off heat) nuclear reactions which can be converted to electric power by means of heat transformation equipment and a turbine-generator unit.

Nuclear power plant A facility in which heat produced in a reactor by the fissioning of nuclear fuel is used to drive a steam turbine.

Nucleation The initial stage in a phase transformation, evidenced by the formation of small particles (nuclei) of the new phase, which are capable of growing.

Nucleotide A monomer of the nucleic acids composed of a five-carbon sugar, a nitrogen-containing base, and phosphoric acid.

Nucleus Core of an atom. The nucleus contains both positive (protons) and neutral (neutrons) subatomic particles.

Null A condition, such as balance, which results in a minimum absolute value of output.

Number 1. A mathematical entity that may indicate quantity or amount of units.

2. Loosely, a numeral. An abstract mathematical symbol for expressing a quantity. In this sense, the manner of representing the number is immaterial. Take 26, for example; this is its decimal form - but it could be expressed as a binary (base 2), octal (base 8), or hexadecimal (base 16) number.

Number representation The representation of numbers by agreed sets of symbols according to agreed rules.

Nutating Moving an antenna feed point in a conical pattern so that the polarization of the beam does not change.

Nyquist frequency If an analogue signal is sampled at a rate more than twice that of its highest frequency component, it can be properly reconstructed when reconverted back to the analogue domain. The required sampling rate is called the Nyquist frequency. Conversely, the analogue bandwidth required to accurately transmit a properly reconstructed sampled image is one half the image sample (pixel clock) rate.

Object linking and embedding (OLE) A Microsoft Windows standard mechanism for embedding one program within another. For example, an Excel spreadsheet can be pasted into a Visual Basic program. If a file is linked to an OLE control, the data stored in that file is displayed in the OLE control.

Obstruction Obstacles such as other buildings and trees can prevent direct insolation depending upon the time of day. Obstruction therefore introduces time-varying shading and insolation which must be taken into consideration when calculating the building thermal and lighting performance.

Occupancy The time during which people are in a building (generally expressed in hours per day)

Occupant behaviour The pattern of activity of occupants in a building, including the number of occupants, their distribution, activities and time spent within the building, and how they interact with the buildings facilities, such as ventilation systems, window opening etc.

Occupied zone The region within an occupied space between 75 and 1800 mm above the floor and more than 600mm from the walls or fixed air-conditioning equipment.

Octal A group of 8 symbols from 0 to 7.

Octal number system A number system based on powers of eight. This system is used extensively in computer work.

Octave Interval between two sounds whose fundamental frequencies differ by a ratio of 2 to 1. 440 Hz. Is one octave above 220 Hz.

OD Outside diameter.

Off line test equipment Equipment that tests and isolates faults in modules or assemblies removed from systems.

Off peak Periods of relatively low system demands.

Off-line UPS An uninterruptible power supply (UPS) which feeds power to the load directly from the utility and then transfers to battery power via an inverter after utility drops below a specified voltage. The delay between utility power loss and inverter startup can be long enough to disrupt the operation of some sensitive loads.

Offset A sustained deviation between the control point and the setpoint of a proportional control system under stable operating conditions.

Offset current A current that comes from a switching card even though no user signals are applied. It comes mostly from the finite coil to contact impedance. It is also generated by triboelectric, piezoelectric, and electrochemical effects present on the card.

Offset null An op amp control pin used to eliminate the effects of internal component voltages on the output of the device.

OFHC Oxygen-free high-conductivity copper. The industrial designation of the pure copper used in a Type T thermocouple.

OHM (Ù) Unit of resistance. One ohm is the electrical resistance between two points of a conductor when a constant potential difference of 1 volt, applied to these points, produces a current of 1 ampere in the conductor.

Ohm's law The current in an electrical circuit is directly proportional to the electromotive force in the circuit. The most common form of the law is E = IR, where E is the electromotive force or voltage across the circuit, I is the current flowing in the circuit, and R is the resistance of the circuit.

Ohmeter An instrument used to measure electrical resistance.

Ohmic contact A resistive contact between a semiconductor material and metal. This type of contact exhibits a straight line I-V characteristic with a small value of resistance and does not significantly rectify.

Ohmic value Resistance in ohms.

Ohmmeter A meter used to measure resistance.

Ohms per square The resistance of any square area of thin film·resistive material as measured between two parallel sides.

Ohms per volt Refers to a value of ohms per volt of full scale defection for a moving coil meter movement. The number of ohms per volt is the reciprocal of the amount of current required to produce full scale deflection of the needle. A meter requiring 50 microamps for full scale deflection has an internal resistance of 20 kw per volt. The higher the ohms per volt rating, the more sensitive the meter.

Oil An inductive device made by looping turns of wire around a core.

Oil filled Describes an enclosure (e.g., a submersible motor) that has been filled with oil to prevent leakage.

Oil mist lubrication dry sump A method for lubricating anti-friction bearings which utilizes oil dispersed on an air stream. The mist is exhausted from the bearing housing so as not to permit oil to accumulate.

Oil mist lubrication wet sump Similar to Oil Mist Lubrication - Dry Sump, except that a pool of oil is developed in the bearing chamber. This oil pool will continue to supply oil to the bearing in the event that the oil mist is interrupted and is fed from a source outside the bearing housing such as a constant level oiler.

Oilcan tube A type of planar tube, similar to the lighthouse tube, which has cooling fins. The oilcan tube is designed to handle large amounts of power at uhf frequencies.

Omnidirectional antenna An antenna that radiates and receives equally in all directions (nondirectional).

On Axis A listening experience which takes place with the ear precisely lined up with the central pressure axis of the emitter. Headphones are a normative example of this effect. Sitting directly in front of a speaker is another.

On off control A special case of two-step control in which one of the output signal values is zero.

On peak energy Energy supplied during periods of relatively high system demand as specified by the supplier.

On/off A method of control that turns the output full on until set point is reached, and then off until the process error exceeds the hysteresis.

One shot Monostable multivibrator.

Online UPS A UPS in which the inverter is on during normal operating conditions supplying conditioned power to the load through an inverter or converter that constantly controls the AC output of the UPS regardless of the line input. In the event of a power failure, there is no delay or transfer time to backup power.

Opaque Those substances that do not transmit (pass) any light rays; that is, the light rays are either absorbed or reflected.

Open (Protected) motor A motor having ventilating openings which permit passage of external cooling air over and around the windings. The term "open machine", when applied to large apparatus without qualification, designates a machine having no restriction to ventilation other than that necessitated by mechanical construction.

Open bearing A ball bearing that does not have a shield, seal or guard on either of the two sides of the bearing casing.

Open circuit 1. The condition of an electrical circuit caused by the breaking of continuity of one or more conductors of the circuit; usually an undesired condition.

2. A circuit that does not provide a complete path for the flow of current.

Open circuit voltage Refers to the voltage difference of two points in a circuit when the two points are electrically disconnected from each other.

Open ended line A transmission fine that has a terminating impedance that is infinitely large.

Open externally ventilated machine A machine which is ventilated with external air by means of a separate motor-driven blower mounted on machine enclosure.

Open frame construction A construction technique where the supply is not provided with an enclosure.

Open loop control A control system in which no monitoring feedback is used. An open-loop system assumes a fixed relationship between a controlled condition and an external condition. It does not take into account changing space conditions from internal heat gains, infiltration/exfiltration, solar gain, or other changing variables in the building. Open-loop control alone does not provide close control and may result in underheating or overheating. For this reason, open-loop systems are not common in residential and commercial buildings.

Open loop mode An amplifier circuit having no means of comparing the output with the input. (No feedback.) .

Open pipe ventilated machine An open machine except that openings for admission of ventilating air are so arranged that inlet ducts or pipes can be connected to them. Air may be circulated by means integral with machine or by means external to and not a part of the machine. In the latter case, this machine is sometimes known as separately- or force-ventilated machine.

Open security Environment that does not provide environment sufficient assurance that applications and equipment are protected against the introduction of malicious logic prior to or during the operation of a system.

Open systems security Provision of tools for the secure internetworking of open systems.

Operating altitude The maximum altitude that is allowed to safely operate a Cosel power supply.

Operating humidity The maximum humidity that is allowed to safely operate a Cosel power supply.

Operating software In a BEMS, the main operating software and program that schedules and controls the execution of all other programs; cf. Application software.

Operating system A collection of programs that controls the overall operation of a computer and performs such tasks as assigning places in memory to programs and data, processing interrupts, scheduling jobs and controlling the overall input/output of the system.

Operating temperature range The range of ambient, base plate, or case temperatures through which a power supply is specified to operate safely and to perform within specified limits.

Operating voltage The value of the voltage under normal conditions at a given instant and at a given point in the system.

Operational amplifier (OP AMP) An amplifier designed to perform computing or transfer operations and that has the following characteristics: 1. very high gain,

2. very high input impedance, and
3. very low output impedance.

Operational data security The protection of data from either accidental or unauthorized, intentional modification, destruction, or disclosure during input, processing, or output operations.

Operational ph The determination of sample ph by relating to ph measurements in a primary standard solution. This relationship assumes that electrode errors such as sensitivity and changes in asymmetry potential can be disregarded or compensated for, provided the liquid junction potential remains constant between standard and sample.

Operator person handling equipment.

Optical coupler A coupler composed of an LED and a photodiode and contained in a lightconducting medium. Suitable for frequencies in the low-megahertz range.

Optical isolation Two networks which are connected only through an LED transmitter and photoelectric receiver with no electrical continuity between the two networks.

Optical radiation Electromagnetic radiation at wavelengths between the region of transition to Xrays (l » 1 nm) and the region of transition to radio waves (l » 1 mm).

Optimum start/stop controller This controller alters the time that the HVAC equipment starts/stops depending on the weather conditions. It works by using an external sensor and, occasionally, an internal sensor, to bring in the heating/cooling plant at the latest possible time to get the building/zone(s) to the required temperature by the start of occupancy.

Optimum working frequency The most practical operating frequency that can be used with the least amount of problems and is roughly 85 percent of the maximum usable frequency.

Opto electronic Materials that can either produce an electric current from light or produce light from a current.

Opto isolated Method of sending a signal from one piece of equipment to another without the usual requirement of common ground potentials.

Optocoupler A combination of an LED and a photodiode to give high isolation between the input and the output.

Optoelectronic devices Devices that either produce or use light in their operation.

Optoelectronics A technology that combines optics and electronics, including many devices based on the action of a pn junction. Examples are LEDs, photodiodes, and optocouplers.

Or gate A gate that performs the logic OR function. It produces an output 1 whenever any or all of its inputs is/are 1.

Order wire circuit A circuit between operators used for operations control and coordination.

Origin The point on a graph where the vertical and horizontal axes cross each other.

Origin of an installation The position at which electrical energy is delivered to an electrical installation.

Oscillate To produce a continuous output waveform without an input signal present.

Oscillator A circuit that produces an alternating waveform as output when the primarily powered by a direct input.

Oscilloscope An instrument for making visible the instantaneous values of one or more rapidly varying electrical quantities as a function of time or of another electrical or mechanical quantity.

Out of circuit meter A meter that is not permanently installed in a circuit. Usually portable and self-contained, these meters are used to check the operation of a circuit or to isolate troubles within a circuit.

Out of phase When the maximum and minimum points of two or more waveshpes do not occur at the same time.

Outage The period during which a generating unit, transmission line, or other facility is out of service.

Outboard rotor A two-journal rotor which has its center of gravity between the journals.

Outdoor air Air taken from the external surroundings and therefore not previously circulated through the system.

Outgassing Evaporation of oil, dirt, or any other substance from a surface after it is placed in a low pressure or vacuum environment.

Outlet A point on the wiring system at which current is taken to supply utilization equipment.

Output bus An output circuit path leading from the output(s) of one or more crosspoint switches arranged in a crosspoint switching array. Typically, only one crosspoint switch at a time can feed a signal to an output bus. Each output connector is fed from an output bus.

Output end The end of a transmission line that is opposite the source; receiving end.

Output impedance Impedance measured across the output terminals of a device without a load connected.

Output noise The RMS, peak-to-peak (as specified) ac component of a transducer's dc output in the absence of a measurand variation.

Output power Amount of power a component, circuit or system can deliver to a load.

Output voltage The voltage measured at the output terminals of a power supply.

Outstation (Also field processing unit, data gathering panel) In a BEMS system, this is the unit (with inputs from the sensors and outputs to the actuators) which controls the plant. It is often situated in the plant room.

Over current protection A protective feature that keeps the output current of a power supply within predetermined limits during overload to prevent damage to the supply or load.

Over damping Aperiodic damping in which the degree of damping is greater than that required for for critical damping.

Over ranged A term used for proportional controllers when the load changes sufficiently large to cause the control point to move above or below the limits of the proportional band.

Over voltage protection A feature or device that senses and responds to a high voltage condition.

Overcurrent Any current in excess of the rated current of equipment or the ampacity of a conductor. It may result from overload, short circuit, or ground fault.

Overcurrent detection A method of establishing that the value of current in a circuit exceeds a predetermined value for a specified length of time.

Overdriven When the input signal amplitude is increased to the point that the transistor goes into saturation and cutoff.

Overlay system (Also, BEMS reset system) A BEMS which is overlaid onto conventional analogueue control systems providing reset signals to change temperature, humidity settings, etc. This contrasts with Integrated DDC. where the BEMS provides direct digital control of temperature, humidity and pressure, eliminating the need for discrete analogueue controller.

Overload Codition that occurs when the load is greater than the system was designed to handle. (Load resistance too small, load current too high.) Overload results in waveform distortion and/or overheating.

Overload current An overcurrent occurring in a circuit which is electrically sound.

Overload protection Protective device such as a fuse or circuit breaker that outomatically disconnects a load when current exceeds a predetermined value.

Overload relay A relay that responds to electric load and operates at a pre-set value of overload.

Overmodulation A condition that exists when the peaks of the modulating signal are limited.

Overpressure An induced pressure above ambient atmospheric pressure or other given reference pressure.

Overshoot A transient change in output voltage I excess of specified output regulation limits, which can occur when a power supply is turned on or off, or when there is a step change in line or load.

Overtone Similar in concept to a harmonic. Overtones are sounds produced by an instrument (or sound source) that are higher in frequency than the fundamental frequency. They may or may not coincide with the frequencies of a harmonic series (harmonics), although they usually do. Harmonics are always musically related to a fundamental in that they are integer multiples of it. Overtones of a sound are often identical to its harmonics except the first overtone is considered the second harmonic because the first harmonic is the fundamental. Overtones are also sometimes called partials.

Overvoltage Similar to a surge but for a longer period of time, over 2.5 second.

Oxidation The removal of one or more electrons from an atom, ion or molecule.

Oxygen trim An energy management function in which an attempt is made to reduce heat losses from a boiler's exhaust gases, i.e., the aim is to try and approach the stochiometric ('ideal') air/fuel ratio.

PA An acronym for pascal; An SI unit of pressure measurement equal to 1 newton per square meter.

Package Protective enclosure for a chip or a sensor, typically made of plastic or ceramic.

Packet A block of data sent over the network transmitting the identities of the sending and receiving stations, error-control information, and message.

Packet filter Inspects each packet for user defined content, such as an IP address but does not track the state of sessions. This is one of the least secure types of firewall.

Pad An electrical circuit used to attenuate or reduce the amplitude of an audio signal by a fixed amount, e.g a -15dB pad reduces the signal by a fixed 15 decibels.

Pan Pot Short for panoramic potentiometer, this is a knob controlling a voltage divider that can send a signal to a combination of two busses, such as left and right. Always found on mixing consoles to set up (pan) a signal within the stereo field, it is also called a 'balance' control on domestic stereo amplifiers.

Panel building or panel type building A building composed of pre-fabricated concrete facade and other building elements. Usually the panels have a low level of thermal insulation, and often the panels are poorly jointed which leads to relatively high levels of infiltration. Consequently many panel buildings suffer from high levels of energy consumption, and often poor comfort conditions inside.

Panel lock A feature that prevents operation of the front panel.

Panelboard A single panel or group of panel units designed for assembly in the form of a single panel, including buses and automatic overcurrent devices, and equipped with or without switches for the control of light, heat, or power circuits; designed to be placed in a cabinet or cutout box placed in or against a wall, partition, or other support; and accessible only from the front.

Paper Normally consists of sheets of cellulose, mainly obtained from wood pulp from which lignin and other non-cellulosic materials have been removed.

Paper capacitor Fixed capacitor using oiled or waxed paper as a dialectric.

Parabolic reflector An antenna reflector in the shape of a parabola. It converts spherical wavefronts from the radiating element into plane wavefronts.

Parallax An optical illusion which occurs in analogue meters and causes reading errors. It occurs when the viewing eye is not in the same plane, perpendicular to the meter face, as the indicating needle.

Parallax error The error in meter readings that results when you look at a meter from some position other than directly in line with the pointer and meter face. A mirror mounted on the meter face aids in eliminating parallax error.

Parallel Circuit having two or more paths for current flow. Also called shunt.

Parallel circuit Two or more electrical devices connected to the same pair of terminals so separate currents flow through each; electrons have more than one path to travel from the negative to the positive terminal.

Parallel connection An electrical circuit with more than one possible path for electronic flow.

Parallel data transmission A transmission method where the data (in the form of 'words', or groups of BITS) are transmitted simultaneously.

Parallel negative limiter A resistor and diode, connected in series with the input signal, in which the output is taken across the diode and the negative alternation is eliminated.

Parallel operation The connection of two or more power sources of the same output voltage which are designed for this type of operation to obtain a higher output current.

Parallel port A port through which two or more data bits are passed simultaneously, such as all the bits of an 8-bit byte, and that requires as many input channels as the number of bits that are to be handled simultaneously.

Parallel positive limiter A resistor and diode connected in series with the input signal, in which the output is taken across the diode and the positive alternation is eliminated.

Parallel resonance A resonance condition usually occurring in parallel RLC circuits, where the voltage becomes a maximum for a given current.

Parallel resonant circuit A resonant circuit in which the source voltage is connected across a parallel circuit (formed by a capacitor and an inductor) to furnish a high impedance to the frequency at which the circuit is resonant. Often referred to as a tank circuit.

Parallel transmission Sending all data bits simultaneously. Commonly used for communications between computers and printer devices.

Paralleling When two or more DC motors are required to operate in parallel - that is, to drive a common load while sharing the load equally among all motors - they should have speed-torque characteristics which are identical. The greater the speed droop with load, the easier it becomes to parallel motors successfully. It follows that series motors will operate in parallel easier than any other type. Compound motors, which also have drooping speed characteristics (high regulation), will generally parallel without special circuits or equalization. It may be difficult to operate shunt or stabilized-shunt motors in parallel because of their nearly constant speed characteristics. Modifications to the motor control must sometimes be made before these motors will parallel within satisfactory limits.

Paramagnetism A relatively weak form of magnetism resulting from the independent alignment of atomic dipoles (magnetic) with an applied magnetic field. Also a type of induced magnetism, associated with unpaired electrons, that causes a substance to be zapped into the inducing magnetic field.

Parameter A value that determines the response of an electronic controller to given inputs.

Parasitic array An antenna array containing one or more elements not connected to the transmission line.

Parasitic element The passive element of an antenna array that is connected to neither the transmission line nor the driven element.

PARD (Periodic and Random Deviation) The sum of all ripple and noise components measured over a specific band width and stated, unless otherwise specified, in peak-to-peak values.

Parity A technique for testing transmitting data. Typically, a binary digit is added to the data to make the sum of all the digits of the binary data either always even (even parity) or always odd (odd parity).

Parity bit An additional bit that is attached to each code group so that the total number of 1s being transmitted is either odd or even.

Part winding start motor Is arranged for starting by first energizing part of the primary winding and subsequently energizing the remainder of this winding in one or more steps. The purpose is to reduce the initial value of the starting current drawn or the starting torque developed by the motor. A standard part winding start induction motor is arranged so that one-half of its primary winding can be energized initially and subsequently the remaining half can be energized, both halves then carrying the same current.

Partial discharge Discharges which are partial and are not flash overs across electrodes.

Particulate matter A state of matter in which solid or liquid substances exist in the form of aggregated molecules or particles. Airborne particulate matter is typically in the size range of 0.01 to 100 micrometers.

Particulate tracer Solid particles of aerosol or bubbles used as a tracer for measuring the rate of air movement. These particles usually have diameters of 2 to 3 microns, and can be detected by using either a) a fluorescent light scattering detector; b) a photomultiplier (P-M) detector, or c)a phosphorescence with a P-M detector.

Particulates Fine liquid or solid particles such as dust, smoke, mist, fumes, and fog found in air and emissions.

Parts per million (PPM) Represents the concentration of gases or vapor in air. For example, 1 ppm means that 1 unit of the gas is present for every 1 million units of air.

Pascal (pa) SI unit of pressure. One pascal is equal to the force of one newton exerted on one square meter.

Pasive filter A filter that contains only passive or non amplifying components.

Pass band The range of frequencies that will be passed and amplified by a tuned amplifier. Also the range of frequencies passed by a band pass filter.

Passive adsorption A process by which a gas or vapour is condensed (out of the air) and held on the surface of a piece of solid material by natural forces only.

Passive attack Attack which does not result in an unauthorized state change, such as an attack that only monitors and/or records data.

Passive component A component that does not provide any amplification or gain such as a resistor, capacitor, LED, volume control, battery, globe. Components that are NOT passive transistor, valve, IC, (these are called active).

Passive element An element which does not generate electricity but either consumes it or stores it.

Passive sampling A method of sampling tracer gas in a building by the process of passive adsorption.

Passive satellite A satellite that reflects radio signals back to earth.

Passive smoker A non smoker who shares the same room, building, or space as a smoker, and thus is exposed to the products of tobacco combustion.

Passive system System that emits no energy. It only receives. It does not transmit or reveal its position.

Passive threat The threat of unauthorized disclosure of information without changing the state of the system. A type of threat that involves the interception, not the alteration, of information.

Password A string of characters from the computer keyboard selected by a computer user to authenticate them when they log on. Passwords prevent unauthorised access to personal accounts, restricted files, subscription services etc.

Patch panel A panel used to tie a receiver or transmitter to its associated equipment.

Path resistance The resistance of a complete signal path, including the switching element's contact resistance, any PC board circuit resistance and connector terminal resistance and or cabling.

Payback period The length of time it takes for the savings received to cover the cost of implementing the technology.

Peak Maximum value of a waveform during a particular cycle or operating time.

Peak amplitude The maximum value above or below the reference line.

Peak clipping Peak clipping is used to reduce a system peak, reducing the need to operate peaking units with relatively high fuel costs. Peak clipping is pursued only when the resources are not expected to be able to meet the impending load requirements.

Peak current The maximum current that flows during a complete cycle.

Peak day The day of highest customer demand for electricity during a year.

Peak demand The maximum load during a specified period of time.

Peak detection Detection that uses the amplitude of pam or the duration of pdm to charge a holding capacitor and restore the original waveform.

Peak factor Ratio of the peak value to the r.m.s. Value of an alternating waveform.

Peak inverse voltage In an electron tube, the maximum negative voltage that can be applied to the plate without danger of arc-over. In a semiconductor diode, the maximum reverse bias voltage that can be applied without reaching the zener (or breakdown) voltage.

Peak output current The maximum current value delivered to a load under specified pulsed conditions.

Peak responding A measurement where the displayed value is equal to the peak value of the input signal.

Peak reverse voltage The peak ac voltage that a rectifier tube will withstand in the reverse direction.

Peak shaving Techniques used by electric utilities to lower the peak demand on the system.

Peak to peak The measure of absolute magnitude of an ac waveform, measured from the greatest positive alternation to the greatest negative alternation.

Peak to peak value The maximum voltage change occurring during one cycle of alternating voltage or current. The total amount of voltage between the positive peak and the negative peak of one cycle or twice the peak value.

Peak torque Maximum torque that can be delivered for even a short period of time.

Peak voltage The maximum value present in a varying or alternating voltage. This value may be positive or negative.

Peak watts (Wp) The maximum power a device produces or consumes.

Peaking capacity Capacity of generating plants normally reserved for operation during the hours of highest daily, weekly, or seasonal loads. This equipment is usually designed to meet the portion of load that is above base load.

Peaking coil An inductor used in an amplifier to provide high-frequency compensation, which extends the high-frequency response of the amplifier.

Peaking plants power plants that operate for a relatively small number of hours, usually during peak demand periods. Such plants usually have high operating costs and low capital costs.

Peltier effect When a current flows through a thermocouple junction, heat will either be absorbed or evolved depending on the direction of current flow. This effect is independent of joule I2 R heating.

Penetration The successful unauthorized access to an automated system.

Penetration signature The description of a situation or set of conditions in which a penetration could occur or of system events which in conjunction can indicate the occurrence of a penetration in progress.

Penetration testing The portion of security testing in which the evaluators attempt to circumvent the security features of a system. The evaluators may be assumed to use all system design and implementation documentation, that may include listings of system source code, manuals, and circuit diagrams. The evaluators work under the same constraints applied to ordinary users.

Pentavalent element Element whose atoms have five valence electrons. Used in doping intrinsic silicon or germanium to produce n-type semiconductor material. Most commonly used pentavalent materials are arsenic and phosphorus.

Pentode tube A five-electrode electron tube containing a plate, a cathode, a control grid, and two grids.

Per channel rate The sample rate for each channel of a scanning A/D system.

Per unit (pu) per unit is a method of expressing the value of a quantity in terms of a reference or base quantity. It is very similar to percentage, except that there is no multiplying constant of 100.

Percent of modulation The degree of modulation defined in terms of the maximum permissible amount of modulation.

Percent of regulation The change in output voltage that occurs between no-load and full-load in a DC voltage source. Dividing this change by the full-load value and multiplying the result by 100 gives percent regulation.

Percent of ripple The ratio of the effective rms value of ripple voltage to the average value of the total voltage. Expressed as a percentage.

Perception Awareness of the effects of stimuli.

Perfectly balanced rotor A rotor is perfectly balanced when its mass distribution is such that it transmits no vibratory force or motion to its bearings as a result of centrifugal forces.

Performance assessment method (PAM) A performance assessment method defines best practice simulation procedures in order to assess the performance of a building.

Performance attributes Performance attributes measure the quality of service and operating efficiency. Loss of load probability, expected energy curtailment, and reserve margin are some of the performance attributes.

Perigee The point in the orbit of a satellite closest to the earth.

Perimeter based security The technique of securing a network by controlling access to all entry and exit points of the network. Usually associated with firewalls and/or filters.

Period Duration between repetitions of a waveform cycle. It is also equal to the inverse of frequency).

Period time The time required to complete one cycle of a waveform.

Periodic function A function which repeats itself after a definite period.

Periodic wave A waveform that undergoes a pattern of changes, returns to its original pattern, and then repeats the same pattern of changes. Examples are square waves, rectangular waves, and sawtooth waves.

Peripheral A device that is external to the CPU and main memory, i.e., printer, modem or terminal, but is connected by the appropriate electrical connections.

Peripheral device In a data processing system, any equipment, distinct from the central processing unit, that may provide the system with additional capabilities.

Periscope An optical instrument, housed in a long tube; used to translate the observer's line of sight in a vertical direction.

Permanence Magnetic equivalent of magnetic inductance and consequently equal to the reciprocal of reluctance, just as conductance is equal to the reciprocal of resistance.

Permanent magnet Magnet normally made of hardened steel that retains its magnetism indefinately.

Permanent magnet motor A PM motor uses a permanent magnet instead of armature windings. The permanent magnets are mounted [embedded] on the rotor. Permanent Magnet [PM] motors are normally small and produce little horsepower.

Permanent magnet speaker A speaker with a permanent magnet mounted on soft iron pole pieces.

Permanent magnet synchronous (PMR) (Hysteresis synchronous) A motor with magnets embedded into the rotor assembly, which enable the rotor to align itself with the rotating magnetic field of the stator. These motors have zero slip (constant speed with load) and provide higher torque, efficiency and draw less current than comparable reluctance synchronous motors.

Permeability [μ] From the relation between magnetic induction and magnetic field (B = μH); for free space, μ0 = 1.26 × 10-6 H/m.

Permeance Inverse of reluctance. (Unit: H)

Permitivity [å] From the relation between polarization charge and electric field; for free space, å0 = 8.85 × 10-12 F/m.

Perpetrator The entity from the external environment that is taken to be the cause of a risk. An entity in the external environment that performs an attack, i.e. hacker.

Persistence The length of time a phosphor dot glows on a CRT before disappearing.

Personnel security The procedures established to ensure that all personnel who have access to any classified information have the required authorizations as well as the appropriate clearances.

Peta (P) Decimal multiple prefix corresponding to 1015.

Petroleum A mixture of hydrocarbons existing in the liquid state found in natural underground reservoirs often associated with gas. Petroleum includes crude oil, fuel oil, kerosene and jet fuel.

Ph junctions The Junction of a reference electrode or combination electrode is a permeable membrane through which the fill solution escapes (called the liquid junction).

Phase Indicates the space relationships of windings and changing values of the recurring cycles of A.C. voltages and currents. Due to the positioning (or the phase relationship) of the windings, the various voltages and currents will not be similar in all aspects at any given instant. Each winding will lead or lag another, in position. Each voltage will lead or lag another voltage, in time. Each current will lead or lag another current, in time. The most common power supplies are either single (10) or three phase (with 120 electrical degrees between the 3 phases).

Phase angle Angle at which the steady state input signal leads the output signal.

Phase conductor A conductor of an a.c. System for the transmission of electrical energy other than a neutral conductor, a protective conductor or a PEN conductor. The term also means the equivalent conductor of a d.c. System unless otherwise specified in the Regulations.

Phase cut signal This is where a sine wave is cut off part way through the cycle to produce a continuously varying output signal from a controller. The signal is then used to directly control a magnetic actuator.

Phase difference Difference in phase angle between two sinusoids or phasors.

Phase locked loop An electronic circuit that controls an oscillator so that it maintains a constant phase angle relative to a reference signal. (PLL).

Phase modulation (PM) Angle modulation in which the phase of the carrier is controlled by the modulating waveform. The amplitude of the modulating wave determines the amount of phase shift, and the frequency of the modulation determines how often the phase shifts.

Phase proportioning A form of temperature control where the power supplied to the process is controlled by limiting the phase angle of the line voltage.

Phase sequence or phase rotation The time order in which the voltages (or currents) pass through their respective maximum values (or any other definable position).

Phase shift Change in phase of a wave form between two points, expressed as degrees of lead or lag.

Phase shift keying Similar to ON-OFF cw keying in AM systems and frequency-shift keying in FM systems. Each time a mark is received, the phase is reversed. No phase reversal takes place when a space is received.

Phase shift oscillator An oscillator that uses three RC networks in its feedback path to produce the 180° phase shift required for oscillation.

Phase splitter A device that provides two output signals from a single input signal. The two output signals differ from each other in phase (usually by 180 degrees).

Phase transformation A change in the number and/or character of the phases that make up the microstructure of an alloy.

Phase voltage The voltage between the phase and neutral of a three phase system is defined as the phase-to-neutral voltage or more commonly as the phase voltage. For a balanced three phase system, the phase voltage is 1/S3 times the line-to-line voltage. The voltage across one arm of either a star-connected load or a delta-connected load is also sometimes referred to as the voltage across a phase or phase voltage. This latter quantity is the same as the phase-to-neutral voltage for a star-connected load, and the line-to-line voltage for a delta-connected load.

Phasor Representation of a sinusoid on the Argand diagram in the form of the magnitude (usually r.m.s.) And phase angle. It may be represented as a complex number in either cartesian co-ordinates or polar co-ordinates.

PHF hack A well-known and vulnerable CGI script which does not filter out special characters (such as a new line) input by a user.

Phonograph Piece of equipment used to reproduce sound stored on a disk called a phonograph record.

Phonon A single quantum of vibrational or elastic energy.

Phosphor Luminescent material applied to the inner face of a cathode ray tube that when bombarded with electrons will emit light of various colours.

Photocell A device that changes resistance when light falls on it. Similar to a light dependent resistor or LDR. This cell does not produce an output current - see Solar cell for output current.

Photoconduction A process by which the conductance of a material is change by incident electromagnetic radiation in the visible light spectrum.

Photoconductive cell Material whose resistance decreases or conductance increases when exposed to light.

Photoconductivity Light shining on the surface of a material increasing the conductivity.

Photodetector Component used to detect or sense light.

Photodiode A light-controlled PN junction. Current flow increases when the PN junction is exposed to an external light source. A semiconductor diode that produces, as a result of the absorption of photons, (a) a photovoltage or (b) free carriers that support the conduction of photocurrent. Photodiodes are used for the detection of optical communication signals and for the conversion of optical power to electrical power.

Photoelectric voltage A voltage produced by light.

Photoetching Chemical process of removing unwanted material in producing printed circuit boards.

Photomicrograph The picture made with a microscope.

Photon A massless particle, the quantum of the electromagnetic field carrying energy, also known as the light quantum.

Photoresistor A device for measuring or detecting electromagnetic radiation. The conductivity of the resistor changes with exposure to light.

Phototransistor A transistor with a window on the top face to allow light to fall on the active surface. Also available as a Darlington photo-transistor to produce a very sensitve light detecting device.

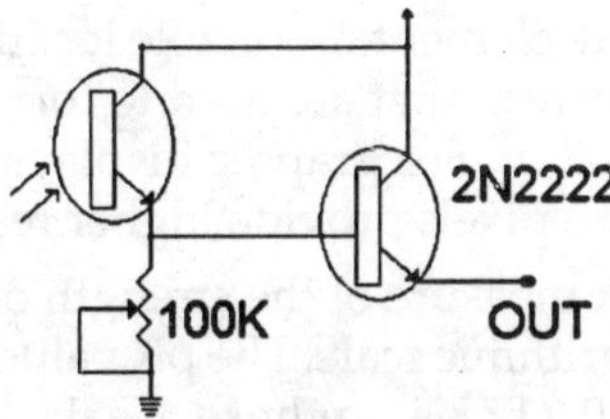

Fig. Phototransistor

Photovoltaic (PV) Refers to any device which produces free electrons when exposed to light. When these electrons are properly gathered, a potential difference (voltage) may be produced by the device, for example, a solar cell produces approximately one half volt in full sun.

Photovoltaic cell (solar cell) A device that acts much like a battery when exposed to light and converts light energy into electrical energy.

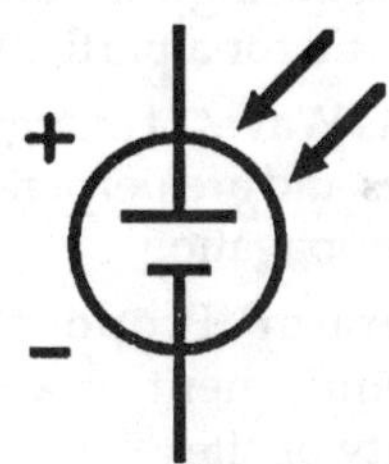

Fig. Photovoltaic cell

Phracker An individual who combines phone phreaking with computer hacking.

Phreak(er) An individual fascinated by the telephone system. Commonly, an individual who uses his knowledge of the telephone system to make calls at the expense of another.

Phreaking The art and science of cracking the phone network.

Physical security The measures used to provide physical protection of resources against deliberate and accidental threats.

Pi Value representing the ratio between the circumference and diameter of a circle and equal to approximately 3.142.

Pi Filter A filter consisting of two line-to-line capacitors and a series inductance used to attenuate noise and ripple. The elements are connected together to resemble the Greek letter pi.

Pico A prefix adopted by the National Bureau of Standards meaning 10-12.

Pierce oscillator A variation of the colpitts oscillator. This oscillator uses a quartz crystal in place of the inductor found in the colpitts oscillator feedback network. The crystal maintains a highly stable output frequency.

Piezo electric effect Vibrtion that occurs when a crystal is excited by an ac signal across its plates.

Piezo sounder Small crystal element that can emit very high sound levels while requiring low current. Also called a Piezo diaphragm as it has no active circuit connected to the diaphragm. A piezo buzzer has an active circuit connected to the diaphragn to emit sound when a DC voltage is applied.

Piezoelectric accelerometer A transducer that produces an electrical charge in direct proportion to the vibratory acceleration.

Piezoelectric crystal Crystal material that will generate a voltage when mechanical pressure is applied and conversely will undergo mechanical stress when subjected to a voltage.

Piezoelectric currents The current caused by mechanical stress to the insulating materials or connectors. To minimize this problem in low current or voltage measurements, the stress must be removed from the insulators, and materials with a low piezoelectric effect must be used.

Piezoelectric effect The effect of producing a voltage by placing a stress, either by compression, expansion, or twisting, on a crystal and, conversely, producing a stress in a crystal by applying a voltage to it.

Piezoelectric material A ferroelectric material in which an electrical potential difference is created due to mechanical deformation, or conversely, in which the application of a voltage causes dimensional changes in the material.

Piggy back The gaining of unauthorized access to a system via another user's legitimate connection.

Pin diode Acronym for positive-intrinsic-negative diode. A photodiode with a large, neutrally doped intrinsic region sandwiched between p-doped and n-doped semiconducting regions. Note: A PIN diode exhibits an increase in its electrical conductivity as a function of the intensity, wavelength, and modulation rate of the incident radiation.

Pin out The pin assignment of a device.

Pinch off region A region on the characteristic curve of a FET in which the gate bias causes the depletion region to extend completely across the channel.

Ping of death The use of Ping with a packet size higher than 65,507. This will cause a denial of service.

Pinhole The term pinhole embraces a wide variety of oxide defects and is used in a broad sense today. Listed in this category are cracks caused by thermal contraction after oxidation or by handling, and regions of oxide with low dielectric strength caused by dust particles, inadequate masking, contamination, or poor resist adhesion.

Pin-out Diagram showing for electronic components the relations between connecting pins and internal components.

Piston flow Also known as plug flow, and displacement flow, and is regarded as the most efficient form of ventilation. The ventilation air acts as a piston, which pushes the "old" air in the room in front of it without actually mixing. Therefore all of the air that reaches an arbitrary point from a small packet of fresh air at the inlet does so at the same time; this time is by definition, the local mean age of air at this point.

Pitch A term used to describe the frequency of a sound heard by the human ear.

Pitting A form of very localized corrosion wherein small pits or holes form, usually in a vertical direction.

Pixel Picture element. Definable locations on a display screen that are used to form images on the screen. For graphic displays, screens with more pixels provide higher resolution.

pK value A measure of the strength of an acid on a logarithmic scale. The pK value is given by log10 (1/.Ka), where Ka is the acid dissociation constant pK values often are used to compare the strengths of different acids.

Plaintext Unencrypted data.

Planar tube An electron tube, constructed with parallel electrodes and a ceramic envelope, that is used at uhf frequencies. It is commonly referred to as a lighthouse tube.

Plane of polarization The plane (vertical or horizontal), with respect to the earth, in which the E field propagates.

Plane separation Of a balancing machine, is the operation of reducing the correction plane interference ratio for a particular rotor.

Plane wavefronts Waves of energy that are flat, parallel planes and are perpendicular to the direction of propagation.

Planned generator Proposal to install generating equipment at an existing or planned facility or site.

Plant A facility containing prime movers, electric generators, and other equipment for producing electric energy.

Plant component Device within a plant network such as a boiler or pump.

Plant connection Topology of plant network.

Plasma An ionized gas containing about equal numbers of positive and negative charges, which is a good conductor of electricity, and is affected by a magnetic field.

Plastic deformation Permanent or nonrecoverable deformation, accompanied by permanent atomic displacement.

Plastic film capacitor Capacitor in which alternate layers of aluminum foil are separated by thin films of plastic dialectric.

Plasticizer A low molecular weight polymer additive that enhances flexibility and workability and reduces stiffness and brittleness.

Plate 1. One of the electrodes in a storage battery.

2. One of the electrodes in a capacitor.

3. The principal electrode to which the electron stream is attracted in an electron tube.

Plate dissipation The amount of power lost as heat in the plate of a vacuum tube.

Plate keying A keying system in which the plate supply is interrupted.

Plate modulator An electron-tube modulator in which the modulating voltage is applied to the plate circuit of the tube.

Plate resistance The plate voltage change divided by the resultant plate current change in a vacuum tube, all other conditions being fixed.

Platinel A non-standard, high temperature platinum thermocouple alloy whose thermoelectric voltage nearly matches a Type K thermocouple (Trademark of Englehard Industries).

Plenum chamber A chamber, at higher/lower pressure than surrounding air, that receives air before/after delivery to a conditioned space or combustion system.

Plug A device, provided with contact pins, which is intended to be attached to a flexible cable, and which can be engaged with a socket outlet or with a connector.

Plug flow A flow regime where the flow is predominately in one direction and contaminants are swept along with the flow.

Plug reversal Reconnecting a motor's winding in reverse to apply a reverse braking torque to its normal direction of rotation while running. Although it is an effective dynamic braking means in many applications, plugging produces more heat than other methods and should be used with caution.

Plume A visible or measurable discharge of a contaminant body from a given point of origin. Can be a visible body of pollution such as smoke coming from a stack or a measured amount such as the heat in water coming from a power plant boiler.

Pn junction A junction between an N-type semiconductor and a p-type semiconductor made by some method of diffusing, fusing or melting.

Pneumatic controller A controller which uses compressed air as the control medium. All the required control terms are readily available, e.g., step control, PID, etc.

PNP Type of bipolar transistor using p-type, n-type, n-type semiconductor material.

Point (in wiring) A termination of the fixed wiring intended for the connection of current using equipment.

Point A physical source or destination for data in the form of analogueue or digital signals.

Point contact diode A diode in which the end of a fine wire is pressed against a semiconductor. Used as a detector or mixer over the microwave region.

Point defect A crystalline defect associated with one or several atomic sites.

Point descriptor The information giving the characteristics of a monitoring point.

Point of common coupling Power supplier-consumer metering point at which power is supplied to the electrical equipment of the consumer.

Point of common coupling PCC: The location of the connection between the network and the Embedded Generator, beyond which other customer loads may be connected on the side. The PCC may be separate from the Point of Supply where a line is dedicated to the connection of an Embedded Generator.

Point of delivery Point for interconnection on the Transmission Provider's System where capacity and/or energy are made available to the end user.

Poisson ratio The ratio between the strain of expansion in the direction of force and the strain of contraction perpendicular to that force v = -Et/E1.

Polar coordinates Either of two numbers that locate a point in a plane by its distance from a fixed point and the angle this line makes with a fixed line.

Polar relay A dc relay in which the direction of movement of the armature depends on the direction of the current flow.

Polarisationchange in the potential of an electrode as the result flow.

Polarised A component or plug etc that must be fitted around a certain way. eg Thick and thin pins spaced apart so that the plug cannot be inserted around the wrong way.

Polarity 1. The condition in an electrical circuit by which the direction of the flow of current can be determined. Usually applied to batteries and other direct voltage sources.

2. Two opposite charges, one positive and one negative.

3. A quality of having two opposite magnetic poles, one north and the other south.

Polarization (P) The total electric dipole moment per unit volume of dielectric material.

Polarization test A ratio of a one-minute meggar test to ten-minute meggar test. Used to detect contaminants in winding insulation done typically on high voltage, V.P.I. motors which are tested by water immersion.

Polarized A component which must be connected in correct polarity to function and/or d/or prevent destruction. Example: Electrolytic capacitor.

Pole piece 1. A piece of ferromagnetic material used to control the istribution of magnetic lines of force; that is, it concentrates the lines of force in a particular place or evenly distributes the lines of force over a wide area.

2. The shaped magnetic material upon which the stator windings of motors and generators are mounted or wound.

Poles In an AC motor, refers to the number of magnetic poles in the stator winding. The number of poles is a determinant of the motor's speed. In a DC motor, refers to the number of magnetic poles in the motor. Creates the magnetic field in which the armature operates. (Speed is not determined by the number of poles).

Polling A method of interrogating a computer to determine whether each device is ready to perform its specific task and/or communicate with the computer.

Pollutant migration The movement of indoor air pollutants throughout the building between rooms or zones.

The concentration within a given portion of air of harmful or unpleasant contaminants such as noxious gases or dust particles. Concentrations are often expressed as time weighted values over 24 hours, a working day or a working week.

Polycrystalline cell (also known as "multicrystalline cell") A wafer of silicon with a multi-grained structure. All grains have the same atomic crystal lattice, however, each grain has a unique orientation in space thereby producing a unique reflection of light, resulting in a 'patchy' appearance.

Polyester A type of capacitor.

Polygon of forces If the forces acting on an object can be represented by the sides of a polygon taken in order, the forces will be in equilibrium.

Polyphase A term that describes systems or units of a system that are activated by or which generate separate out-of-phase voltages. Typical polyphase systems are 2-phase and 3-phase; their voltages are 90- and 120-degrees out of phase, respectively. This term means the same as multiphase.

Polyphase motor Two or three-phase induction motors have their windings, one for each phase, evenly divided by the same number of electrical degrees. Reversal of the two-phase motor is accomplished by reversing the current through either winding. Reversal of a three-phase motor is accomplished by interchanging any two of its connections to the line. Polyphase motors are used where a polyphase (3-phase) power supply is available and is limited primarily to industrial applications. Starting and reversing torque characteristics of polyphase motors are exceptionally good. This is due to the fact that the different windings are identical and, unlike the capacitor motor, the currents are balanced. They have an ideal phase relation which results in a true rotating field over the full range of operation from locked rotor to full speed.

Polysilicon Polycrystalline silicon used as conductor in integrated circuits, and especially FETs.

Porcelain Hard, white material made by the firing of a mixture of pure kaolin (china clay) with felspar and quartz, or with other materials containing silica.

Port A pair of terminals through which a single current may enter or leave a network.

Portable equipment Electrical equipment which is moved while in operation or which can easily be moved from one place to another while connected to the supply.

Position adjust type signal A controller output using two binary signals, working as a pair. Normally, they drive a reversing actuator by sending either an increase, hold, or decrease comand. When used with DDC the controller may calculate the percentage opening of the controlled device based on the actuator running time to go from fully closed to fully open.

Position sensor A component in a servosystem that measures position and converts the measurement into a form convenient for transmission as a feedback signal.

Position servosystem A servosystem whose end function is to control the position of the load it is driving.

Positive Polarity of point that attracts electrons as opposed to negative which supplies electrons.

Positive charge A charge that exists in a body that has fewer electrons than protons.

Positive clamper A circuit that clamps the lower extremity of the output waveshape to a dc potential of 0 volts.

Positive feedback A feedback signal that is in phase with an amplifier input signal. Positive feedback is necessary for oscillation to occur.

Positive ground A system whereby the positive terminal of the source is connected to the system's conducting chassis.

Positive ion Atom that has lost one or more valence electrons resulting in a net positive charge.

Positive logic The form of logic in which the more positive logic level represents 1 and the more negative level represents 0.

Positive pressure A condition that exists when more air is supplied to a space than is exhausted, so the air pressure within that space is greater than that in surrounding areas.

Positive sequence A balanced set of three phase components which have the same magnitude, same sequence as the original unbalanced set, and phase angle diferring from each other by 120o. The frequency is of course the same as the original unbalanced three phase system.

Positive temperature coefficient An increase in resistance due to an increase in temperature.

Potential difference Potential difference is the work done in moving a unit positive electric charge from one point to another. (Unit: volt or V).

Potential divider A combination of impedances which allows a fraction of the input voltage to be taken as output.

Potential energy Energy caused by the position of one body with respect to another body or to the relative parts of the same body.

Potential equalization conductor Conductor providing a connection between equipment and the potential equalization busbar of the electrical installation.

Potential transformer An instrument transformer specifically designed to give an accurate voltage ratio for measurement and/ or control purposes. They are always connected in parallel with the circuit (like a voltmeter).

Potential value The limiting value of the controlled condition that tends to be attained following a particular adjustment of the corrector unit, all other factors which may effect the value of the controlled condition being maintained constant.

Potentiometer A variable resistor with three terhree terminals. Mechanical turning of a shaft can be used to produce variable resistance and potential. Example: A volume control is usually a potentiometer.

Power(W) Amount of energy converted by a circuit or component in a unit of time, normally seconds. Measured in units of watts. (joules/second).

Power amplifier An amplifier designed to deliver maximum power output to a load. Example: In an audio system, it is the power amplifier that drives the loudspeaker.

Power architecture The technical description of a power supply devices and power usage devices that represent a Power System.

Power code Identifies the type of power supply providing power to a DC motor. Frequency, voltage, and type of rectifier configuration.

Power conditioning systems A broad class of equipment that includes filters, isolation transformers, and voltage regulators. Generally, these types of equipment offer no protection against power outages.

Power density 1) The ratio of the power available from a power source to its weight. E.g., watts/pound) The ratio of the power available from a power source to its volume. E.g., watts/cubic inch.

Power derating factor A transistor rating that tells how much the maximum allowable value of PD decreased for each 1°C rise in ambient temperature.

Power dissipation Amount of heat energy generated by a device in one second when current flows through it.

Power distribution unit A portable electrical distribution unit that provides an easily expandable and flexible electrical environment for an equipment and its associated peripherals.

Power exchange This is a commercial entity responsible for facilitating the development of transparent spot prices for energy capacity, and/or ancillary services.

Power factor The ratio of the total power in watts (resistive load) to the total apparent power in voltamperes (VA) (reactive load). The difference between watts and VA is due to reactive load impedance. Apparent power equals watts only for a purely resistive load (i.e., zero degrees phase shift between the applied voltage and the resultant current). Power factor is best thought of intuitively as the multiplier (ranging between 0 and 1) that you must use to obtain the real power from the apparent power. For example if the rms voltage and current of a circuit is measured and multiplied together, the apparent power is obtained, but this value must be multiplied by the power factor to obtain the real power. If the load is purely resistive then the phase difference between the voltage and current will be zero and the power factor will be one, and the apparent power will equal the true power — but only for a resistive load. For a reactive load (any load with inductive and/or capacitive reactance, i.e., any real world load) there will be a phase difference between the voltage and the current due to the phase delay introduced by the reactive elements. Simply put, since the maximum voltage and current do not occur at the same instant of time the amount of power developed is less than the measured rms voltage and current multiplied together.

Power factor correction capacitor This is a device that helps improve the efficiency of the flow of electricity through distribution lines by reducing energy losses. It is installed in substations and on poles. Usually it is installed to correct an unwanted condition in an electrical system.

Power failure signal A logic signal from a power supply that provides advance notice that the output voltage is about to fall out of specifications due to loss of line.

Power gain In an antenna, the ratio of its radiated power to that of a reference.

Power good signal A logic signal from a power supply indicating that outputs are within predetermined specifications.

Power grid (National Grid) A network of power lines and associated equipment used to transmit and distribute electricity over a geographic area (or country).

Power handling The wattage level that a driver can handle without sustaining damage due to over-heating the voice coil or over-extending the cone excursion.

Power loss 1. The electrical power, supplied to a circuit, that does no work and is usually dissipated as heat.

2. The heat loss in a conductor as current flows through it.

Power marketers Entities engaged in buying and selling electricity.

Power pentode A special purpose tube used to provide high-current gain or power amplification. Each grid wire is directly in line with the one before and after it, a fact which allows more electrons to reach the plate.

Power plant A generating station where electricity is produced.

Power pool An association of two or more interconnected electric systems having an agreement to coordinate operations and planning for improved reliability and efficiencies.

Power rating Power available at the output terminals of a power source based on the manufacturers specifications.

Power standing wave ratio (PSWR) The ratio of the square of the maximum and minimum voltages of a transmission line.

Power supply A device for the conversion of available power of one set of characteristics to another set of characteristics to meet specified requirements. Typical application of power supplies include to convert raw input power to a controlled or stabilized voltage and/or current.

PPM Abbreviation for "parts per million," sometimes used to express temperature coefficients. For instance, 100 ppm is identical to 0.01%.

Preamplifier (Preamp) An amplifier that raises the output of a low-level source for further processing without appreciable degradation of the signal-to-noise ratio.

Precedence Effect Also known as the Haas effect, this phenomenon identifies the tendency for the ear to attribute all perceived sound to the nearest emitter, even if a more distant speaker is actually louder. Thus, drums that yield 90 percent or more of their energy in the non-directional lower frequencies are perceived as located in the space created by the tiny amount of higher frequency overtones. It is also for this reason that sounds from the rear speakers of a surround system are delayed by 10 to 30 milliseconds, so that they can be experienced as coming from the rear direction.

Precision A term used about information retrieval - the number of relevant documents retrieved divided by the total number of documents retrieved.

Preconditioning period One or more days of simulation before the requested simulation period. These are required to remove any effects that may result from the assumed initial conditions of the simulation. Calculated values during the preconditioning period (or start-up period) are not saved.

Predicted mean vote (PMV) Comfort rating derived from the work of Fanger. PMV is derived from the physics of heat transfer combined with an empirical fit to sensation. PMV* represents a new temperature index that incorporates skin wettedness into the PMV equation using SET* or ET* to characterize the environment.

Predicted percent dissatisfied due to draft PD is a fit to data of persons expressing thermal discomfort due to drafts. The inputs to PD are air temperature, air velocity , and turbulence intensity.

Predictive control A control system which attempts to predict the effects of a disturbance on the future output. Predictive control forms the basis of many self-tuning control methods; cf. adaptive control.

Preset A collection of system settings stored in a memory that is virtually instantly recallable, typically at the touch of a button.

Present value The amount of money required to secure a specified cash flow on a future date at a given rate of return.

Present worth factor The adjustment factor that discounts a sum of future dollars back to the current year.

Pressure The force per unit area acting on a surface.

Pressure attenuation technique A method of estimating the leakage of a building by releasing air inside the building causing instant pressurization, the pressure returning to normal as the air leaks out. The rate of reduction of the pressure is proportional to the leakage.

Pressure coefficient A dimensionless coefficient relating the velocity pressure on the outer surface of the building to the velocity pressure derived from the mean wind velocity at a reference point.

Pressure compensator Device that equalizes the internal pressure of the motor with the external pressure of the environment.

Pressure differential The difference in pressure across a building envelope or component whether caused by natural or artificial means.

Pressure, total In flowing air, the sum of the static pressure and the velocity pressure.

Pressure, velocity .In flowing air, the pressure due to the velocity and density of the air.

Pressurization A method of testing air leakage of a building or component by installing a fan in the building envelope, for example through a door or window, and creating a static pressure excess inside the building. The airflow rate through the fan and the pressure difference across the envelope are measured from which the air leakage is assessed.

Primary First winding of a transformer. Winding that is connected to the source as opposed to secondary which is a winding connected to a load.

Primary cell An electrochemical cell in which the chemical action eats away one of the electrodes, usually the negative electrode.

Primary circuit A circuit electrically connected to the input or source of power to the device.

Primary device Part of a flowmeter which is mounted internally or externally to the fluid conduit and produces a signal corresponding to the flowrate and from which the flow may be determined.

Primary standards Aqueous ph buffer solutions established by the National Bureau of Standards within the 2.5 to 11.5 ph range of ionic strength less than 0.1 and which provide stable liquid junction potential and uniformity of electrode sensitivity.

Primary winding That winding of a motor, transformer or other electrical device which is connected to the power source.

Prime mover The source of the turning force applied to the rotor of a generator. This may be an electric motor, a gasoline engine, a steam turbine, and so forth.

Principal axes The axes of maximum and minimum normal stress.

Printed circuit A method of manufacturing parts of electronic equipment in which the wiring between components, and certain fixed components themselves, are printed on to an insulating board.

Printed circuit board A flat, insulating surface upon which printed wiring and miniaturized components are connected in a predetermined design and attached to a common base.

Printer A device for producing paper copies of data.

Printout Refers to anything printed out by a peripheral, or any computer-generated hardcopy.

Priority A rank assigned to a task that determines its precedence in receiving system resources.

Private key cryptography An encryption methodology in which the encryptor and decryptor use the same key, which must be kept secret. This methodology is usually only used by a small group.

Probe An oscilloscope input device, usually having a pointed metal tip for making electrical contact with a circuit element and a flexible cable for transmitting the signal to the oscilloscope.

Probe coupler A resonant conductor placed in a waveguide or cavity to insert or extract energy.

Problem configuration This is the core description of a problem which contains references to the various geometry, scheduling, mass flow network, plant and thermophysical property files which constitute the description of a problem.

Process A general term that describes a change in a measurable variable (e.g., the mixing of return and outdoor air streams in a mixed-air control loop and heat transfer between cold water and hot air in a cooling coil). Usually considered separately from the sensing element, control element, and controller.

Process control characteristic This gives information on how the controlled variable is affected by small changes in the intermediate position of the final control element.

Process reaction rate (PRR) The rate at which the plant responds to a disturbance. Usually plotted with the control variable on y-axis and time on the x-axis.

Production costing A method used to determine the most economical way to operate a given system of power resources under given load conditions.

Productivity The efficiency with which a person performing a specific function does a job, or the output of a worker under specific environments and conditions.

Profile Patterns of a user's activity which can detect changes in normal routines.

Program A collection of instructions needed to solve a particular problem or to guide the computer in its operation.

Programmable point A control or monotoring point for which the user may program an associated control scheme.

Programmable UJT Unijunction transistor with a variable intrinsic stand-off ratio.

Programmed I/O A standard method of accessing an I/O device - the CPU reads each byte of data from or writes each byte of data to the device.

Programmed start/stop An energy management function which operates to selectively shutdown electrically-operated equipment. This is accomplished on a predetermined time-schedule, usually paralleling occupancy schedules.

Project financing Most commonly used method to finance the construction of independent power facilities. Typically, the developer pledges the value of the plant and part or all of its expected revenues as collateral to secure financing from private lenders.

Prom Programmable read-only memory. A semiconductor memory whose contents cannot be changed by the computer after it has been programmed.

Promiscuous mode Normally an Ethernet interface reads all address information and accepts follow-on packets only destined for itself, but when the interface is in promiscuous mode, it reads all information (sniffer), regardless of its destination.

Proof pressure The specified pressure which may be applied to the sensing element of a transducer without causing a permanent change in the output characteristics.

Propagation Travelling of electromagnetic, electrical or sound waves through a medium.

Propagation delay The specified amount of time for a signal to pass through a previously closed signal path. The delay must be considered, for example, when the signal is used to synchronize other signals, or is being used in a Clock / Data configuration. This is due to both the electrical length of the signal path, and any active components in the signal path.

Propagation time Time required for a wave to travel between two points.

Proportional action In this type of control action, the output of the controller is proportional to the error. If the error is large, the signal output to the actuator is large, and if the error gets smaller, the output signal gets smaller proportionally. The relationship between the two is determined by a constant called the proportional gain. The error band within which the output is between 0% and 100% is called the proportional band. The higher the gain, the higher the proportional band. The main problem with proportional control is offset. This can be reduced by the addition of integral action, although proportional action alone is used successfully in many situations where the degree of offset is tolerable. A sluggish response to sudden, or very large, load changes may be improved by incorporating derivative action.

Proportional band (PB) That range of values of deviation corresponding to the full operating range of output signal of the controlling unit resulting from proportional action only. The PB can be expressed as a percentage of the range if the controlled condition which the measuring unit of the controller is 'designed' to measure. With HVAC control systems, the proportional band is expressed in absolute units (for example: K); with industrial control, the proportional band is expressed in percentages (of the input value).

Proportional control factor The broken loop amplification at an infinite period with any integral action removed from the controller.

Proportional controller A controller with proportional action only.

Proportional gain The inverse of the proportional band.

Proportional-speed controller A variation of floating control is proportional-speed control. In this type of controller, the farther the control point moves beyond the deadband, the faster the actuator moves to correct the deviation.

Proportioning band A temperature band expressed in degrees within which a temperature controller's time proportioning function is active.

Proportioning control mode A time proportioning controller where the amount of time that the relay is energized is dependent upon the system's temperature.

Proportioning control plus derivative function A time proportioning controller with derivative function. The derivative function senses the rate at which a system's temperature is either increasing or decreasing and adjusts the cycle time of the controller to minimize overshoot or undershoot.

Proportioning control plus integral A two-mode controller with time proportioning and integral (auto reset) action. The integral function automatically adjusts the temperature at which a system has stabilized back to the setpoint temperature, thereby eliminating droop in the system.

Proportioning control with integral and derivative functions Three mode PID controller. A time proportioning controller with integral and derivative functions. The integral function automatically adjusts the system temperature to the set point temperature to eliminate droop due to the time proportioning function. The derivative function senses the rate of rise or fall of the system temperature and automatically adjusts the cycle time of the controller to minimize overshoot or undershoot.

Prorated bills The computation of a bill based upon proportionate distribution of the applicable billing schedule.

Prospective fault current The value of overcurrent at a given point in a circuit resulting from a fault of negligible impedance between live conductors having a difference of potential under normal operating conditions, or between a live conductor and an exposed conductive part.

Protection head An enclosure usually made out of metal at the end of a heater or probe where connections are made.

Protection tube A metal or ceramic tube, closed at one end into which a temperature sensor is inserted. The tube protects the sensor from the medium into which it is inserted.

Protective conductor A conductor used for some measures of protection against electric shock and intended for connecting together any of the following parts: Exposed conductive parts, Extraneous conductive parts, The main earthing terminal, Earth electrode(s), The earthed point of the source, or an artificial neutral.

Protective conductor current Electric current which flows in a prospective conductor under normal operating conditions:

Protective earth conductor Conductor to be connected between the protective earth terminal and an external protective earthing system.

Protective earth terminal Terminal connected to conductive parts of Class I equipment for safety purposes. This terminal is intended to be connected to an external earthing system by a protective earth conductor.

Protective extra low voltage PELV An extra low voltage system which is not electrically separated from earth, but which otherwise satisfies all the requirements for SELV.

Protective relay A relay, the principal function of which is to protect service from interruption, or to prevent or limit damage to apparatus.

Protoboard Board with provision for attatching components without solder. Also called a breadboard. Primarily used for constructing experimental circuits.

Protocol A formal set of conventions governing the format and control of interaction among communicating functional units.

Proton Sub atomic particle within the nucleus of an atom. Has a positive charge.

Prowler A daemon that is run periodically to seek out and erase core files, truncate administrative logfiles, nuke lost+found directories, and otherwise clean up.

Proximity sensor or proximity switch A sensor or switch with the ability to detect it's relationship to a metal target without making physical contact.

Proxy A firewall mechanism that replaces the IP address of a host on the internal (protected) network with its own IP address for all traffic passing through it. A software agent that acts on behalf of a user, typical proxies accept a connection from a user, make a decision as to whether or not the user or client IP address is permitted to use the proxy, perhaps does additional authentication, and then completes a connection on behalf of the user to a remote destination.

P-Type semiconductor A semiconductor for which the predominant charge carriers responsible for electrical conduction are holes.

Pull down period The time taken for a measured temperature to reduce from some given higher value to a required set-point.

Pull in torque The maximum constant torque which a synchronous motor will accelerate into synchronism at rated voltage and frequency.

Pulse A step rise, a level, and a step fall of voltage or current. Characteristics of a pulse are: rise, time, duration and fall time.

Pulse amplitude modulation (PAM) Pulse modulation in which the amplitude of the pulses is varied by the modulating signal.

Pulse code modulation (PCM) A modulation system in which the standard values of a quantized wave are indicated by a series of coded pulses.

Pulse duration (PD) The period of time during which a pulse is present.

Pulse duration modulation (PDM) Pulse modulation in which the time duration of the pulses is changed by the modulating signal.

Pulse fall time Time for a pulse to decrease from 90% of its peak value to 10% of its peak value.

Pulse frequency modulation (PFM) Pulse modulation in which the modulating voltage varies the repetition rate of a pulse train.

Pulse modulation A form of modulation in which one of the characteristics of a pulse train is varied.

Pulse oscillator A sine-wave oscillator that is turned on and off at specific times. Also known as a ringing oscillator.

Pulse position modulation (PPM) Pulse modulation in which the position of the pulses is varied by the modulating voltage.

Pulse rate Frequency of pulses applied to the drive.

Pulse repetition frequency (PRF) The rate, in pulses per second, at which the pulses occur.

Pulse repetition time (PRT) Interval between the start of one pulse and the start of the next pulse; reciprocal of pulse-repetition frequency.

Pulse rise time Time required for a pulse to increase from 10% of its peak value to 90% of its peak value.

Pulse time modulation (PTM) Pulse modulation that varies one of the time characteristics of a pulse train (pwm, pdm, ppm, or pfm).

Pulse width Duration of time between the leading and trailing edges of a pulse.

Pulse width modulation (PWM) Pulse modulation in which the duration of the pulses is varied by the modulating voltage.

Pulsed input The representation of a value by a series of abrupt and relatively short cyclic changes in a signal.

Pumped storage A facility designed to generate electric power during peak load periods with a hydroelectric plant using water pumped into a storage reservoir during off-peak periods.

Pumped storage hydroelectric plant A plant that usually generates electric energy during peak-load periods by using water previously pumped into an elevated storage reservoir during off-peak periods when excess generating capacity is available.

Push pull amplifier An amplifier that uses two transistors (or electron tubes) whose output signals are in phase opposition.

Pushbutton part of an electrical device, consisting of a button that must be pressed to effect an operation.

PV Array Two or more photovoltaic modules wired in series and/or parallel.

Pyroelectricity The property of certain crystals, such as tourmaline, of acquiring opposite electrical charges on opposite faces when heated.

Pyrophoric A chemical that will ignite spontaneously in air at a temperature less than or equal to 54.4°C (130°F).

Pythagorean theorem A theorem in geometry: The square of the hypotenuse of a right triangle equals the sum of the squares of the other two sides. In electronics used for vector analysis of AC circuits.

Q Quality factor of an inductor or capacitor. It is the ratio of a component's reactance (energy stored) to its effective series resistance (energy dissipated). For a tuned circuit, a figure of merrit used in bandwidth calculations. Q is the ratio of reactive power to resistive power in a tuned circuit. Also the symbol for charge in coulombs (Q for quantity).

Q-factor (quality factor) A measure of the "quality" of a resonant system. Resonant systems respond to frequencies close to the natural frequency much more strongly than they respond to other frequencies. On a graph of response versus frequency, the bandwidth is defined as the 3 dB change in level besides the center frequency.

Quadrature amplitude modulation (QAM) Quadrature modulation in which the two carriers are amplitude modulated. In analogue communications, the representation (i.e., transmission) of digital information by encoding bit sequences of fixed, specified length (number of bits), and representing these bit sequences as a function of (a) the amplitude of an analogue carrier; or (b) a phase shift of the analogue carrier with respect to the phase that represented the preceding bit sequence, and where the permissible phase shift is an integral multiple of /2 radians (90°, or one-quarter unit interval); or (c) both.

Quadrature circuit A circuit designed to provide a 900 signal phase displacement.

Quadrupole Let a collection of electric or magnetic charges be distributed around a point; for example, the center of mass of a system of atoms, molecules, or nuclei. The potential at a distance r from this point may be represented by an infinite series of terms in inverse powers of r. The term in the inverse first power is the Coulomb potential, the inverse second power term is the dipole potential, the inverse third power term is the quadrupole potential, etc. A typical example is an array of four charges of equal magnitude so spaced that they coincide with the vertices of a parallelogram. Charges located on opposite vertices are of the same sign; the distance of separation between charges is taken to be of the order of molecular or infinitesimal dimensions.

Quadrupole radiation Radiation emitted by a quadrupole. Since the selection rules were deduced by analoguey between the behavior of a classical electric dipole and the emission of radiation by quantum transitions, then there are arrangements of two dipoles (e.g., a linear arrangement in which the positive charges coincide and are located at the center of two negative charges separated by a distance 2x (x = distance from center), so Óe i x i = 0) when the quadrupole moment (Óå i ·x i 2) not zero but varies as 1/r3, and therefore acts as a source of radiation.

Quad-shield Four layers of shielding.

Qualification testing Testing performed one time only on a sample basis to determine the suitability of a product or product family for usage against certain specifications or for certain programs.

Qualifying activity The government agency or original equipment manufacturer (OEM) responsible for determining vendor qualification status with respect to a given specification or series of specifications. In many cases the qualifying activity will be different from the procuring activity.

Quality conformance inspection Ongoing sample testing on a periodic basis to determine conformance with the quality and reliability standards established in Qualification.

Quality Conformance to a set of predetermined design and workmanship standards. Quality and reliability are not synonymous.

Quantization A process whereby the continuous range of input-signal values is divided into nonoverlapping subranges. Each of these subranges has a discrete value of the output uniquely assigned. Once a signal value falls within a given subrange, the output provides the corresponding discrete value.

Quantization error The inherent uncertainty in digitizing an analogue value that is caused by the finite resolution of the conversion process; increasing the resolution of an ADC reduces the uncertainty.

Quantizing noise The noise (deviation of a signal from its original or correct value) that results from the quantization process. In serial digital video, it is a granular type of noise that occurs only in the presence of a signal.

Quantum dots* A class of nanocrystals that emit varying colours of light depending on their size. They can be used to label different biological structures.

Quantum vacuum Also known as quantum-mechanical vacuum. Quantum physics reveals that even an ideal vacuum, with a measured pressure of zero Pascals (Pa), isn't really empty. One reason is that the walls of the vacuum chamber emit light in the form of black-body radiation: visible light if they are at a temperature of thousands of degrees, infrared light if they are cooler. This soup of photons will be in thermodynamic equilibrium with the walls, and the vacuum can consequently be said to have a particular temperature. More fundamentally, there are quantum-mechanical fluctuations in the vacuum. This may be responsible for the observed value of the cosmological constant.

Quartz Silicon dioxide. A material used in the construction of equipment, wafer carriers and masks.

Quartz clock A clock containing a quartz oscillator that determines the accuracy and precision of the clock.

Quartz crystal Synthetic quartz is composed of Silicon and Oxygen (Silicon Dioxide SiO2) and is cultured in autoclaves under high pressure and temperature. Quartz exhibits piezoelectric properties that generate an electrical potential when pressure is applied on the surfaces of the crystal. Conversely, when an electrical potential is applied to the surfaces of a crystal, mechanical deformation or vibration is generated. These vibrations occur at a frequency determined by the crystal design and oscillator circuit. Under proper conditions, quartz can be used to stabilize the frequency of an oscillator circuit.

Quartz crystal microbalance (QCM) A piezoelectric quartz crystal that utilizes the Converse Piezoelectric Effect to determine mass changes as a result of frequency change of the crystal. Material is coated onto the crystal that attracts a certain other material. When this other material becomes attached, it lowers the crystal frequency, the amount being directly related to its mass. The Sauerbrey equation relates the frequency change to the mass change. The mass change can be in the nanogram range.

Sauerbrey equation: Df = -Dmdf2/_qvq

where:

Df = change in frequency

Dmd = change in mass density (md, mass per surface area)

f2 = frequency squared

_q = density of quartz = 2650 kg m-3

vq = propagation of sound in quartz = 3340 m s-1

Quartz crystal oscillator A timing device that consists of a crystal and an oscillator circuit, providing an output waveform at a specified reference frequency.

Quartz crystal unit A completed quartz crystal, including a resonator plate with electrodes, a holder with suitable mounting structures, and a permanently sealed cover. Usually called a crystal.

Quartz oscillator An oscillator in which a quartz crystal is used to stabilize the frequency.

Quick break/quick make Switches designed to make or break circuits in less than 5 milliseconds to make or break. Recommended for use on DC circuits.

Quick change This feature allows you to exchange CDs on a carousel player, while a CD is playing.

Quick disconnect A type of connector shell that permits rapid locking and unlocking of two connector halves.

Quick disconnect coupling A design feature, apparent in the quick disconnect connector; it permits relatively rapid joining and separation.

Quick set record (QSR, QTR, XPR) A one-touch button that allows instant recording for 30, 60 or 90 minutes. This feature eliminates the need to use the programming mode.

Quick start A refined start mechanism that results in less time spent waiting for the picture to appear on the screen.

Quiescent At rest. For an amplifier the term is used to describe a condition with no active input signal.

Quiescent current The standing current that flows in a circuit when the signal is not applied. The quiescent current is usually very low or lower than when processing a signal.

Quiescent point (Q point) A point on the DC load line of a given amplifier that represents the quiescent (no signal) value of output voltage and current for the circuit.

R

R face One of the three larger faces which occur at the ends of the natural quartz crystal.

Raceway An enclosed channel designed expressly for holding conductors and cables, including conduit and tubing, wireways, and busways.

Rad (Si) The quantity of any type of ionizing radiation that will impart 100 ergs of energy per gram of silicon.

Radar Radar (coined from radio directing and ranging) is a system for locating reflecting objects by means of radio signals. Anything which will reflect enough energy of a radio signal of high frequency to actuate the receiver may be detected and accurately located in space over distances approximating line-of-sight distances from the radar stations. This will be over 100 miles for high-flying planes. Since the radio signals penetrate fog, darkness, rain, and haze, this method of locating planes, ships, and ground objects does not have many of the limitations of the older optical systems. In addition, the ranging by radar means may be much more accurate than that of the optical methods. Although radar was originated a number of years ago, it was not until just before and during World War II that it was brought to the high degree of perfection which permitted its use in detecting enemy equipment and then controlling the gunfire which destroyed that equipment, often without the enemy being visible. Because of the varied reflecting characteristics of many ground installations, bodies of water, types of earth, terrain formations, etc., certain types of radar were used for blind bombing when visual sighting was impossible. These wardeveloped applications point the way to many obvious peacetime applications such as blind navigation of ships in heavy fogs, blind flying, blind landings, and storm tracking.

Radio A form of wireless communications in which the output of the transmitter takes the form of dissipating electromagnetic radiation which spreads outward from the antenna through free space. The signal strength drops off as the square of the distance from the source of radiation. Distant radio receivers have to be very sensitive to detect signals that can measure only a few microvolts per meter in strength.

Radio broadcast Transmission of music, voice and other information on radio carrier waves that can be received by the general public.

Radio communication Communication between distant points not directly connected by an electrical conductor may be accomplished by electromagnetic waves radiated through space. Thus radio communication utilizes radiated energy instead of the conducted energy of the wired methods. Obvious benefits of the radio method are the elimination of the expense of installation and upkeep of a wire communication system, and the communication between points difficult of access by wired systems. Furthermore, since energy may be radiated in all directions in space, broadcasting to large numbers of persons is simpler. Radiated electromagnetic waves can be used for communication purposes in three ways. First, they may be used to create a monotone signal of dots and dashes comprising a code. This is radio telegraphy. It has certain important commercial and governmental uses, but is not suitable for broadcasting, as it must be interpreted by trained operators. Secondly, electromagnetic waves may be modulated so that they carry the electrical equivalent of sound waves. At the receiver they are caused to reproduce the original sound, whether it was voice or music. This is the field of radio telephony and broadcasting. Thirdly, it is possible to transmit, by means of these waves, electrical pulses which will re-create at the point of reception, a scene which originated at the transmitter, with accompanying sound. This is television.

Radio control The remote control of an apparatus by signals conveyed by electromagnetic waves.

Radio frequency (RF) A term that refers to alternating current (AC) having characteristics such that, if the current is input to an antenna, an electromagnetic (EM) field is generated suitable for wireless broadcasting and/or communications.

Radio frequency identification A method of identifying unique items using radio waves. Typically, a reader communicates with a tag, which holds digital information in a microchip. But there are chipless forms of RFID tags that use material to reflect back a portion of the radio waves beamed at them.

Radio frequency interference (RFI) Any electromagnetic disturbance that interrupts, obstructs, or otherwise degrades or limits the effective performance of electronics/ electrical equipment.

Radio frequency transponder The core of TIRIS is a tag, which can be attached to or embedded in an object. In a transponder system, a reader sends a signal to a tag via an antenna, charging the transponder and allowing it to return a signal carrying the data stored within it. The data collected from the transponder can be sent directly to a host computer or stored and later uploaded.

Radio telephony The use of radio to communicate sounds.

Radio wave An electromagnetic wave of a frequency arbitrarily lower than 3000 GHz.

Radioastronomy Branch of astronomy that studies the radio waves generated by celestial bodies and uses these emissions to obtain information about them.

Radiocommunication Term used to describe the transfer of information between two or more points by use of radio or electromagnetic waves.

Radio-frequency amplifier Amplifier having one or more active devices to amplify radio signals.

Radio-frequency generator Generator capable of supplying RF energy at any desired frequency in the radio-frequency spectrum.

Radio-frequency probe Probe used in conjunction with an AC meter to measure radio-frequency signals.

Radiotelegraphy The use of radio (instead of wire) to communicate a message(s) over a distance.

Radix Refers to the number of digits in a numbering system. For example, the decimal numbering system is said to be radix-10. May also be referred to as the "base".

Radome A typically rigid dielectric cover over the radiating portion of an antenna, and nearly always separated from the radiator by an air gap. A radome (the merger of radar and dome) has the purpose of protecting the radiator from natural weather phenomena and contamination by dirt. It usually includes aerodynamic shaping to minimize wind loading.

Ragone plot The graphical illustration of the specific energy of a cell as a function of its specific power.

Rail-to-rail input or output The allowable input and output voltage ranges include the power-supply rails.

Rail-to-rail The span from the negative power supply to the positive power supply.

Rainproof So constructed, projected, or treated as to prevent rain from interfering with the successful operation of the apparatus under specified test conditions.

Raintight So constructed or protected that exposure to a beating rain will not result in the entrance of water.

Random jitter (RJ) Includes all jitter components not defined as deterministic jitter (i.e., the jitter that is not related to the signal and known noise sources).

Random noise Also known as "white noise." A noise signal that never repeats and has a flat frequency spectrum. Random noise can also be a digital pattern that never repeats. Random noise is generally considered to have a Gaussian amplitude distribution, but numerically generated noise can also have a flat amplitude distribution. The amplitude of random noise is normally measured as the rms value.

Range 1.The difference between the minimum and maximum values of a parameter, such as voltage;

2. a statement of the minimum and maximum values of a parameter.

Range bin Discrete element along a single radial of radar data at which the received signals are sampled. Range bins from Poldirad data are spaced at multipes from 150 m intervals.

Rangeability The ratio of the maximum flowrate to the minimum flowrate of a meter.

Rankine (°R) An absolute temperature scale based upon the Fahrenheit scale with 180° between the ice point and boiling point of water. 459.67°R = 0°F.

Rapid prototyping, rapid prototyping platform A platform of programmable technologies for rapid prototyping.

Rapid thermal processing (RTP) A method of rapidly heating up a wafer by exposing it to bright lamps. Wafers can be raised from room temperature to up to 1100°C in seconds, and cooled in a similar length of time.

Raster A raster is the series of scan lines that make up a TV picture or a computer's display. The term raster line is the same as scan line. All of the scan lines that make up a frame of video form a raster. Lines and rows of dots such as those on the illuminated face of a video screen. A matrix of pixels or the scan lines on a CRT.

Ratchet Ratchet is a term applied to any component of a customer's bill that is based on a previous level of use.

Rate action The derivative function of a temperature controller.

Rate time the time interval over which the system temperature is sampled for the derivative function.

Rated capacity (battery) The number of Amp-Hours a battery can deliver under specific conditions (rate of discharge, end voltage, temperature).

Rated lamp life With regard to lighting, the point in time when 50% of a statistically significant number of lamps has failed.

Rated output The output at standard calibration.

Rated output current The maximum continuous load current a power supply is designed to provide under specified operating conditions.

Rated power Nominal power output of an inverter; some units cannot produce rated power continuously.

Rated temperature The maximum temperature at which an electric component can operate for extended periods without loss of its basic properties.

Rated voltage The maximum voltage at which an electric component can operate for extended periods without undue degradation or safety hazard.

Ratio detector A dual-diode frequency-modulation detector.

Ratiometric measurement A measurement technique where an external signal is used to provide the voltage reference for the dual-slope A/D converter. The external signal can be derived from the voltage excitation applied to a bridge circuit or pick-off supply, thereby eliminating errors due to power supply fluctuations.

Reactance The opposition to an alternating current presented by inductance, capacitance, or a combination of the two. Reactance is measured in ohms and is represented by the letter X.

Reactance circuit A circuit designed to behave as either an inductive or capacitive reactance, and hence, as an equivalent inductor or capacitor.

Reactive power Also called imaginary power or wattless power. It is the power value in "volt amps" obtained from the product of source voltage and source current in a reactive circuit.

Read The process of retrieving data stored on an RFID tag by sending radio waves to the tag and converting the waves the tag sends back into data.

Read only memory (ROM) Memory that contains fixed data. The computer can read the data, but cannot change it in any way.

Read range The distance from which a reader can communicate with a tag. Active tags have a longer read range than passive tags because they use their own power source (usually a battery) to transmit signals to the reader. With passive tags, the read range is influenced by frequency, reader output power, antenna design, and method of powering up the tag. Low-frequency tags use inductive coupling, which requires the tag to be within a few feet of the reader.

Read rate A term usually used to describe the number of tags that can be read within a given period or the number of times a single tag can be read within a given period. The read rate can also mean the maximum rate at which data can be read from a tag expressed in bits or bytes per second.

Read-and-record The recording, by individual device serial number, of the actual parametric values measured for that device at a specific electrical test point. Read-and-record can be done for device characterization, drift (delta) measurement, or temperature coefficient computation.

Reader A device used to communicate with RFID tags. The reader has one or more antennas, which emit radio waves and receive signals back from the tag. The reader is also sometimes called an interrogator because it "interrogates" the tag.

Reader field The area of coverage. Tags outside the reader field do not receive radio waves and can't be read. This is also sometimes referred to as the read field.

Reader module The electronics of a reader, including a digital signal processor, on a circuit board. Modules can be put in an RFID label printer or other device, as opposed to a standalone reader.

Real power The average value of the instataneous product of volts and amps over a fixed period of time in an AC circuit.

Real time The time interval over which the system temperature is sampled for the derivative function.

Real time counter A digital tape counter in VCRs, DVDs, DVD/VCR combos and camcorders that indicates elapsed time in terms of actual hours, minutes and seconds.

Rear channel preamp output This feature allows you to connect the sound from the rear channel to a separate power amplifier, for more amplification or further processing.

Rear lobe A region of pickup at the rear of a super-cardioid or hypercardioid microphone polar pattern. A bidirectional microphone has a rear lobe equal to its front pickup.

Rebonding The placement of a new bond on the same pad or post as a previously attempted bond. If the original bond has been removed, the new bond is still considered a rebond. A bond-off, that is, an extra bond placed at the edge of the post (never the pad) for the purpose of clearing the bonding machine, is not considered a rebond.

Receiver Any circuit that intercepts a signal, processes the signal, and converts it to a form useful to a person. The signal may be in any form such as electric currents in a wire, radio waves, modulated light, or ultrasound. The receiver converts signals into audio information, video information, or both.

Receptivity Measure of a material's resistance to current flow.

Recloser A switching device that rapidly recloses a power switch after it has been opened by an overload. In reclosing the power feed to the line, the device tests the circuit to determine if the problem is still there. If not, power is not unnecessarily interrupted on the circuit.

Reclosure A reclosure is a switch that functions like a circuit breaker and protects primary circuits from faulty conditions. A reclosure will automatically restore the circuit in the event of a temporary fault. Temporary faults can be caused by things such as a tree branch falling onto the lines or a squirrel on the lines.

Recombinant system Sealed secondary cells in which gaseous products of the electrochemical charging cycle are made to recombine to recover the active chemicals. A closed cycle system preventing loss of active chemicals. Used in Ni-Cads and SLA batteries.

Recombination Process by which a conduction band electron gives up energy (in the form of heat or light) and falls into a valence band hole.

Reconditioning One or more deep discharges below 1.0 V/cell with a very low controlled current, causing a change to the molecular structure of the cell and a rebuilding of its chemical composition. Reconditioning helps break down large crystals to a more desirable small size, often restoring the battery to its full capacity. Applies to nickel-based batteries.

Record A collection of unrelated information that is treated as a single unit.

Recovery The lowering of the polarization of a cell during rest periods.

Recovery time The time for a sensor to return to baseline value after the step removal of the measured variable. Usually specified as time to fall to 10% of final value after step removal of measured variable.

Recrystalling The removal of one crystal and the replacement by a crystal of a different frequency and the realignment of the equipment.

Rectangular coordinates A Cartesian coordinate of a Cartesian coordinate system whose straight-line axes or coordinate planes are perpendicular.

Rectangular wave Also known as a pulse wave. A repeating wave that only operates between two levels or values and remains at one of these values for a small amount of time relative to the other value.

Rectification The process of changing an alternating current to a unidirectional current.

Rectifier A component that passes current only in one direction, e.g., a diode.

Recycling Reclamation of materials without endangering human health and the environment.

Red eye reduction Some cameras with red eye reduction have a small bright red light, located on the front of the camera, that turns on when the shutter button is first depressed. The eye's iris will close in response to this light, making the opening in the eye smaller. The regular cameraflash then fires with less chance of being reflected back from the rear of the eye.

Redox A contraction of the words "reduction" and "oxidation". The two chemical reactions on which cell chemistries depend.

Redox battery A battery in which the chemical energy is stored in two dissolved ionic reactants separated by a membrane.

Redox potential The potential developed by a metallic electrode when placed in a solution containing a species in two different oxidation states.

Reduced instruction set computer (RISC) Is a computer architecture that has reduced chip complexity by using simpler processing instructions. It is faster than its more complex counterparts, thanks to its simplicity, and is also designed and built more economically.

Reduction stepper Steppers that reduce the image contained on the reticle nX times, usually 5X, but sometimes 4X, 2.5X or 2X, before exposing the wafer.

Reduction The gain of electrons by a chemical species.

Redundant power supply A second power supply circuit sometimes specified for systems used in critical applications. Redundancy is useful where unexpected power failures can cause a major system to fail, often at great expense. Redundant power supplies could be fed from different AC power (mains) circuits for maximum system reliability. Power supplies are usually "diode or'ed" and should be hot swappable. A "redundant capacity" supply is usually a single supply with built-in redundancy included and is not hot swappable, but each supply voltage is duplicated allowing the user to schedule down time to replace the supply.

Reed relay Relay consisting of two thin magnetic strips within a glass envelope. When a coil around the envelope is energized, the relay's contacts snap together making a connection between leads attached to the reed strips.

Reel A cylinder device used to hold wire and cable until installed. There are standard reel sizes that are used in the electrical industry that are either wood (non-returnable) or steel (returnable).

Reference designator The name of a component on a printed circuit by convention beginning with one or two letters followed by a numeric value. The letter designates the class of component; eg. "Q" is commonly used as a prefix for transistors. Reference designators appear as usually white or yellow epoxy ink (the "silkscreen") on a circuit board. They are placed close to their respective components but not underneath them, so that they are visible on the assembled board. By contrast, on an assembly drawing a reference designator is often placed within the boundaries of a footprint – a very useful technique for eliminating ambiguity on a crowded board where reference designators in the silkscreeen may be near more than one component.

Reference edge The edge of a blank or wafer identified for use in orienting the blank or wafer on an x-ray chuck for making orientation measurements.

Reference flat The flat edge on an otherwise circular blank for use as a reference edge. Called an X-flat when perpendicular to the X-axis.

Reference frequency Frequency having a fixed and specified position with respect to the assigned frequency.

Reference ground Defined point in a circuit or system from which potential measurements shall be made.

Reference junction The cold junction in a thermocouple circuit which is held at a stable known temperature. The standard reference temperature is 0°C (32°F). However, other temperatures can be used.

Reference mark Any diagnostic point or mark which can be used to relate a position during rotation of a part to its location when stopped.

Reference plane Any plane perpendicular to the shaft axis to which an amount of unbalance is referred.

Reference voltage The defined or specified voltage to which other voltages are compared.

Reflection loss The part of a signal which is lost due to reflection of power at a line discontinuity.

Reflections With video signals, reflections can be caused by energy that is not absorbed by the load (or a termination) and is reflected and possibly combined with the original signal. Reflected signals can occur when the impedance does not match (as a result of wrong termination or mixing of cable impedance). Some of the undesirable results of reflection include Y/C delays, colour smearing, ringing on luma (but not on colour), and ghosts.

Reflective lens A lens that throws back or bends light from a surface. AKA a mirror.

Reflectivity factor (Z) Integral over the backscatter cross-section of the particles in a pulse volume. For particles small compared to the wavelength the scatter cross-section is D6, where D is the diameter of the particle. Radars are calibrated in the way to give directly (assuming the dielectric constant of water) the reflectivity factor from the received backscattered energy. Units for the reflectivity factor are mm6 m-3 or the logarithmic value of this in dBZ.

Reflow oven An oven employing infrared radiation or hot air.

Reflow soldering A process for joining surface mount parts into a solder paste for permanent interconnection via passage through various stages including preheat, stabilization/drying, reflow spike and cooldown.

Reflow spike The portion of the reflow soldering process during which the temperature of the solder is raised to a value sufficient to cause the solder to melt.

Refraction The bending of sound waves by a change in the density of the transmission medium, such as tem-perature gradients in air due to wind.

Refractive lens A lens (composed of glass lens components) that bends or deflects the path of light as it passes between different media, for example, air or gas.

Refractor A device used to re-direct the luminous flux from a light source by the process of refraction.

Refractory metal Metals such as tungsten, titanium, and molybdenum which are capable of withstanding extremely high, or refractory, temperatures.

Refractory metal thermocouple A class of thermocouples with melting points above 3600°F. The most common are made from tungsten and tungsten/rhenium alloys Types G and C. They can be used for measuring high temperatures up to 4000°F (2200°C) in non-oxidizing, inert, or vacuum environments.

Refurbishing The repair of worn out or damaged batteries. This is not the same as reconditioning.

Regeneration The process for providing an internal positive feedback within an electronic system for increasing signal amplitudes. Excessive regeneration produces oscillation and, hence, signal generation.

Regenerative braking This uses the electrical drive motor in an electric vehicle to act as a generator returning energy to the battery when overdriven mechanically by the vehicle wheels. This provides a powerful braking effect and at the same time captures energy which would otherwise be wasted or dissipated in the brakes.

Regenerative feedback Positive feedback. Feedback from the output of an amplifier to the input such that the feedback signal is in phase with the input signal. Used to produce oscillation.

Regenerative receiver An AM radio receiver in which positive feedback is used in order to increase the sensitivity and selectivity of the receiver.

Regenesys A high power Sodium Polysulfide Bromine "Flow Battery".

Register In printed board manufacture, many terms are borrowed from the subject of printing. Proper alignment of various plates, stones, or screens to assure clear and accurate reproduction, as of colour. In printed circuit design, the designer gets his photoplot files in register before he views them with his Gerber file viewer. The board manufacturer produces film from the Gerber files and uses them in register with respect to the panels of material from which he will build the boards. He is going to want the pads on both sides and on internal layers to be in register before he drills holes in the panel. The term registration is often used in the printed circuit industry for this sense of the noun register .

Regulated power supply A device that maintains within specified limits a constant output voltage or current for specified changes in line, load temperature or time.

Regulating transformer A transformer used to vary the voltage, or phase angle, of an output circuit. It controls the output within specified limits and compensates for fluctuations of load and input voltage.

Regulation The process of holding constant selected parameters, the extent of which is expressed as a percent.

Reject number For a sample test, the number of failed devices which will cause lot rejection. This will normally be one higher than the accept number.

Relative Not independent. Compared with or with respect to some other measured quantity.

Relative antenna power gain The ratio of the average radiation intensity of the test antenna to the average radiation of a reference antenna with all other conditions remaining equal.

Relative bechmann angle (RBA) Relative Bechmann Angle is the apparent angle of orientation of an AT crystal blank, which is derived from a plot of the temperature vs. frequency characteristic of a crystal resonator.

Relative permeability The ratio of permeability of a medium to that of a vacuum. In the cgs system, the permeability is equal to 1 in a vacuum by definition. The permeability of air is also for all practical purposes equal to 1 in the cgs system.

Relaxation oscillator An oscillator circuit that generates a signal by periodic conduction-nonconduction periods. A nonsinusoidal waveform is produced and, generally, no resonant circuits are employed.

Relay (mechanical) An electromechanical device that completes or interrupts a circuit by physically moving electrical contacts into contact with each other.

Relay (solid state) A solid state switching device which completes or interrupts a circuit electrically with no moving parts.

Relays distance Relays used on transmission lines that use a variety of sensors and measurements to determine when an unusual condition exists at some distance, out on the transmission circuit.

Relays over-current Protective relays used on power systems that detect excessive currents and send signals to protective devices, such as power circuit breakers.

Relays voltage Protective relays used on power systems that detect when line voltage has gone outside of an acceptable range, either up or down, and send a signal to a protective device or system.

Reliability The assurance that a component will perform in a specified manner for a specified time under a set of specified conditions that include electrical, mechanical, thermal, and environmental stresses. The concept of reliability encompasses the elements of both quality and longevity.

Relief valve A relief valve is designed to reduce pressure quickly.

Reluctance, R Analogueous to resistance in an electrical circuit, reluctance is related to the magnetomotive force, F, and the magnetic flux by the equation R = F/(Magnetic Flux), paralleling Ohm's Law where F is the magnetomotive force (in cgs units).

Remanence, BD The magnetic induction that remains in a magnetic circuit after the removal of an applied magnetizing force. If there is an air gap in the circuit, the remanence will be less than the residual induction, Br.

Remote control 1) (general) Control of an operation from a distance: this involves a link, usually electrical, between the control device and the apparatus to be operated. Note: Remote control may be over (A) direct wire, (B) other types of interconnecting channels such as carrier-current or microwave, (C) supervisory control, or (D) mechanical means.

2. (programmable instrumentation) A method whereby a device is programmable via its electrical interface connection in order to enable the device to perform different tasks.

Remote cut-off tube Sometimes known as a variable-mu tube. It is characterised by a wide negative-bias latitude during which current flows.

Remote diode A diode or diode-connected bipolar transistor used as a temperature-sensing element, often integrated onto an integrated circuit whose temperature is to be measured.

Remote sensing A technique for regulating the output voltage of a power supply at the load by connecting the regulator error-sensing leads directly to the load. Remote sensing compensates for specified maximum voltage drops in the load leads. Care should be exercised to avoid opening load handling leads to avoid damaging the power supply. Polarity must be observed when connecting sense leads to avoid damaging the system.

Remote temperature sensor A remotely located PN junction used as a temperature sensing device, usually located on an integrated circuit other than the one doing the measurement.

Remote turn on lead The lead from the head unit which supplies a signal (12V+) to the "remote turn on" lead of the amplifier turning the amplifier on when the head unit is turned on, and allowing the amplifier to be mounted in a location out of reach of the user. This is NOT the amplifier's main source of power.

Repair An operation that restores a part or assembly to a condition in which it can be used.

Repeatability The ability of an instrument to give the same output or reading when repeatedly operated under identical conditions by the same user.

Repeater Devices that receive a radio signal, amplify it and re-transmit it in a new direction. Used in wireless networks to extend the range of base station signals, thereby expanding coverage – within limits – more economically than by building additional base stations. They are typically used for buildings, tunnels or difficult terrain.

Reproducibility The ability of an instrument to give the same output or reading when operated under identical conditions by different users.

Resealable cap (battery) A safety vent valve which is capable of closing after each pressure release from within a cell.

Resealable safety vent The resealable vent internal to a cell to release excessive internal pressure.

Reserve battery Batteries which are stored in an inactive state without their electrolyte. They are only activated when needed by the introduction of the electrolyte.

Reserve capacity The number of minutes at which the battery can be discharged at 25 Amps and maintain a terminal voltage higher than 1.75 volts per cell, on a new, fully charged battery at 80degrees Fahrenheit(27°C). Defines a battery's ability to power a vehicle with an inoperative alternator or fan belt. Used for comparing automotive SLI batteries.

Reserved word A word that has a defined function in the language, and cannot be used as a variable name.

Reset signal A signal used to return a circuit to a desired state.

Resettable fuse A fuse which protects against excessive current and temperature by interrupting the flow of current. After opening it will reset after the fault conditions have been removed but only after it has cooled. It requires no manual resetting or replacement. The "Polyswitch" is an example of this.

Residential (lighting) A residential development, or a mixture of residential and small commercial establishments, characterized by few pedestrians during nighttime hours. This definition includes area with single-family homes, townhouses, and/or small apartment buildings.

Residual (final) unbalance Residual unbalance is that unbalance of any kind that remains after balancing.

Residual induction, Br This is the point at which the hysteresis loop crosses the B axis at zero magnetizing force, and represents the maximum flux output from the given magnet material. By definition, this point occurs at zero air gap, and therefore cannot be seen in practical use of magnet materials.

Residual magnetism Magnetism remaining in the core of an electromagnet after the coil current is removed.

Residue Any undesirable material that remains on a substrate after a process step.

Residue removal Removal of all remaining residues left on the wafer after the implant or etch process.

Resin An organic polymer which, when mixed with a curing agent, crosslinks to form a thermosetting plastic.

Resin flux A resin and small amounts of organic activators in an organic solvent.

Resin impregnation The process of coating a glass fabric by resin using metering rolls to control the fabric to resin ratio.

Resist A material which is used to coat the substrate and is then selectively cured to form an impervious layer.

Resist coat Used as a template for etching pattern features into underlaying layer surface.

Resist processing Coating wafer with photosensitive chemicals prior to exposure.

Resist thickness The typical thickness of photoresist spun onto a wafer surface ranges from one to two microns. Resist thickness used in thin-film head processing can be much greater, ranging up to 10 microns.

Resistance(R) Resistance is the opposition that a substance offers to the flow of electric current. It is represented by the uppercase letter R. The standard unit of resistance is the ohm, sometimes written out as a word, and sometimes symbolized by the uppercase Greek letter omega. When an electric current of one ampere passes through a component across which a potential difference (voltage) of one volt exists, then the resistance of that component is one ohm. In general, when the applied voltage is held constant, the current in a direct-current (DC) electrical circuit is inversely proportional to the resistance. If the resistance is doubled, the current is cut in half; if the resistance is halved, the current is doubled. This rule also holds true for most low-frequency alternating-current (AC) systems, such as household circuits. In some AC circuits, especially at high frequencies, the situation is more complex, because some components in these systems can store and release energy, as well as dissipating or converting it. The electrical resistance per unit length, area, or volume of a substance is known as resistivity. Resistivity figures are often specified for copper and aluminum wire, in ohms per kilometer. Opposition to AC, but not to DC, is a property known as reactance. In an AC circuit, the resistance and reactance combine vectorially to yield impedance.

Resistance ratio characteristic For thermistors, the ratio of the resistance of the thermistor at 25°C to the resistance at 125°C.

Resistance temperature characteristic A relationship between a thermistor's resistance and the temperature.

Resistance to solvents A test which requires immersion of sample devices in such solvents as trichlorotrifluoroethane and methylene chloride, followed by brushing to determine the durability of unit marking.

Resistance weld Procedure involving pressure sealing with electricity and backfilling with nitrogen to force out oxygen and moisture. This results in superior aging characteristics.

Resistance welding Resistance welding is a process used to join metallic parts with electric current. There are several forms of resistance welding, including spot welding, seam welding, projection welding, and butt welding.

Resistive load Load that offers resistance to the flow of current. Examples of resistive loads: electric heaters, ranges and ovens, toasters, and irons. If the device is supposed to get hot and doesn't move, it's mostly likely a resistive load.

Resistive power Amount of power dissipated as heat in a circuit containing resistive and reactive components. True power as opposed to reactive power.

Resistive temperature detector (RTD) Temperature detector consisting of a fine coil of conducting wire (such as platinum) that will produce a relatively linear increase in resistance as temperature increases.

Resistivity (volume) The resistance that a unit volume of a material offers to the passage of electricity, the electric current being perpendicular to two parallel faces. More generally, the volume resistivity is the ratio of the potential gradient parallel with the current in the material to the current density.

Resistor An electrical component that limits or regulates the flow of electrical current in an electronic circuit. Resistors can also be used to provide a specific voltage for an active device such as a transistor. A component used to introduce resistance into a circuit.

Resistor colour code Coding system of coloured stripes on a resistor to indicate the resistor's value and tolerance.

Resistor-transistor logic (RTL) Logic gates implemented using particular configurations of resistors and bipolar junction transistors. For the majority of today's designers, resistor-transistor logic is of historical interest only.

Resolution (horizontal) The amount of detail in a horizontal direction in a video image. It is expressed as the number of distinct vertical lines, alternately black and white, that can be seen in the width of the picture. This information is usually derived from observation of the vertical wedge of the test pattern. Horizontal resolution depends on the high frequency amplitude and phase response of the pick-up equipment, as well as the transmission medium and the monitor itself.

Resolution (vertical) The amount of resolvable detail in a vertical direction in a video image. It is expressed as the number of distinct horizontal lines, alternately black and white, that can be seen in a test pattern. Vertical resolution is primarily determined by the number of horizontal scanning lines in a frame.

Resonance 1. The state in which the natural response frequency of a circuit coincides with the frequency of an applied signal, or vice versa, yielding intensified response.

2. The state in which the natural vibration frequency of a body coincides with an applied vibration force, or vice versa, yielding reinforced vibration of the body.

Resonance pitch When you tighten a drum, you raise it's resonance.

Resonance resistance The resistance of the crystal unit alone at the resonance frequency.

Resonant circuit A circuit in which inductive and capacitive elements are in resonance at an operating frequency.

Resonant converter A class of converters that uses a resonant circuit as part of the regulation loop.

Resonant frequency The natural frequency at which a circuit oscillates or a device vibrates. In an L-C circuit, inductive and capacitive reactances are equal at the resonant frequency.

Resonator A body that is capable of being set into resonance by the application of a periodic force.

Resonator plate The quartz blank or resonator.

Response time or time constant The amount of time (in seconds) in which the sensor operates after being subjected to a step temperature increase where the difference between the initial soak temperature and actuation temperature equals 63% of the step temperature. The response time is expressed for a designated flow (feet per second), media and system pressure.

Rest periods Interruptions to the charging process to allow the chemical reactions in the battery to stabilise.

Retentivity The degree by which a material retains magnetism.

Reticle An etched, chrome-coated quartz glass plate used to transfer the pattern for a single layer of device circuitry of an integrated circuit.

Retrace During the scanning of a picture onto a screen, scan lines are produced from left to right. Before scanning the next line, the electron beam must get back to the left side of the screen. This is called retrace. The beam must be turned off (blanked) during retrace time.

Retrace time The time required to move the scanning beam from the right side to the left side of the screen.

Retro A rear-projection video display.

Return The name for the common terminal of the output of a power supply; it carries the return current for the outputs.

Return loss A measure of reflected energy in decibels at a specific frequency and cable length.

Return path Conduction elements in a magnetic circuit which provide a low reluctance path for the magnetic flux..

Reuse The assignment of frequencies or channels to cells so that adjoining cells do not use the same frequencies and cause interference whereas more distant cells can use the same frequencies. Reuse expands the capacity of a cellular network by enabling the use of the same channels throughout the network.

Reverberation The persistence of sound in an enclosed space, as a result of multiple reflections after the sound source has stopped. The decaying residual signal that remains after a sound occurs, created by multiple reflections as the original sound wave bounces off walls, furniture, and other non-absorbing barriers within a room or other acoustical environment. A room with very little reverberation is called a "dead" room, which is the opposite of a "live" acoustic space which is very reflective.

Reverberation time The length of time required for the sound field to collapse, after the sound source has stopped.

Reverse bias Bias on a PN junction that allows only leakage current (minority carriers) to flow. Positive polarity on the n-type material and negative polarity to the p-type material.

Reverse breakdown voltage Amount of reverse bias that will cause a PN junction to break down and conduct in the reverse direction.

Reverse channel The path through which energy travels from the RFID tag to the interrogator, or reader. It is also sometimes called the back channel.

Reverse current Current through a diode when reverse biased. An extremely small current also referred to as leakage.

Reverse saturation current Reverse current through a diode caused by thermal activity. This current is not affected by the amount of reverse bias on the component, but does vary with temperature.

Reverse voltage protection A circuit or circuit element that protects a power supply from damage caused by a voltage of reverse polarity applied at the input or output terminals.

Reversible reaction A chemical reaction which can be reversed to reconstitute the original components.

Reversible temperature coefficient A measure of the reversible changes in flux caused by temperature variations.

Rework A manufacturing operation that restores a part or an assembly to an operable condition. The reworked part/assembly should meet or surpass specifications.

Reynolds number The ratio of inertial and viscous forces in a fluid defined by the formula Re = rVD/μ, where: r = Density of fluid, μ = Viscosity in centipoise (CP), V = Velocity, and D = Inside diameter of pipe.

RF (radio frequency) An electromagnetic disturbance is a radio frequency if the wavelength falls within the range of 30 km to 1 mm. This represents a frequency rang of 10 kHz to 3000 GHz. The input signal from the antenna is an RF signal.

RF adapter A device that allows video and audio signals from a video tape recorder (VTR) or computer to be shown on a standard TV receiver. This device produces comparatively poor resolution and picture quality. Also called "RF converter."

RF amplifier An amplifier designed to increase RF signal amplitudes.

RF choke An inductor designed to have a high reactance when used in series with a signal-carrying lead.

RF control A medium of remote control from which signals are sent to the controlled equipment via data pulses modulated on an RF carrier signal.

RF modulator A device that converts a signal (typically audio and/or video) into a radio frequency.

RFID tag A microchip attached to an antenna that is packaged in a way that it can be applied to an object. The tag picks up signals from and sends signals to a reader. The tag contains a unique serial number, but may have other information, such as a customers' account number. Tags come in many forms, such smart labels that can have a barcode printed on it, or the tag can simply be mounted inside a carton or embedded in plastic. RFID tags can be active, passive or semi-passive.

Rheostat A variable resistor having two terminals, as distinguished from the potentiometer where three terminals are used.

Ridge pin A device that allows the mounting of a pin type insulator to a pole. The ridge pin is bolted to the top of the pole and the insulator is screwed onto the threads at its top.

Rigid flex Hybrid constructions which combine standard rigid circuit boards with flexible printed circuits, thereby reducing the component count, weight, and susceptibility to vibration of the circuit, and greatly increasing its reliability.

Rigid rotor A rotor is considered rigid when it can be corrected in any two (arbitrarily selected) planes and after that correction, its unbalance does not significantly exceed the balancing tolerances (relative to the shaft axis) at any speed up to maximum operating speed and when running under conditions which approximate closely to those of the final supporting system.

Rinse The removal of cleaning solutions, etchants or developers, etc. from the wafer using water. Rinsing stops the processes by removing the active chemical from the surface. There are several different methods of rinsing including overflow rinsing, dump rinsing and spin rinsing.

Ripple The periodic ac component at the power source output harmonically related to source or switching frequencies. Unless specified otherwise, it is expressed in peak-to-peak units over a specified band width.

Ripple frequency Frequency of the ripple present in the output of a DC source.

Ripple voltage The periodic ac component of the dc output of a power supply.

Rise time The time required for an output voltage of a digital cirucit to change from low voltage level (0) to high voltage leve 1., after the change has started. Very short rise times, not high clock speeds, are the primary cause of cross-talk in PCBs. Rise times are charactericstic of the technology being used in a circuit. Gallium Arsenide components can have rise times around 100-picoseconds (millionths of millionths of seconds), 30 to 50 times faster than some CMOS components.

Riser Risers are the cables between underground and overhead flow of electricity.

Riser pole A pole used to transition from overhead and underground cables.

Rising-edge A transition from a logic 0 to a logic 1. Also known as a positive edge.

Roadmap An international reference document of requirements, potential solutions, and their timing for the semiconductor industry. It identifies needs and encourages innovative solutions to meet future technical challenges, and provides an ongoing emphasis on obtaining consensus industry drivers, requirements, and technology timelines.

Roaming A method that enables subscribers of one wireless operator to use their handset in another carrier's service area. Customers cannot roam on a network unless their home carrier and the visited operator have a roaming agreement and a user has compatible equipment.

Rocking chair cell A lithium ion cell

Rolled annealed copper A type of copper used in manufacturing TAB tape. The copper is passed through a series of rollers to achieve a desired thickness, typically 1 ounce or 1.4 mils thick, and then annealed to remove the internal stresses created by the rolling process.

Rolling blackouts A controlled and temporary interruption of electrical service.

Rolloff A gradual decrease in response below or above some specified frequency.

Rolloff frequency A filter which has a reduced output as the frequency is increased. A tone control is a rolloff filter when turned down. For example, microphones usually have a bass rolloff filter to remove wind noise and/ or excessive breath pops.

Roll-off rate Rate of change in gain when an amplifier is operated outside of its bandwidth.

Rooming With large matrix switching systems, specific outputs can be assigned to a room. That room sees only those outputs, even though they are part of a total switching system. The outputs seen by a room have virtual numbers (may be different from the physical numbers). Room presets using those outputs can be saved and recalled without affecting other switcher outputs.

Rotary switch Electromechanical device that has a rotating shaft connected to one terminal capable of making or breaking a connection to one or more other terminals.

Rotor A rotor is a rotating body whose journals are supported by bearings.

Roughness Lack of planarity of solid surface at the atomic level; a parameter that measures lack of planarity; in high quality Si wafers better than 0.1 nm.

Route 1. A layout or wiring of a connection. 2. The action of creating such a wiring.

Router A data switch that handles connections between different networks. A router identifies the addresses on data passing through the switch, determines which route the transmission should take and collects data in so-called packets that are then sent to their destinations.

S

Safe operating area (SOA) A manufacturer specified power/time relationship that must be observed to prevent damage to power bipolar semiconductors.

Safety ground A conductive path from a chassis, panel or case to earth to help prevent injury or damage to personnel and equipment.

Safety vent A safety mechanism that is activated when the internal gas pressure rises above a normal level.

Sag The amount of vertical displacement of an overhead conductor between support points. Sag is a consideration when designing a pole or tower line and will be a determining consideration in the overall height of the structure. Sag varies with the temperature of the conductor and charts are available to determine it.

Salt atmosphere Exposure of sample devices to a salt rich environment to determine long-term durability of the package materials

Salt bridge The salt bridge of a reference electrode is that part of the electrode which contains the filling solution to establish the electrolytic connection between reference internal cell and the test solution. Auxiliary Salt Bridge: A glass tube open at oneend to receive intermediate electrolyte filling solution, and the reference electrode tip and a junction at the other end to make contact with the sample.

Salt effect (fx) The effect on the activity coefficient due to salts in the solution.

Sample A device or devices randomly chosen from a lot of material. Sampling assumes that randomly selected devices will exhibit characteristics during testing that are typical of the lot as a whole.

Sample and hold (S/H) circuit A circuit that acquires an analogue voltage and stores it temporarily in a capacitor; also referred to as a sample-and-hold amplifier (SHA).

Sample rate The rate at which an analogue signal is sampled. It is frequently expressed as kilosamples/sec (kS/s) or Megasamples/sec (MS/s. For example, 44.1 kHz is the standard sample rate for compact disks; 48 kHz is often used with digital audio tape (DAT) recording. A higher sample rate allows a higher frequency response. In order to accurately reconstruct a sound, the sample rate must be at least twice the highest frequency in the sound.

Sample volume or pulse volume The volume in which the radar data for one **range bin** are measured. Defined by the width of the radar beam (app. 1°) and half the length of the transmitted pulse. 1 μs pulses are 150 m deep, 2 μs pulses are 300 m deep.

Sampling Is the process of converting continuous signals into discrete values.

Sampling rate An A/D converter converts an analogue signal into a stream of digital numbers, each representing the analogue signal's amplitude at a moment in time. Each number is called a "sample." The number sample per second is called the sampling rate, measured in samples per second.

Saponifier An alkaline chemical added to water to improve its ability to dissolve rosin flux residues.

Sapphire Single-crystal Al2O3; can be synthesized and processed into various shapes; highly resistant chemically; transparent to UV radiation.

Satellite A radio relay station that orbits the Earth. A complete satellite communications system also includes earth stations that communicate with each other via the satellite.

Satellite communications A telecommunications service provided via one or more satellite relays and their associated uplinks and downlinks.

Satellite dish A kind of antenna used to pick up transmissions broadcast from a satellite.

Satellite link A radio link between a transmitting Earth station and a receiving Earth station through one satellite.

Saturable reactor A saturable reactor is an adjustable inductor in which the current versus voltage relationship is adjusted by control magnetomotive forces applied to the core.

Saturation Condition in which a further increase in one variable produces no further increase in the resultant effect. In a bipolar junction transistor, the condition when the emitter to collector voltage is less than the emitter to base voltage. This condition puts forward bias on the base to collector junction.

Savants A term used to describe distributed middleware designed by the Auto-ID Center to filter data from EPC readers and pass it on to enterprise systems. It was envisioned that Savants would reside on servers across the EPC Network and pass data to one another and act as a kind of nervous system for the network. The term is being phase out by EPCglobal and many of the functions of Savants are being incorporated in commercial middleware products.

Sawtooth voltage A waveform characterised by a gradual incline and an abrupt decline of amplitude, so that successive signals produce a sawtooth pattern.

Sawtooth wave Repeating waveform that rises from zero to maximum value linearly drops back to zero and repeats. A ramp waveform.

Scalar notation A notation in which each signal is assigned a unique name; for example, a3, a2, a1, and a0.

Scalar potential Mathematically, a scalar field whose negative gradient is a given vector field. If the scalar potential is denoted by the Greek letter ö and the vector field it generates as **v**, then **v** = -sö. The vector field can be a velocity field or a force field. Equation 1. therefore means a movement or acceleration towards the direction in which there will be a decrease of potential. Physically, the scalar potential is similar or identical to potential energy. Any conservative force field can be represented as the negative gradient of some scalar potential. Any lamellar field can be represented as having a scalar potential, but a solenoidal field generally does not have a scalar potential (except the degenerate case when it is Laplacian)

Scale The array of calibrated marks from which the input quantity may be read and interpreted.

Scale division The interval between the centers of two consecutive scale marks.

Scale length The length of the path of motion from one end of the scale as indicated by the tip of the pointer. In the case of knife-edge pointers and others, extending beyond the scale division mark, the pointer shall be considered as ending at the outer end of the shortest scale division marks. In multi-scale instruments the longest scale shall be used to determine the scale length. The case of "antiparallax" instruments of the step-scale type with graduations on a raised step into the plane of and adjacent to the pointer tip, the scale length shall be determined by the end of the scale divisions adjacent to the pointer tip.

Scaler A device that takes a standard video signal, decodes it, and uses advanced digital signal processing technology to scale the image to the optimal or native resolution of a display device. (Usually at a higher rate)

Scaling Scaling is changing the size of an image to fit the native rate (or pixel size) of a display device, without changing its shape. For example, to fit a 720 x 480 resolution TV image on a 1024 x 768 XGA resolution display, the TV image has to be scaled "up"; pixels need to be created in order for the original image to fill the screen. Alternately, to fit a 1280 x 1024 SXGA resolution image on an XGA resolution screen, the image will need to be scaled "down"; pixels need to be removed from the original image in order for it to fit on the screen. There are many different methods for image scaling, and some produce better results than others.

Scaling resistor A resistor added to an output circuit of measurement equipment to provide a scaled voltage output. The output is not a "true" voltage output and may be susceptible to loading errors.

Scan 1. In video, to move an electron beam across the raster in a camera or monitor.

2. To feed visual information into a computer by means of an optical device called a scanner.

Scan converter Also called "video converter" or "TV converter," a scan converter is a device that changes the scan rate of a source video signal to fit the needs of a display device. Examples: computer-video to NTSC (TV), or NTSC to computer-video.

Scan doubler A device used to change composite interlaced video to non-interlaced component video, thereby increasing brightness and picture quality. Also called "line doubler."

Scan-doubling The process of making the scan lines less visible by doubling the number of lines and filling in the blank spaces. Also called "line-doubling."

Scanner An electronic device that can send and receive radio waves. When combined with a digital signal processor that turns the waves into bits of information, the scanner is called a reader or interrogator.

Scanning electron microscope (SEM) A photo produced by a microscope that uses a beam of electrons to form a large image of a very small object.

Scart Also known as Euroconnector or Peritel, a 21-pin connector commonly used in Europe to interconnect satellite receivers, television sets, and other audiovisual equipment (e.g. videocassette recorders). A single connector combines audio and video signals. The name comes from "Syndicat des Constructeurs d'Appareils Radiorécepteurs et Téléviseurs."

Peritel is an abbreviation for "péritélévision." Peri is a prefix that means around or surrounding – in this case, it suggests the connection between the television and its electronic environment.

Scattering A change in the light wave passing through an optical fiber caused by an impurity or change of density in the fiber. This effect produces losses in the fiber.

Schematic diagram A diagram which shows, by means of graphic symbols, the electrical connections and functions of a curcuit.

Schmitt trigger A building block in which the input must be taken to about 70% of rail voltage before the output will change. The lower level for change is about 30%. This produces a gap and thus noise is prevented from entering.

Schottky barrier A potential barrier formed between a metal and a semiconductor, frequently used in the creation of Schottky diodes.

Schottky diode Hot carrier diode. A diode using an aluminium-silicon junction in which carrier storage is negligible, leading to very fast on and off states and thus very fast switching speeds. The forward voltages are 0.3v.

Scientific notation Numbers entered as a number from one to ten multiplied by a power of ten. Example: 8765 = 8.765 × 103.

Screen grid The electron-beam accelerating electrode in a tube. This grid is usually placed at signal ground to act as an electrostatic shield between input and output sections.

Screen printing The transfer of a pattern onto a surface by forcing a suitable material through a screen with a squeegee.

Screening 100% testing of a device, as opposed to sampling.

Scribe grid The area separating two adjacent dice on a wafer through which the scribing tool will pass during the separation of the wafer into individual dice.

Scrolling The displayed image (or interfering noise on the image) rolling constantly on the screen.

Scrubbing The process of vibrating two pieces of metal, or metal coated materials, at ultrasonic frequencies to create a friction weld.

SDTV compatible Televisions that are SDTV (Standard TV) compatible enable you to hook up digital component video inputs (Y, Pr, Pb) to the unit for viewing of standard definition programming broadcast in either 480i or 480p formats. In order to view SDTV programming, the TV must be connected to a digital TV set-up box and switched to DTV mode. A DTV signal must be present in your area and the set-up box must be tuned to the proper channel. An external over-the-air antenna may be required. If your cable company "passes through" a network DTV signal unchanged, a set-top box can decode and tune the program. However, standards for cable connectivity have not yet been finalized. Consequently, this product may not be compatible with your cable system.

Sealed cells A cell which remains closed and does not release gas or liquid when operated within the limits of charge and temperature specified by the manufacturer. An essential component in recombinant cells.

Sealed enclosure A type of speaker enclosure that does not allow the pressure generated by the back wave of the speaker to leave the enclosure.

Sealing The process whereby the lid is fastened onto a cavity (or hermetic) type semiconductor device. Sealing methods include solder (whereby a metal lid is soldered to a metal seal ring), weld (where a metal lid is welded to a metal package base), and glass (or ceramic) seal (where a ceramic lid is fastened to a ceramic base with reflowed glass).

Seamless switching A feature found on many video switchers. This feature causes the switcher to wait until the vertical interval to switch. This avoids a glitch (temporary scrambling) which normally is seen when switching between sources.

Sea-of-cells Popular name for a channel-less gate array.

Sea-of-gates Popular name for a channel-less gate array.

Search coil A coil conductor, usually of known area and number of turns that is used with a fluxmeter to measure the change of flux linkage with the coil.

Second The basic unit of measure of time, equivalent to "the duration of 9,192,631,770 periods of the radiation corresponding to the transition between the two hyperfine levels of the ground state of the cesium-133 atom." For our purposes, one "second" is 1/60th of a minute.

Second harmonic distortion Second harmonic distortion (HD2): Ratio of second-order harmonic to the input signal (carrier). Often measured as dBc.

Secondary Output winding of a transformer. Winding that is connected to a load.

Secondary battery A battery which can be recharged and used repeatedly.

Secondary cell Electrolytic cell used to store electricity. Once discharged may be restored by recharging by putting current through the cell in the direction opposite to that of discharge current.

Secondary circuit A circuit electrically isolated from the input or source of power to the device.

Secondary device A part of the flowmeter which receives a signal proportional to the flowrate, from the primary device, and displays, records and/or transmits the signal.

Secondary emission Electron emission from a nonheated electrode when struck by a high-velocity electron beam.

Secondary frequency standard A frequency standard that does not have inherent accuracy, and therefore must be calibrated against a primary frequency standard.

Secondary output An output of a switching power supply that is not sensed by the control loop.

Secondary standard pH buffer solutions which do not meet the requirements of primary standard solutions but provide coverage of the pH range not covered by primary standards. Used when the pH value of the primary standard is not close to the sample pH value.

Sectionalizer A sectionalizer is similar to a reclosure, but only opens when a line is "dead" due to the operations of a reclosure or breaker upstream. It serves to isolate the section in fault and allow the remainder of the circuit to remain energized.

Sectoral antenna A directive antenna with a radiation pattern aperture (3 dB beamwidth) larger than 45°. Sectoral antennas are generally used for point-to-multipoint systems or combined with several antennas to create a base station.

Secure hash standard This standard specifies a Secure Hash Algorithm, SHA-1, for computing a condensed representation of a message or a data file.

SED Surface-conduction electron-emitter display. SED is a new type of flat-panel display technology developed by Canon and Toshiba. SED is capable of high levels of brightness and colour performance, as well as a wide angle of visibility, that on a par with a CRT. Large screens can be produced by simply increasing the number of electron emitters in accordance with the required number of pixels.

Seebeck effect When a circuit is formed by a junction of two dissimilar metals and the junctions are held at different temperatures, a current will flow in the circuit caused by the difference in temperature between the two junctions.

Seebeck EMF The open circuit voltage caused by the difference in temperature between the hot and cold junctions of a circuit made from two dissimilar metals.

Seed crystals Crystal starting material introduced into a crucible of molten semiconducting material onto which atoms are attached resulting in the formation of a cylinder-shaped solid having the same crystal structure as the seed.

Seed value An initial value loaded into a linear feedback shift register or random number generator.

Segmented display A pattern of a specific number of segments arranged in a rectangular form to provide an alphanumeric display when energized.

Selected item drawing A drawing prepared for the purpose of procuring material which has been selected for parameters or characteristics differing from those to which the part is normally manufactured or tested.

Selectivity Selectivity refers to the ability of an etch process to only etch the layer which is to be patterned and not attack the layer underneath. A high selectivity is a good thing.

Self-alignment The tendency of certain slightly misaligned components (during placement) to self-align with respect to their land patterns during reflow soldering. It occurs due to the surface tension of molten solder.

Self-assembly* The ability of objects to assemble themselves into an orderly structure. Routinely seen in living cells, this is a property that nanotechnology may extend to inanimate matter.

Self biasing Gate bias for a field effect transistor in which source current through a resistor produces the voltage for gate to source bias.

Self discharge (battery) The decrease in the state of charge of a battery or cell, over a period of time, due to internal electro-chemical losses.

Self heating Internal heating of a transducer as a result of power dissipation.

Self inductance Property that causes a counter electromotive force to be produced in a conductor when the magnetic field expands or collapses with a change of current.

Self-planarizing head A mechanism integrated into the head of an outer lead bonder that allows the bottom surface of the thermode blades to adjust to the plane defined by the surface of the substrate.

Self-replication* The ability of an entity such as a living cell to make a copy of itself. Theoretically, nanotechnologists could invent self- replicating devices.

Self-terminating A switching configuration which automatically terminates a signal path when it is not connected to any other signal path. It is usually most important to terminate unused inputs to a unit to assist in reducing noise and improve crosstalk isolation.

Self timer Operates the camcorder for a specific time then stops recording, allowing the camera operator to appear in their own video.

Semi-con The semi-conducting material extruded over the insulation on medium voltage insulated cables. It is extruded onto the cable, normally at the same time as the insulation.

Semiconductor An electronic conductor whose resistivity at room temperature is in the range 10-2 to 109 ohm-em (which is between metals and insulators), in which the electrical charge carrier concentration increases with increasing temperature over some temperature range. Certain semiconductors possess two types of carriers, namely negative electrons and positive holes. The essential difference between a semiconductor and a metal is that, according to the band theory of solids, the number of free electrons in the former is very small, the energy bands being either entirely full or entirely empty, except for a few electrons and holes created by thermal excitation (intrinsic semiconductor) or by the presence of impurities. Important semiconducting materials include Si, Ge, Se, Cu 2 O, PbTe, PbS, SiC, etc., and their uses are manifold in such devices as rectifiers, modulators, detectors, thermistors, photocells and transistors.

Semiconductor device, multiple unit A semiconductor device having two or more sets of electrodes associated with independent carrier streams. It is implied that the device has two or more output functions which are independently derived from separate inputs, e.g., a duo-triode transistor.

Semiconductor device, single unit A semiconductor device having one set of electrodes associated with a single carrier stream. It is implied that the device has a single output function related to a single input.

Semiconductor, compensated A semiconductor in which one type of impurity or imperfection (e.g., donor) partially cancels the electrical effects of the other type of impurity or imperfection (e.g., acceptor).

Semiconductor, degenerate A semiconductor in which the number of electrons in the conduction band is so high that they must be described by Fermi-Dirac statistics, as in a metal.

Semiconductor, extrinsic A semiconductor with electrical properties dependent upon impurities.

Semiconductor, impurity A semiconductor whose properties are due to the presence of foreign atoms, giving rise to impurity levels. Such materials are very useful, since their electrical properties can be profoundly altered by changing the concentration and type of impurity.

Semiconductor, intrinsic A material which is naturally semiconducting, in the form of its pure, ideal crystal even when entirely free of impurities. All semiconductors have this property, but it may be very small compared with the impurity semiconductor, and only to be observed at high temperatures. It is to be described, according to the band theory of solids, by the thermal excitation of electrons from the filled band the whole width of the energy gap to the conduction band.

Semiconductor, n-type An extrinsic semiconductor in which the conduction electron density exceeds the hole density. It is implied that the net ionized impurity concentration is donor type.

Semiconductor, p-type An extrinsic semiconductor in which the hole density exceeds the conduction electron density. It is implied that the net ionized impurity concentration is acceptor type.

Semiconductors, III-V III-V Semiconductors are fabricated using elements from 3rd and 5th group of periodic table; e.g. GaAs, GaP, GaN, GaAlAs.

Semi-cutoff (lighting) Luminaire light distribution is classified as semi-cutoff when the candlepower per 1000 lamp lumens does not numerically exceed 50 (5.0%) at an angle of 90 degrees above nadir (horizontal), and 200 (20%) at a vertical angle of 80 degrees above nadir. This applies to any lateral angle around the luminaire.

Semi passive tag Similar to active tags, but the battery is used to run the microchip's circuitry but not to broadcast a signal to the reader. Some semi-passive tags sleep until they are woken up by a signal from the reader, which conserves battery life. Semi-passive tags can cost a dollar or more. These tags are sometimes called battery-assisted tags.

Sense amplifier An amplifier which is connected to the output voltage divider to determine, or sense, the output voltage. (bertan high voltage).

Sense line return The conductor which routes the voltage on the output return to the control loop.

Sense line The conductor which routes output voltage to the control loop.

Sense resistor A resistor placed in a current path to allow the current to be measured. The voltage across the sense resistor is proportional to the current that is being measured and an amplifier produces a voltage or current that drives the measurement.

Sensing element That part of the transducer which reacts directly in response to the input.

Sensitivity A standard way to rate audio devices like microphones, loudspeakers, and amplifiers. For loudspeakers, sensitivity is rated by applying 1 watt of power to the speaker and measuring sound pressure level (SPL) at one meter. For microphones, sensitivity is expressed as the minimum input signal required to produce a standard output level. For a power amplifier, sensitivity is rated as the input level required to produce 1 watt of power output into a specified load impedance, typically 4 ohms or 8 ohms.

Sensitivity shift A change in slope of the calibration curve due to a change in sensitivity.

Sensor A device that responds to a physical stimulus and produces an electronic signal. Sensors are increasingly being combined with RFID tags to detect the presence of a stimulus at an identifiable location.

Separately derived system A premises wiring system whose power is derived from a battery, a solar photovoltaic system, or from a generator, transformer, or converter windings, and that has no direct electrical connection, including solidly connected grounded circuit conductor, to supply conductors originating in another system.

Separates A speaker system with more than one type of driver. The most common type of separates system is a set containing two high frequency drivers (tweeters), two lower frequency drivers (mids or woofers), and two crossover networks (filters).

Separation The degree to which left and right channels in a stereo signal can be kept apart.

Separator A film or grid to separate 2 electrodes to prevent short-circuiting and retain electrolyte.

Sequencing The process that forces the order of turn on and turn off of individual outputs of a multiple output power supply.

Sequential A function whose output value depends not only on its current input values, but also on previous input values. That is, the output value depends on a sequence of input values.

Sequential access An access mode in which records are retrieved in the same order in which they were written. Each successive access to the file refers to the next record in the file.

Serial data A way to transfer information by breaking the characters of a word into bits, which are then transmitted sequentially along a single line. Compare to parallel, which uses more than one line.

Serial port An input/output connection on the computer that allows it to communicate with other devices in a serial fashion — data bits flowing on a single pair of wires. The serial port is used with RS-232 protocol.

Serial transmission Sending one bit at a time on a single transmission line. Compare with parallel transmission.

Serial-in parallel-out (SIPO) Refers to a shift register in which the data is loaded in serially and read out in parallel.

Serial-in serial-out (SISO) Refers to a shift register in which the data is both loaded in and read out serially.

Serialization Application of a unique alphanumeric identifier to each unit of a lot of devices to afford traceability to variables data, individual radiographs, etc.

Series 1. The interconnection of two or more power sources such that alternate polarity terminals are connected so their voltages sum at a load.

2. The connection of circuit components end to end to form a single current path.

Series circuit A circuit configuration in which a single current path is arranged among all components. Connecting the positive speaker output of an amplifier channel to the positive terminal of speaker # 1 , connecting the negative terminal of # 1 to the positive terminal of speaker # 2, and the negative terminal of # 2 to the negative output of the same amplifier channel is a series connection.

Series connection A wiring configuration used to increase voltage. Series wiring is positive to negative (+ to -) or negative to positive (- to +). Opposite of parallel connection.

Series parallel network Network that contains components connected in both series and parallel.

Series pass A controlled active element in series with a load that is used to regulate voltage.

Series regulator A regulator in which the active control element is in series with the dc source and the load.

Series resistor (impedance) A resistor (impedance) intended to be connected in series with an associated instrument in order to produce a higher voltage range that can be obtained by the instrument alone.

Series resonance Condition that occurs in a series LC circuit at the frequency where inductive reactance equals capacitive reactance. Impedance is minmum, current is maximum limited only by resistance in the circuit.

Series resonant circuit A circuit in which the components are connected in series and capable of oscillating at a particular frequency. It exhibits minimum impedance at resonance.

Series vs. parallel load resonance A crystal can be used in an oscillator circuit to operate in either of two resonant modes: Series Resonance or Parallel Load Resonance (also known as antiresonance). The crystals used in these two types of modes are physically the same crystal, but calibrated to slightly different frequencies.

When a crystal is placed into an oscillator circuit, they oscillate together at a tuned frequency. This frequency is dependent upon the crystal design and the amount of load capacitance, if any, the oscillator circuit presents to the crystal. Specified in picoFarads (pF), load capacitance is comprised of a combination of the circuit's discrete load capacitance, stray board capacitance, and capacitance from semiconductor Miller effects.

Series wiring A group of electrical devices, such as batteries or PV modules, wired together to increase voltage, while ampacity remains constant. Two 100 amp hour 12 Vdc batteries wired in series form a 100 amp hour 24 Vdc battery bank.

Serration pulse A vertical synchronizing pulse divided into a number of small pulses, each acting for the duration of half a line in a television system. Serration pulses are used to keep the horizontal oscillator synchronized during the vertical sync pulse interval.

Serration pulse removal switch A switch on certain interfaces that removes the incoming serration pulses from the computer video source.

Service conductors The supply conductors that extend from the street main or transformers to the service equipment of the premises being supplied

Service drop A service drop is the lines running to the customer's house. Usually, a service drop is made up of two 120 volt lines and a neutral line, from which the customer can obtain either 120 or 240 volts of power. When these lines are insulated and twisted together, the installation is called triplex cable.

Service drop Run of cables from the power company's aerial power lines to the point of connection to a customer's premises.

Service entrance cable The conductors that connect the service conductors (drop or lateral) to the service equipment of the building.

Service entrance conductors (overhead) The service conductors between the terminals of the service equipment and a point usually outside the building, clear of building walls, where joined by tap or splice to the service drop.

Service entrance conductors (underground) The service conductors between the terminals of the service equipment and the point of connection to the service lateral.

Service entrance equipment Service entrance equipment is the main control and means of disconnect for the supply of electricity to a building. Usually it consists of circuit breakers, switches, and fuses.

Developers and builders of residential houses will frequently include 200 amp service in their advertisements. This refers to the size of the service entrance equipment. It is listed as a feature to indicate that the house is adequately supplied in terms of electric energy. Older homes generally have 100 amp service.

Service equipment The necessary equipment, usually consisting of a circuit breaker or switch and fuses and their accessories, located near the point entrance of supply conductors to a building and intended to constitute the main control and cutoff means for the supply to the building.

Service lateral The underground service conductors between the street main, including any risers at a pole or other structure or from transformers, and the first point of connection to the service-entrance conductors in a terminal box, meter, or other enclosure with adequate space, inside or outside the building wall. Where there is no terminal box, meter, or other enclosure with adequate space, the point of connection is the entrance point of the service conductors into the building.

Service life (battery) The total period of usefull life of a battery, normally expressed in the total number of Charge/Discharge cycles.

Set point The temperature at which a controller is set to control a system.

Set-and-forget digital recording level Since a digital compact cassette recorder (DCC) can receive pure digital signals from CD players or other digital sources via coaxial and fibre optic digital inputs, there is no need to adjust an incoming record level for these sources.

Setting range The range over which the value of the stabilized output quantity may be adjusted.

Settling time The time required for a device or circuit's output voltage to settle and remain within a specified tolerance around the final value every time the output changes.

Seven segment display Device made of several light emitting diodes arranged in a numeric or alphanumeric pattern. By lighting selected segments numeric or alphabet characters can be displayed.

Sexagesimal Base-60 numbering system.

Shadow mask A metal plate with holes or vertical lines that is used to determine exactly where the electron beam strikes the CRT screen.

Shadowing The failure of molten solder to wet the leads of surface mount components due to their location on the board during wave soldering or the cause for insufficient heating of surface mount components due to their location on the board during infrared reflow soldering.

Shallow trench isolation A method for electrically isolating semiconductor devices that uses a dielectric filled trough.

Shaped response A frequency response that exhibits significant variation from flat within its range. It is usu-ally designed to enhance the sound for a particular application.

Sharpness control Same as Peaking control.

Shear modulus The ratio of the shear stress and the angular shear distortion.

Shear strength The force required to shear apart adhesive-bonded and cured materials and/or components.

Shear stress Where normal stress is perpendicular to the designated plane, shear stress is parallel to the plane.

Shearing strain A measure of angular distortion also directly measurable, but not as easily as axial strain.

Sheath The outer covering or jacket of a multiconductor cable.

Sheath thermocouple A thermocouple made out of mineral-insulated thermocouple cable which has an outer metal sheath.

Shedding The loss of material from the plates of Lead Acid batteries.

Shelf life The duration a cell can be kept in storage and still retain its ability to give a specified performance.

Shells or bands Orbital path containing a group of electrons having a common energy level.

Shield A metallic foil or braided wire layer surrounding conductors which is designed to prevent electrostatic or electromagnetic interference from external sources.

Shield effectiveness A measurement of how well the shielding material (braid, solid tape, etc.) protects the external environment from radiation produced by the center conductor.

Shield effectiveness The relative ability of a shield to screen out undesirable radiation. Frequently confused with the term shield percentage, which it is not.

Shielding A metal enclosure or gasket for a circuit, or a metal shield surrounding wire conductors (coaxial or triaxial cable) to lessen interference, interaction, or current leakage. The shield is usually grounded.

Shift register Two or more bistable elements (flip-flops) connected in series. With each tick of the clock, the output of stage n is shifted to stage n+1. Applications include clock or signal delays, delay lines, linear-feedback shift registers.

Shipping tube In reference to molded carrier ring or TAB components in slide carriers, it is the tube in which components are vertically stacked for shipping and feeding into excise and form feeders.

Ship-to-stock A system whereby a component user develops a statistical evaluation of the quality of his vendors' products, enabling him to use the products of those vendors whose quality is acceptable without performing additional product testing upon receipt at his plant.

Shmoo plot 1.A plot of pass/fail test results that involves pairs of test parameters such as frequency vs. voltage, or voltage vs. temperature;

2. a one-axis or two-axis pass/fail plot of a series of measurements.

Shock hazard A potentially dangerous electrical condition that may be further defined by various industry or agency specifications.

Shock sensor An acceleration sensor, generally a piezoelectric type, that can measure high acceleration but cannot measure static g forces.

Short An unwanted connection between conductor paths.

Short circuit A short circuit is an accidentally established connection between two points in an electric circuit such as when a tree limb or an animal bridge the gap between two conductors. This will cause an overload of current on the line causing melting of lines, blown fuses, and the faulty operation of protective devices such as reclosures and circuit breakers.

Short circuit current (SCC) The initial value of the current obtained from a power source in a circuit of negligible resistance.

Short circuit protection A protective feature that limits the output current of a power supply to prevent damage.

Short circuit test A test in which the output is shorted to ensure that the short circuit current is within its specified limits.

Short distribution (lighting) A luminary is classified as having a short light distribution when its max candlepower point falls between 1.0MH - 2.25MH TRL. The maximum luminaire spacing-to-mounting height ratio is generally 4.5 or less.

Short short circuit 1. An abnormal connection of relatively low resistance between two points of a circuit. The result is excess (often damaging) current between these points. Such a connection is considered to have occurred in a printed wiring CAD database or artwork anytime conductors from different nets either touch or come closer than the minimum spacing allowed for the design rules being use.

Short wavelength red laser Supplies the shorter wavelength needed to match the tiny DVD pits. The laser has a noise reduction function to suppress interference and it exhibits stable oscillation at high temperature (60 degrees Celsius).

Shotgun stick A specialized hot stick that allows the capture of certain types of clamps and devices in its hook. It is also called a "Grip All" stick.

Shrink The reduction of die size through conversion to a process within the same basic process family, but with smaller feature sizes.

Shrouded terminals Terminals surrounded by an insulating shroud which prevents accidental contact with the terminal.

Shuffle play The CD player will play the songs from one CD in a random order. Some CD players offer another type of shuffle play, where the songs from all the CDs in the deck are played in random order.

Shunt 1. A parallel conducting path in a circuit. 2. A low value precision resistor used to monitor current.

Shunt capacitance Abbreviated as C0. A parameter associated with a quartz crystal unit, used to identify the capacitance resulting from the presence of the electrodes plus stray capacitance associated with the holder. The static capacitance between the crystal terminals. Measured in picoFarads (pF), shunt capacitance is present whether the device is oscillating or not (unrelated to the piezoelectric effect of the quartz). Shunt capacitance is derived from the dielectric of the quartz, the area of the crystal electrodes, and the capacitance presented by the crystal holder.

Shunt leads Leads that connect a circuit of an instrument to an external instrument shunt. The resistance of these leads is taken into account in the adjustment of the instrument.

Shunt regulator A voltage regulator which uses a transistor or FET, in parallel with the load, which shorts out the excess voltage when the applied input voltage exceeds a specified limit producing a regulated output voltage. It is a simple but lossy design.

Shunt resistor Resistor connected in parallel or in shunt with another component or circuit.

Shutdown The termination of the running of an operating system.

Shutter speeds Variable speeds at which the camcorder captures images. The faster the speed, the less blurred the playback image will be. Ideal for sporting events where slow motion or stop action is important.

Shuttlecock cell A lithium ion cell.

SI (System internationale) The name given to the standard metric system of units.

Side brazed dip A dual-in-line package with leads brazed to feedthroughs on the side of the package. Leads will normally come straight down rather than at angles as on other dual-in-line packages. No molded versions of this package are available.

Sideband The bands of frequency above and below a carrier frequency that are produced by its modulation.

Side-emitting laser diode A laser diode constricted at the edge of an integrated circuit's substrate such that, when power is applied, the resulting laser beam is emitted horizontally; that is, parallel to the surface of the substrate.

Sidewalk (lighting) Paved or otherwise improved areas for pedestrian use, located within public street rights-of-way also containing roadways for vehicular traffic.

Sidewall pressure The force exerted on a cable as it is dragged around a bend. The longer the pull and the tighter the bend radius, the higher the sidewall pressure will become. High sidewall pressure damages cable. There is a higher chance of destroying cable by high sidewall pressure than by high tensile tension. Electrical cable manufacturers' specifications limit sidewall pressure to a range of 300 - 1500 lbs/ft, depending on cable type. There are three ways to reduce sidewall pressure while pulling cable with bends: shorten the pull, enlarge the bend radius, or use a high performance lubricant. Polywater's "Pull Planner 2000" software package automatically calculates sidewall pressure for each segment of a cable pull (a "segment" contains one straight section and one bend). This allows a designer to play "what if," easily changing pull parameters and instantly seeing the affect on sidewall pressure.

SiGe Abbreviation for silicon-germanium.

Sigma delta An ADC architecture consisting of a 1-bit ADC and filtering circuitry which over-samples the input signal and performs noise-shaping to achieve a high-resolution digital output. The architecture is relatively inexpensive compared to other ADC architectures.

Sigma-delta converter An ADC that uses a single-bit quantizer with a very high sampling rate combined with a single-bit DAC in a feedback loop.

Sign bit The most significant binary digit, or bit, of a signed binary number. If set to a logic 1, this bit represents a negative quantity.

Signal An electrical transmittance (either input or output) that conveys information.

Signal attenuation The weakening of RF energy from an RFID tag or reader. The energy emitted by the reader naturally decreases with distance. The rate of decrease is proportional to the inverse square of the distance. Passive UHF RFID tags reflect back a signal at very low power levels. A tag's reflected signal decreases as the inverse fourth power of the distance between tag and reader. Attenuation can be increased by external factors as well. For instance, water absorbs UHF energy, causing signal attenuation.

Signal conditioner A device placed between a signal source and a readout instrument to change the signal. Examples are attenuators, preamplifiers, charge amplifiers, and sophisticated level-translating deivices that can compensate for non-linearities in the sensor or amplifier.

Signal conditioning Amplifying, filtering, or otherwise processing a signal.

Signal frequency shift Any change in frequency. Any change in the frequency of a radio transmitter or oscillator.

Signal generator A circuit that produces a variable and controllable signal.

Signal ground The common return or reference point for analogue signals.

Signal layer A layer carrying tracks in a circuit board, hybrid, or multichip module.

Signal loss A video problem that shows up as a faint picture for lack of video information.

Signal processing Electronic functions that enhance the representations of physical or electrical phenomena. Temperature, pressure, vibration, acceleration and flow are examples of physical properties that rely on signal processing enhancements. The detection and conversion of RF, X-ray or ultrasonic energy into images and sound is another form of signal processing.

Signal to noise ratio (S/N) Relative power of the signal to the noise. As the ratio decreases on a line, it becomes more difficult to distinguish between information and non-information (noise). S/N deteriorates with distance because the noise builds every time the signal is repeated.

Signature analysis A guided-probe functional-test technique based on signatures.

Signature Refers to the checksum value from a cyclic-redundancy-check when used in the guided-probe form of functional test.

Signed binary number A binary number in which the most-significant bit is used to represent a negative quantity. Thus, a signed binary number can be used to represent both positive and negative values.

Silicide A compound formed when silicon or polysilicon reacts with a metal.

Silicon A brittle, grey, crystalline chemical element which, in its pure state, serves as a semiconductor substrate in microelectronics. It is naturally found in compounds such as silicon dioxide.

Silicon bumping The process of depositing additional metalization on a die's pads to raise them fractionally above the level of the Barrier Layer.

Silicon chip Although a variety of semiconductor materials are available, the most commonly used is silicon and integrated circuits are popularly known as silicon chips, or simply chips.

Silicon compiler The program used in compiled cell technology to generate the masks used to create components and interconnections. May also be used to create data-path functions and memory functions.

Silicon controlled switch An SCR with an added terminal called an anode gate. A positive pulse either at the anode gate or the cathode gate will turn the device on.

Silicon dioxide, SiO2 Silica; native oxide of silicon; the most common insulator in semiconductor device technology; high quality films are obtained by thermal oxidation of silicon; thermal SiO2 forms smooth, low-defect interface with Si; can be also readily deposited by CVD; Key parameters: energy gap Eg ~ 8eV; dielectric strength 5-15 x 106 V/cm; dielectric constant k = 3.9; density 2.3 g/cm3; refractive index n =1.46; melting point ~ 1700°C; prone to contamination with alkali ions and sensitive to high energy radiation (i.e. X-rays); single crystal SiO2 is known as quartz.

Silicon gate An MOS process fabricated with a polysilicon material as the gate electrode.

Silicon germanium A compound of the elements of Silicon and Germanium. SiGe is used for high speed applications such as communications transitions as well as an enhancement for the transistors in advanced logic devices.

Silicon nitride Typically used as a passivation, masking or insulating layer.

Silicon tetrachloride A colourless, corrosive, nonflammable liquid that forms hydrogen chloride when combined with water. It is used for epitaxial deposition of single-crystal silicon and for high-temperature silicon dioxide CVD.

Silicon tetrafluoride A colourless, corrosive, gas used in its pure state and in oxygen mixtures to etch silicides, silicon, adn polysilicon films. Silicon Tetrafluoride disassociates in the presence of an RF field to produce reactive fluoride ions which react with and etch silicon films.

Silicon transistor A bipolar junction transistor using silicon as the semiconducting material.

Silicon wafer A thin, iridescent, silvery disk of silicon which contains a set of integrated circuits, prior to their being cut free and packaged. A silicon wafer will diffract reflected light into rainbow patterns and, being a similar size, looks so much like a music CD that it could be mistaken for one (except that it has no label or hole in the middle). On closer inspection, one can see the individual (usually rectangular- or square-shaped) integrated circuits which form a uniform patchwork quite unlike the surface of a music CD. When cut or etched from the wafer these circuits are then called chips or dice.

Silicon-controlled rectifier (SCR) Three terminal active device that acts as a gated diode. The gate terminal is used to turn the device on allowing current to pass from cathode to anode.

Silicon-on-insulator (SOI) A sandwich-like substrate in which the silicon surface is electrically isolated from the substrate by an insulator.

Silver (Ag) Precious metal that does not easily corrode and is more conductive than copper.

Silver chromate test A qualitative check for the presence of ionic halides in RMA fluxes.

Silver mica capacitor Mica capacitor with silver deposited directly onto the mica sheets instead of using conductive metal foil.

Silver solder Solder composed of silver, copper and zinc. Has a melting point lower than pure silver, but higher than lead-tin solder.

Simple hybrid A hybrid device having an inner seal perimeter (i.e., a cavity perimeter) of less than 2 inches.

Simplex Communication in only one direction at a time.

Simplexed circuit A circuit used to transmit two signals simultaneously.

Simulated stereo surround mode With the Simulated Stereo Surround Mode, one of several different modes in audio receivers and systems shape sound to more closely resemble a live performance. Simulated settings provide a simulated spatial surround sound when playing mono source material.

Simulation The use of a CAD tool that mimics the behaviour of the actual circuit through a collection of sub-circuit models enabling designers to test, verify, and debug circuits before layout and fabrication.

Simulcast Broadcasting a program simultaneously in two different forms, for example a program broadcast in both AM and FM.

Simultaneous record and play Users can watch previously recorded material on a DVD recorder while recording current material at the same time.

Sine Sine of an angle of a right angle triangle is equal to the opposite side divided by the hypotenuse.

Sine wave A smooth, continuously moving waveshape that has no break in its appearance. It has positive and negative half-cycles that are generally symmetrical with respect to a reference. The cyclical repetition of these waves produces a waveshape that has a specified frequency in hertz (number of cycles per second) and a specified amplitude.

Single break/single make Contacts that open and close a circuit at only one place.

Single chip package (SCP) A package that supports a single microelectronics device so that its electrical, mechanical, thermal and chemical performance needs are adequately served.

Single in-line package Package containing several electronic components (generally resistors) with a single row of connecting pins.

Single layer TAB tape Tape constructed with a single conductive metal layer, typically 70 micron thick copper. This type of tape is generally used in low lead count applications, and is not commonly seen.

Single phase line A single phase line carries electrical loads capable of serving the needs of residential customers, small commercial customers, and streetlights. It carries a relatively light load as compared to heavy duty three phase constructs.

Single pole - SP (switches) A switch device that opens, closes, or changes the connection of a single conductor in an electrical circuit.

Single pole double throw A SPDT switching element has one normally open, one normally closed and one common terminal. Three terminals mean that the switch can be wired with the circuit either normally open (N/O) or normally closed (N/C).

Single pole single throw (SPST) Two terminal switch or relay thet can open or close one circuit.

Single pole switch A switch in which only one circuit is controlled.

Single precision The degree of numeric accuracy that requires the use of one computer word. In single precision, seven digits are stored, and up to seven digits are printed. Contrast with double precision.

Single sideband (SSB) AM radio communication technique in which the transmitter suppresses one sideband and therefore transmits only a single sideband.

Single throw (ST) A switch that opens, closes or completes a circuit at only one of the extreme positions of its actuator.

Single throw switch Switch containing only one set of contacts which can be either opened or closed.

Single wafer process Only one wafer is processed at the time; tools that are designed specifically for single-wafer processing become more common as wafer diameter increases.

Single-crystal silicon Silicon material with a single crystal structure. A common material for the construction of solar PV cells.

Single-crystal structure A material having a crystalline structure such that a repeatable or periodic molecular pattern exists in all three dimensions.

Single ended An unbalanced circuit where one side of the circuit or transmission line is grounded. Single-ended audio is unbalanced audio.

Single ended input A signal-input circuit where SIG LO (or sometimes SIG HI) is tied to METER GND. Ground loops are normally not a problem in AC-powered meters, since METER GND is transformer-isolated from AC GND.

Single frequency interference Interference caused by a single-frequency source.

Single in line package A package (either hermetic or molded) with all its leads emanating from one side of the package.

Single layer board A printed circuit board that contains metallized conductors on one side of the board only.

Single phase This implies a power supply or a load that uses only two wires for power. Some "grounded" single phase devices also have a third wire used only for a safety ground, but not connected to the electrical supply or load in any other way except for safety grounding.

Single plane (static) balancing machine A single plane balancing machine is a gravitational or centrifugal balancing machine that provides information for accomplishing single plane balancing.

Single sided A printed circuit board with tracks on only one side.

Sink Device such as a load that consumes power or conducts away heat.

Sink current The amount of conventional current flow into a DUT pin with a pull-up load.

Sink voltage The voltage at a DUT pin acting as a current sink.

Sintered plate (battery) The plate of a alkaline cell, the support of which is made of sintered metal powder, and into which the active material is introduced.

Sintering Heating a mixture of powdered metals, sometimes under pressure, to the melting-point of the metal in the mixture which has the lowest melting-point, the melted metal binding together the harder particles.

Sinusoidal Varying in proportion to the sine of an angle or time function. AC voltage in which the instantaneous value is equal to the sine of the phase angle times the peak value.

Skew 1.An observed difference in time between two events that occur simultaneously;

2. the effect of different propagation delays along an electrical path or channel.

Skew-free A reference to special twisted pair cable in which the length differences between cables reduced to a minimum, thus reducing cable skew.

Skin effect In the case of high frequency signals, electrons are only conducted on the outer surface, or skin, of a conductor. This phenomenon is known as the skin effect.

Skip bond A bond placed between two metallization traces on a hybrid substrate.

Skip search Enables you to skip through unwanted tape segments at 30-second increments.

Sleep timer Used to set a time period after which the unit automatically turns off.

Slew rate The ability of audio equipment to reproduce fast changes in amplitude. Measured in volts per microsecond, this specification is most commonly associated with amplifiers, but applies to most types of audio products. In amplifiers, a low slew rate "softens" the attack of a signal, "smearing" the transients and sounding "mushy." Since high frequencies change in amplitude the fastest, this is where slew rate is most critical. An amp with a higher slew rate will sound "tighter" and more dynamic.

Slide carrier A carrier for handling singulated TAB components. The slide carrier looks and functions similarly to the outer ring of a molded carrier ring component. It serves to facilitate component test, burn-in, and mechanical registration in a set of tooling. Slide carriers come in three basic sizes, 35mm, 48mm, and 70mm.

Slide switch Switch having a sliding button, bar or knob.

Slotted antenna An antenna that consists only of a narrow slot cut into an electrical conductor connected to the transponder. Slotted antennas exhibit the same orientation sensitivity as dipoles.

Slow acting relay Slow operating relay that when energized may not pull up the armature for several seconds.

Slow break/slow make Switches designed to make or break circuits within e 8-12 milliseconds. Typically used for AC applications.

Slow charge Charging overnight in 14 to 16 hours at about 0.1C rate. Safe and simple.

Slow start A feature that ensures the smooth, controlled rise of the output voltage, and protects the switching transistors from transients when the power supply is turned on.

Slow blow fuse Fust that can withstand a heavy current (up to ten times its rated value) for a small period of time before it opens.

Slow wave helical resonator A quarter wavelength- or one-half wavelength-mode resonant circuit constructed as a single-layer cylindrical inductor wound in the form of a helix (called a solenoid), with capacitive end-loading. The velocity of a wave disturbance moving along the helix axis is significantly less than that of an electromagnetic wave in free space, making it an inefficient structure for launching space waves. A Tesla coil's secondary winding and a magnifying transmitter's base-driven extra coil both behave as slow-wave helical resonators. For a close-wound coil, the disturbance propagates at an axial velocity of about 0.1% the speed of light in free space. The axial velocity of the disturbance is established by the pitch of the coil and the speed at which the exciting electrical charge propagates through the wire itself.

Slump A spreading of material (solder paste, adhesive,thick film, etc.) after stencil printing (or dispence) but before curing.

Slurry A highly engineered chemical formulation – typically comprised of fumed silica abrasives – that smoothes and flattens the wafer surface and prepares it for subsequent photolithography steps in which the pattern of the IC is imaged onto the wafer.

SMA A small type of threaded coaxial signal connector typically used in higher frequency applications. This connector is typically usable to 26GHz.

Small outline integrated circuit (SOIC) An integrated circuit surface mount package with two parallel rows of gull-wing leads.

Small outline j-leaded (SOJ) An integrated circuit surface mount package with two parallel rows of J-leads.

Small outline transistor (SOT) Discrete surface mount transistors with a molded plastic outline that serve small and medium power applications.

Smallest bending radius The smallest radius that a strain gage can withstand in one direction, without special treatment, without suffering visible damage.

Small-scale integration (SSI) Refers to the number of logic gates in a device. By one convention, small-scale integration represents a device containing 1 to 12 gates.

Smart antenna An antenna system whose technology enables it to focus its beam on a desired signal to reduce interference. A wireless network would employ smart antennas at its base stations in an effort to reduce the number of dropped calls, improve call quality and improve channel capacity.

Smart battery An intelligent battery which contains information about its specification, its status and its usage profile which can be read by its charger or the application in which it is used.

Smart cards A credit card or other kind of card with an embedded microchip. When the card uses RFID technology to send and receive data it is called a contactless smart card.

Smart cut Process used to fabricate bonded SOI substrates by cleaving the top wafer close to the desired thickness of the active layer; before bonding one wafer is implanted with hydrogen to a depth that will determine the thickness of an active layer in the future SOI wafer; following bonding, the wafer is annealed (at ~500 oC) at which time the wafer splits along the plane stressed with implanted hydrogen. The result is a very thin layer of Si forming a SOI substrate. .

Smart label A generic term that usually refers to a bar code label that contains an RFID transponder. It's considered "smart" because it can store information, such as a unique serial number, and communicate with a reader.

Smart phone A phone with a microprocessor, memory, screen, and built-in modem. The smart phone combines some of the capabilities of a PC in a handset device and typically include Internet connectivity.

Smart signal conditioner Signal conditioner that is programmable or has a flexible architecture to allow it to accomplish sophisticated signal transformations and corrections.

Smartphone A combination of mobile phone and personal digital assistant.

Soft bake The lower-temperature bake step, prior to the expose step, that evaporates the solvent from the photoresist layer.

Soft copy An electronic form of a document; a data file in computer memory or stored on storage media. When one is looking at a soft copy he is viewing the document as displayed on a computer monitor.

Soft error An error of a nonpermanent nature introduced into a cell or cells of a memory device as a result of either voltage or current transients or radiation exposures. Although soft errors may be corrected, the data previously stored in the cell may not be retrievable.

Soft macro (macro function) A logic function defined by the manufacturer of an application-specific integrated circuit. The function is described in terms of the simple functions provided in the cell library and connections between them. The assignment of cells to basic cells and the routing of the tracks is determined at the same time, and using the same tools, as for the other cells specified by the designer.

Soft magnetic material Ferromagnetic material that is easily demagnetized.

Solar cell Photovoltaic cell that converts light into electric energy. Especially useful as a power source for space vehicles.

Solar energy Energy from the sun.

Solar heating Solar heating is heat created by the energy of the sun. Solar heat can be either active or passive. Passive solar heating, as the name suggests, takes advantage of the heat created through natural means (heat created when sunlight passes through a window and becomes trapped inside a building). Active solar heating systems are made up of three components: the solar collector, energy storage, and distribution pipes or ducts. Sunlight is collected by absorber panels, conditioned if required, and distributed through the building by a heat transfer fluid or by air.

Solar hot water Solar hot water is a similar setup to an active solar heating system except it is used to preheat water used principally for normal domestic hot water use. Collectors trap sunlight creating heat. Transfer fluid moves heat from collectors to water holding tanks. This system requires a supplemental heating source for days when there is inadequate sunlight to heat water to required temperatures.

Solar panel A collection of solar modules connected in series, in parallel, or in series-parallel combination to provide greater voltage, current, or power than can be furnished by a single solar module. Solar panels can be provided to furnish any desired voltage, current, or power. They are made up as a complete assembly. Larger collections of solar panels are usually called solar arrays.

Solar power Electricity generated from sunlight.

Solder An alloy of tin and lead with a comparatively low melting point used to join less fusible metals. Typical solder contains 60% tin and 40% lead - increasing the proportion of lead results in a softer solder with a lower melting point, while decreasing the proportion of lead results in a harder solder with a higher melting point. Note that the solder used in a brazing process is of a different type, being a hard solder with a comparatively high melting point composed of an alloy of copper and zinc (brass).

Solder balls Small spheres of solder adhering to the laminate, mask, or conductor surfaces usually after wave or reflow soldering.

Solder bridging Solder paste or solder on two or more adjacent pads that come into contact to form a conductive path (forming a bridge).

Solder bumping A flipped chip technique in which spheres of solder are formed on the die's pads. The die is flipped and the solder bumps are brought into contact with corresponding pads on the substrate. When all the chips have been mounted on the substrate, the solder bumps are melted using reflow soldering or vapor-phase soldering.

Solder bumps Round solder balls bonded to the pads of components and subsequently used for face-down bonding techniques.

Solder connection The joining of two or more metal parts by means of an electrical or mechanical connection.

Solder dip Dipping of the leads of devices into molten solder in order to later facilitate soldering of the devices to circuit boards. Sometimes inaccurately referred to as "tin dip.

Solder mask A dielectric material used to cover the entire surface (except where the joints are to be formed) of the PCB primarily to protect the circuitry from environmental damage. Solder mask also helps to reduce bridging.

Solder mask over bare copper (SMOBC) A technique in which the solder mask is applied in advance of the tin-lead plating. This results in lighter circuit boards because the tin-lead alloy is only used to plate the pads.

Solder Metallic alloy used to join two metal surfaces.

Solder paste A homogenous and kinetically stable mixture of minute spherical solder particles, flux, solvents and binder that is screen printed onto the printed circuit board and then reflowed to form the solder joints.

Solder powder The solder alloy in solder paste exists in the form of powder. Solder powder is the major ingredient that affects the printability of the paste and the quality of the solder joint.

Solder seal Semiconductor device sealing accomplished by soldering a metal lid to a metal seal ring.

Solder thickness The amount of solder deposited on a pad for reflow. Optimum thickness will vary with pad size and pitch, but must be consistent across a single bonding site.

Solder wicking The capillary movement of molten solder onto a pad or component lead or between metal surfaces, such as strands of wire.

Solderability The ability of a conductor to be wetted by solder and to form a strong bond with the solder.

Solderability testing Immersion of the leads of sample devices in solder, followed by visual inspection to determine that the quality of the finish is such that it will accept an even coating of solder.

Soldering A process of joining metallic surfaces with solder, without melting the base material.

Soldering iron Tool with an internal heating element used to heat surfaces being soldered to the point where the solder becomes molten.

Solenoid An air core coil. Equipped with a movable iron core the solenoid will produce motion. As a result of current through the coil the iron core is pulled into the center of the winding. When the coil is deenergized, a spring pulls the movable core away from the center of the winding. Machanical devices connected to the movable core are made to move as a result of current through the coil. Example: Electric door locks on some automobiles.

Solid A physical state of matter in which the motion of molecules is restricted. A solid has a definite shape and volume.

Solid conductor Conductor having a single solid wire instead of strands of fine wire twisted together.

Solid state battery Cells with solid electrolytes. Lithium polymer cells are examples of this technology.

Solid state Refers to the electronic properties of crystalline material, as opposed to vacuum and gas-filled tubes that transmit electricity. Compared with earlier vacuum-tube devices, solid-state components are smaller, less expensive, more reliable, use less power and generate less heat.

Solid state relay A relay that switches electric circuits by use of semiconductor elements without moving parts or conventional contacts.

Solid state sensor The use of chip technology (MOS or Metal Oxide Semi-Conductor and CCD or Charged Couple Device) to produce images. Pixels are the individual sensors making up the solid state device.

Solids content The metal powder content as a percentage of the mass of the wet solder paste, or the percentage by weight of rosin in a flux formulation.

Solid-state device An electronic device that is built on or within a solid material, usually a semiconductor material. Contrast this to a vacuum tube which requires electron movement between electrodes in a vacuum/ low pressure environment.

Solvation Ions in solution are normally combined with at least one molecule of solvent. This phenomenon is termed solvation.

Solvent A solution capable of dissolving a solute.

Solvent cleaning A cleaning method employing chlorinated and fluorinated hydrocarbon liquids.

Solvent extraction The removal of one or more components from a liquid mixture by intimate contact with a secondary liquid that is nearly insoluble in the first liquid and which dissolves the impurities and not the substance to be purified.

Sonar Acronym for "sound navigation and ranging." A system using reflected sound waves to determine the position of some target.

SONET synchronous optical network A North American standard for transmission in synchronous optical networks. It defines a family of rates, formats, interfaces, transport options, and maintenance capabilities. The minimum rate for SDH is 155Mbps.

Sonic Pertaining to sound.

Sonotrode Horn which couples the ultrasonic energy from the generator to the tool.

SOP system-on-package A single component, multi-function, multi-chip package providing all the needed system-level functions. Functions include analogue, digital, optical, RF and MEMS.

SOS (silicon-on-sapphire) Special case of SOI where an active Si layer is formed on top of a saphire substrate (an insulator) by means of epitaxial deposition; due to a slight lattice mismatch between Si and sapphire, Si epitaxial layers larger than the critical thickness have a high defect density.

SOT Small outline transistor

Sound A wave propagated in air producing an auditory sensation in the ear by the change of pressure at the ear.

Sound chain The series of interconnected audio equip-ment used for recording or PA.

Sound level meter An instrument designed to measure sound pressure level.

Sound pressure level (SPL) An acoustic measurement of sound energy, typically expressed in dB SPL.

Sound reinforcement Amplification of live sound sources.

Sound wave Pressure waves propagated through air or other plastic media. Sound waves are generally audible to the human ear if the frequency is between approximately 20 and 20,000 vibrations per second. (hertz)

Soundstage A sonic recreation of an environment in three dimensions: front to back, side to side and up and down. In a home theatre, the soundstage's sonics should match the visuals that appear on the TV screen in terms of placement, size and imaging.

Source 1. Any device that produces electrical energy, or current.

2. Electrical contact region of an MOS transistor device to which electricity is supplied.

Source code A non-executable program written in a high-level language. A compiler or assembler must translate the source code into object code (machine language) that the computer can understand and process.

Source control drawing Similar to a specification control drawing except that purchase of the specified devices from manufacturing sources other than those listed on the drawing is prohibited.

Source current The amount of conventional current flow out of a DUT pin with the pull-down load.

Source follower Amplifier in which signal is applied between gate and drain with output taken between source and drain. Also called "common drain."

Source impedance Impedance through which output current is taken from a source.

Source inspection Surveillance or inspection by a customer's quality representative or by a government inspector at the vendor's facility of material being assembled or screened by that vendor.

Source voltage The voltage at a DUT pin that is acting as current source.

Source voltage effect The change in stabilized output produced by a specified primary source voltage change.

South pole Pole of a magnet into which magnetic lines of force are assumed to enter.

SP Service Provider

Space Used to refer to the width of the gap between adjacent tracks.

Space charge Electrons, protons, and ions moved around in the dielectric of a capacitor by the applied voltage. Charge tends to build up at "discontinuities" in the dielectric, such as the dielectric-electrode interface in film capacitors, at the grain boundaries in crystalline dielectrics, and at various molecular sites simply called "charge traps". Space charge is a major factor in dielectric behavior, such as leakage, high-voltage reliability, and dissipation factor vs. frequency and vs. temperature. Its role in dielectric aging is still the subject of research. Researchers have shown a special interest in cross-linked polyethylene (XLPE) because of its widespread use as a high-voltage insulation in power transmission, but work is also being done with polyethylene napthalate, PMMA, silicon dioxide, and other common dielectrics. Space charge has been extensively studied for many years, but I can´t remember any books or journal articles specifically relating space charge phenomena to capacitor behavior. It may be that capacitor people and dielectric theorists don´t talk to each other much. When a metal object is placed in a vacuum and is heated to incandescence, the energy is sufficient to cause electrons to "boil" away from the surface atoms and surround the metal object in a cloud of free electrons. This is called thermionic emission. The resulting cloud is negatively charged, and can be attracted to any nearby positively charged object, thus producing an electrical current which passes through the vacuum. Space charges can also occur within a solid, liquid, or gas dielectric. For example, when gas near a high voltage electrode begins to undergo dielectric breakdown, electrical charges are injected into the region near the electrode, forming space charge regions in the surrounding gas. Space charges can also occur within solid or liquid dielectrics that are stressed by high electric fields. Trapped space charges in solid dielectrics is often a contributing factor for dielectric failures within high voltage capacitors and power cables. In a plasma, a net charge which is distributed through some volume. Most plasma are electrically neutral or at least quasineutral, because any charge usually creates electric fields which rapidly move surplus charge out of the plasma. However, in some applications one wishes to apply external electric fields to the plasma, and a net space charge can be produced as a result. The resulting space charge must often be accounted for in the physics of these sorts of devices.

Space transformer Abbreviated ST. A major component of certain high-density probe cards . It provides pitch reduction, high routing density and localized mid-frequency decoupling. A major developer of ATE systems which use space transformers is Wentworth Labs. .

Space waves Radio waves or electromagnetic radiation emitted by a 1/2-wave electric dipole antenna in free space. They can reach the receiver either by ground-wave or sky-wave propagation.

Spacing to mounting height ratio Ratio specification used to insure that fixtures are adequately spaded, thus preventing "hotspots"

Span adjustment The ability to adjust the gain of a process or strain meter so that a specified display span in engineering units corresponds to a specified signal span. For instance, a display span of 200°F may correspond to the 16 mA span of a 4-20 mA transmitter signal.

Span The difference between the upper and lower limits of a range expressed in the same units as the range.

S-parameters The reflection and transmission coefficients used in impedance matching between high-speed (RF) devices and transmission lines/traces.

Spare A connector point reserved for options, specials, or other configurations. The point is identified by an (E#) for location on the electrical schematic.

Spark An electric spark is a sudden breakdown of the insulating strength of the dielectric separating two electrodes, due to the formation of ions by an intense electric field, accompanied by a rush of electricity across the "spark gap," and a flash of light indicating very high temperature. Unlike the are, glow, and brush discharges, the spark is of very short duration. It may be oscillatory or intermittent, several discharges taking place in quick succession. In gases, the spark takes place only at appreciable pressures, such as normal atmospheric pressure.

Spark test A test designed to locate pin-holes in the insulation of a wire or cable by application of a voltage for a very short period of time while the wire is being drawn through the electrode field.

Spatial resolution Spatial resolution is a measurement of the total number of pixels displayed in an entire image, usually noted in terms of horizontal by vertical (640 x 480).

Spatial sound Gives a wider stereo effect in some tabletop systems. This feature is especially handy for listening to music from monaural sources and making the sound stage seem wider when listening to the system in a small room.

Speaker Also called "loudspeaker." Transducer that converts electrical energy into mechanical energy at audio frequencies.

Speaker (conical) coverage Ceiling speakers, much like a spot light, generally project audio in a conical coverage pattern. As with any device featuring a conical coverage pattern, the higher it's mounted, the larger the circle of coverage. As the area of coverage increases, audio amplification wattage must also increase to maintain the same sound pressure level per unit of area of coverage.

Speaker level The signal output after the amplifier stage. This is the signal that is fed to the speakers.

Special cut crystals Special cut crystals are those crystals that have been cut at a different angle of orientation to enhance performance for specific needs. Typical special cuts would include BT cut crystals, SC cut crystals, IT cut crystals, and FC cut crystals.

Special effects Unusual sounds that are contained in a program's soundtrack.

Special effects generator Special effects generator. A video-mixing device that allows switching among several cameras and a variety of special effects such as dissolves, fades, wipes and inserts.

Special emergency radio service A Private Land Mobile Radio Service employed by persons or organizations engaged in emergency medical and rescue service, health care or similar activity.

Specialized mobile radio (SMR) A two-way radio service provided within a designated portion of the 800 and 900 MHz frequency bands.

Specific energy Same as Gravimetric Energy Density (Wh/Kg)

Specific gravity SG The ratio of the weight of a solution compared with the weight of an equal volume of water at a specified temperature. It is used to determine the charge condition in lead acid batteries.

Specific gravity The ratio of mass of any material to the mass of the same volume of pure water at 4°C.

Specific heat The ratio of thermal energy required to raise the temperature of a body 1° to the thermal energy required to raise an equal mass of water 1°.

Specific power Same as Gravimetric Power Density (W/Kg)

Specific-gravity (battery) The weight of the electrolyte compared to the weight of an equal volume of pure water. It is used to measure the strength or percentage of sulfuric acid in th electrolyte.

Specified frequency The frequency specified by the customer.

Spectral filter A filter which allows only a specific band width of the electromagnetic spectrum to pass, i.e., 4 to 8 micron infrared radiation.

Spectral width (w) A measure of the dispersion of velocities within the pulse volume. Standard deviation of the velocity spectrum. Spectral width depends among others from the turbulence within the pulse volume.

Spectrum The resolving of overall vibration into amplitude components as a function of frequency.

Spectrum analysis Utilizing frequency components of a vibration signal to determine the source and cause of vibration.

Spectrum analyzer An electronic device for automatically displaying the spectrum of the electromagnetic radiation from one or more devices. A cathode ray tube display is commonly used to display this power-versus frequency spectrum.

Spectrum assignment Federal government authorization for use of specific frequencies or frequency pairs within a given allocation, usually at stated a geographic location(s). Mobile communications authorizations are typically granted to private users, such as oil companies.

Speed up capacitor Capacitor added to the base circuit of a BJT switching circuit to improve the switching time of the device.

SPFP Signal power functional part

Spherically compliant suspension A patented mechanism based on a four-bar linkage that can be integrated into an outer lead bonder to have self-planarization of the thermode head to the substrate. This mechanism has been developed and patented by Universal Instruments Corporation.

SPI serial peripheral interface A 3-wire serial interface developed by Motorola.

SPICE Simulation program with integrated circuit emphasis

Spike A transient disturbance of an electrical circuit caused by, for example, load variations on the AC power line.

Spill light Unwanted light directed onto a neighboring property. Also referred to Light Trespass.

Spinner / spin coater Device for holding a substrate and spinning it at a controlled rate of speed for applying extremely thin coating as with resist.

Spinning A technique in which wafers are coated by spinning liquid photoresist on a rotating wafer surface.

Spin-rinse-dryer (SRD) / spin tool This is a machine for etching or cleaning wafers in wet chemicals. A cassette holding its wafers is turned round and round as chemicals are dispensed over the wafers. Wafers are rinsed in pure (de-ionized) water, and dried by spinning at high speeds.

Spiral wound Battery construction in which the electrodes with the electrolyte and separator between them are rolled into a spiral like a jelly roll (Swiss roll).

Spiral wrap The helical wrap of a tape or thread over a core.

SPL sound pressure level Is the loudness of sound relative to a reference level of 0.0002 microbars.

Splice A non-separable junction joining two fiber optic conductors. A splice can be accomplished either by fusing the two optical conductors with heat or by using an epoxy adhesive.

Split screen A video effect where portions of images from two sources divide the screen.

Split-screen display Lets you view two programs at once by dividing the screen down the middle. Each program is shown at full height, with one program on the left side and the other on the right. The feature is similar to picture-in-picture, but at a much larger size. Audio is played from the program on the left. To listen to audio from the program on the right, simply switch the pictures using the SWAP button.

Spooling The process of queuing tasks or jobs such as background printing.

Spot market The same as any commodity; a "spot" opportunity for immediate delivery or sale.

Spot size The diameter of the circle formed by the cross section of the field of view of an optical instrument at a given distance.

Spray tool This is a machine for etching or cleaning wafers in wet chemicals. Chemicals are sprayed onto the wafers as they are rotated in the chamber.

Spread spectrum A technology that modulates a signal over many carrier frequencies at once. This method can be used to make transmissions more secure, reduce interference, and improve bandwidth-sharing.

Spread The distance a substance (e.g., an adhesive) moves after it has been applied at ambient conditions.

Sprocket pitch The center to center distance of two adjacent sprocket holes.

Spur A substitution for the term spurious frequency response. The word "spur" is used to refer to a frequency occurring at some point higher than the desired mode, but lower than the next overtone.

Spurious emission or radiation Far-field electromagnetic radiation transmitted on a frequency outside the bandwidth required for satisfactory transmission of the required waveform. Spurious emissions include harmonics, parasitic emissions, and intermodulation products, but exclude necessary modulation sidebands of the fundamental carrier frequency.

Spurious error Random or erratic malfunction.

Spurious frequency response In radio reception, a response in the receiver intermediate frequency (IF) stage produced by an undesired emission in which the fundamental frequency (or harmonics above the fundamental frequency) of the undesired emission mixes with the fundamental or harmonic of the receiver local oscillator.

Spurious-free Unwanted frequencies are not present.

Sputter etch The use of ion bombardment to remove a film from the wafer surface.

Square wave A rectangular-shaped (step-function) periodic wave with a positive and negative half-cycle of equal lengths of time or duration. A square wave consists of a sine wave's fundamental frequency combined with the odd harmonics (multiples) of its fundamental frequency.

Squeegee A rubber or metal blade used in printing to wipe solder paste (or glue) across the stencil's face, forcing the material through the patterned apertures and onto the PCB.

Squelch A circuit which mutes the signal when it is below a certain level. Typically used to quiet the signal when only noise is present.

Stabilization Exposure of a magnet to demagnetizing influences expected to be encountered in use in order to prevent irreversible losses during actual operation. Demagnetizing influences can be caused by high or low temperatures, or by external magnetic fields.

Stabilization bake Placement of devices in a chamber at elevated temperature (normally 150°C) without electrical bias in order to test the construction of the devices. Sometimes this is performed prior to seal in order to bake out impurities on the die surface. Stabilization bake differs from high temperature storage only in that it is typically of shorter duration.

Stable datum A datum along which all other data align. From any confusion, order and sanity can emerge providing one merely selects a datum, assigns it importance or seniority and then begins to align other data against it.

Stacked bond A rebond placed directly over a previous bond.

Stadium surround mode One of several different modes in audio receivers and systems to shape sound to more closely resemble a live performance. Stadium Mode recreates the reverberation and open air atmosphere of a rock concert or sporting event when playing stereo (but not mono) source material.

Stage 3 emergency In the state of California, if power reserves ever fall below 1.5 percent, Cal-ISO, the independent system operator in California, will declare a Stage 3 emergency and the state's investor-owned utilities, may be ordered to immediately reduce the demand for electricity. At that point, the utilities will implement a series of temporary, controlled rotating power outages.

Staggered leads: Leads emanating from the same side of a package but bent in alternating directions such that they will insert into two or more parallel lines of mounting holes.

Stagnation pressure The sum of the static and dynamic pressure.

Stand alone A solar system that operates without connection to a grid or another supply of electricity. A battery bank stores unused daylight production for nighttime power. Commonly used in remote regions such as mountains, ocean platforms or communication towers.

Stand off height Height of component above the PCB.

Standard calibration The nominal point at which a measurement device is adjusted.

Standard calibration tolerance The allowable deviation from nominal, expressed in parts per million – ppm, at a specific temperature at +25 °C.

Standard cell An application-specific integrated circuit which, unlike a gate array, does not use the concept of a basic cell and does not have any pre-fabricated components. The manufacturer creates custom masks for every stage of the device's fabrication allowing each logic function to be created using the minimum number of transistors.

Standard charge The normal C/10 charge used to recharge a cell or battery in 10 hours. Other definitions (charging periods) also apply.

Standard electrode potential (E0) The standard potential E0 of an electrode is the reversible emf between the normal hydrogen electrode and the electrode with all components at unit activity.

Standard impedance The nominal impedance associated with the transmission line and test equipment.

Standardization A process of equalizing electrode potentials in one standardizing solution (buffer) so that potentials developed in unknown solutions can be converted to pH values.

Standardized mechanical interface (SMIF) A wafer manufacturing concept in which wafers are kept in sealed pods when they are not being processed in machines.

Standby power A fully charged battery ready to take over supplying a load in case of emergency.

Standby state A state in which the main functions of a circuit have been powered down to save energy, but power remains applied to the circuit ready to make a rapid restart.

Standby time The length of time a battery can power a mobile phone when it is switched on but not making or receiving calls.

Standing wave A stationary sound wave that is rein-forced by reflection between two parallel surfaces that are spaced a wavelength apart.

Standing waves Created when two waves in opposite directions interfere. When a reflected wave reinforces a reflection of the original waveform, the sound waves reinforce themselves, increasing in altitude.

Standoff A mechanical support, which may be an insulator, used to connect and support a wire or device away from the mounting surface.

Star ground A pcb layout technique in which all components connect to ground at a single point. The traces make in a "star" pattern, emanating from the central ground.

Star point A point from which all traces leave in a "star" pattern in pcb layout.

Stark effect The effect of a strong, transverse electric field upon the spectrum lines of a gas subjected to its influence. In many respects it resembles the more complicated types of Zeeman effect, but is subject to different laws, and may change radically in character and in multiplicity of component lines with increasing field intensity. The phenomenon, first observed by Stark in 1913, is conveniently studied by means of a canalray tube having behind the cathode a third electrode which may be given a high positive potential, in order to impose the desired field upon the radiating canal-ray particles.

Starter A device used in conjuction with a ballast for the purpose of starting an electric discharge lamp.

Starting current Current required by the ballast during initial arc tube ignition. Current changes as lamp reaches normal operating light level.

Starting resistance The non-linear change in the resistance of a crystal as a function of the power level used to drive the crystal.

Start-up time The specified time from oscillator power-up to the time the oscillator reaches steady state oscillation.

Starved cell (battery) A cell containing little or no free fluid electrolyte solution. This enables gasses to reach electrode surfaces readily, and permits relitive high rates of recombination.

Stat A prefix used to indicate that electrical quantities are expressed in the electrostatic system of units. This prefix is employed before the names of units in the practical system, e.g., statvolt, statoersted.

State assignment The process by which the states in a state machine are assigned to the binary patterns that are to be stored in the state variables.

State diagram A graphical representation of the operation of a state machine.

State machine The actual implementation (in hardware or software) of a function that can be considered to consist of a set of states through which it sequences.

State of charge (battery) The available amp-hours in a battery at any point of time. State of Charge is determined by the amount of sulfuric acid remaining in the electrolyte at the time of testing or by the stabilized open circuit voltage.

State of health (SOH) A measurement that reflects the general condition of a battery and its ability to deliver the specified performance compared with a fresh battery. It takes into account such factors as charge acceptance, internal resistance, voltage and self-discharge. It is not as precise as the SOC determination.

State transition An arc connecting two states in a state diagram.

State variable One of a set of registers whose values represent the current state occupied by a state Static Flex.

State, steady A condition of dynamic balance, as in an equilibrium reaction, where at equilibrium the concentration of each of the reactants remains constant. In such cases the loss of reactants to form products just balances the formation of reactants from the products in the reverse reaction. A physical system is said to be in a steady state if the various quantities describing the system are either independent of time or are periodic functions of time. Thus an alternating current circuit is in a steady state after all transient effects of a disturbance have disappeared.

Statement A sentence that asserts or denies an attribute about an object or group of objects. For example, "Your face resembles a cabbage."

Static Crackling noise heard on AM radio receivers. Caused by electric storms or electric devices.

Static calibration A calibration recording pressure versus output at fixed points at room temperature.

Static electricity Stationary electric charges.

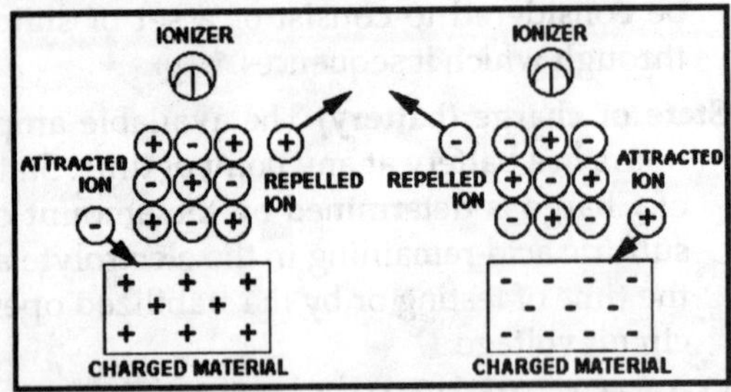

Fig. **Static electricity**

Static error band The error band applicable at room temperature.

Static fields Electric and magnetic fields that do not vary in intensity or strength with time.

Static flex A type of flexible printed circuit which can be manipulated into permanent three-dimensional shapes for applications such as calculators and high-tech cameras which require efficient use of volume and not just area.

Static mesh A basic de-interlacing process used in scalers for video content that contains no movement. This type of processing results in a sharp image with crisp details but will cause images to tear when motion occurs.

Static pressure Pressure of a fluid whether in motion or at rest. It can be sensed in a small hole drilled perpendicular to and flush with the flow boundaries so as not to disturb the fluid in any way.

Static RAM (SRAM) A memory device in which each cell is formed from four or six transistors configured as a latch or a flip-flop. The term static is used because, once a value has been loaded into an SRAM cell, it will remain unchanged until it is explicitly altered or until power is removed from the device.

Static reverse current Reverse current through a zener diode when the reverse voltage across the diode is less than the zener voltage rating of the device.

Static unbalance Static unbalance is that condition of unbalance for which the central principal axis is displayed only parallel to the shaft axis

Static wire A wire placed above the phase wires of a distribution of transmission circuit to protect against lightning. It is normally galvanized or aluminized steel.

Statistical quality control A quality control system utilizing statistical analysis of process defect data to detect quality trends on a real time basis.

Stator Stationary part of some rotary device such as a variable capacitor.

Steady flow A flow rate in the measuring section of a flow line that does not vary significantly with time.

Steady state A condition in which nothing is changing or happening.

Steady state vibration That condition of vibration induced by an unchanging continuing periodic force.

Steady-state dc voltage A fixed polarity of positive or negative voltage with respect to a reference. This form of voltage is used as the power source for electronic circuits.

Steering The critical part of the Dolby® Pro Logic®. Most soundtracks are mixed for playback on multi-speaker systems, using "cues" that steer specific sounds to specific speakers. In a home theatre system, the DPL processor typically does the steering to five or six speakers.

Stencil A metal mask in which patterns or apertures matching the component locations on the PCB are made so a suitable material can be forced through the apertures by a squeegee onto a substrate. Common materials are stainless steel and brass.

Stencil printing Deposition of a specific material, such as solder paste, using a stencil.

Step-down transformer (power and distribution transformer) A transformer in which the power transfer is from a higher voltage source circuit to a lower voltage circuit.

Step-index fiber The simplest form of optical fiber that is based on the principle of total internal reflection within the fiber strand or rod. When a light beam is applied at one end, the light will bounce through the fiber in a zig-zag manner and emerge at the other end with the emerging angle equal to the incident angle.

Stepnp platform System-level exploration platform for multi-processors A multiprocessor based simulation environment.

Stepper A photolithography tool used in the production of semiconductor devices, thin film recording heads, micromachined devices and so forth. Also referred to as a step-and-repeat projection aligner, stepper works by transferring the image of a circuit or component from a master photomask image onto a small portion of the wafer surface. The substrate is then moved or stepped, and the image is exposed once again onto another area of the wafer. This process is repeated until the entire wafer surface is exposed.

Step up DC-DC A switch-mode voltage regulator in which output voltage is higher than its input voltage.

Step up transformer (power and distribution transformer) A transformer in which the power transfer is from a lower voltage source circuit to a higher voltage circuit.

Stereo A process of using separate audio signals on separate channels for the left and right audio, thereby giving depth, or dimension to the sound.

Stereo audio Plays hi-fi/stereo rental tapes and records stereo TV broadcasts when used with a stereo TV or decoder using two sound tracks at the top edge of the videotape.

Stereo headphone jack Allows you to connect optional stereo headphones to the unit, so you can listen to your favourite TV shows without disturbing others or hear TV audio in an unusually noisy environment. The mini-jack configuration accepts either 8-ohm or 16-ohm mini-plug headphones.

Stereo sound System in which reproduced sound is delivered through two or more channels to give a sense of direction to the source.

Stereo/SAP reception with dbx noise reduction This feature has a built-in stereo decoder that can receive both TV stereo and separate audio program (SAP) telecasts where available. A dbx noise reduction system helps provide clear, crisp sound with proper stereo separation.

Stereolithography (SLA) A Rapid Prototyping (RP) system for creating plastic parts directly from 3D CAD files. The RP model speeds design validation and is also finds use as a master pattern.

Stereophonic (stereo) Consisting of two or more audio channels in an audio system during recording and playback to give a more natural distribution of sound.

Stereophonic Sound reproduction utilising separate amplification channels feeding respective loudspeaker systems to provide separation of output signals for various instrumental or vocal groupings.

Sticking The condition caused by physical interference with the normal motion of the moving element.

Stiffness The ratio of the force required to create a certain deflection or movement of a part expressed as (Force/deflection) lbs/in or grams/cm.

Stimulus An input signal that initiates action or reaction in a circuit; an input signal used for circuit excitation.

Stinger Slang for the wire connecting a fused cutout or switch to a transformer bushing.

Stitch bonding A bonding technique whereby the tip of the bonding wire is fed under the bonding head, which then applies heat and pressure to "stitch" the wire to the pad or post.

Frequently several stitches will be employed on the same bond. Stitch bonding can be visually differentiated from ultrasonic bonding because the impressions normally run across the bond rather than along its length.

Stitch length Overall length between the first and second wire bond, generally restricted to 100 the wire diameter.

Stoichiometry The branch of chemistry that deals with the numerical proportions in which substances react.

Stone (quartz) A term used to describe a quartz crystal before any machining operations have been done. The synthetic single crystals grown in the Hydrothermal Process are referred to as a "Stone".

Stop band Range of frequencies outside the pass band of a tuned amplifier.

Stop bit A signal following a character or block that prepares the receiving device to receive the next character or block.

Stopband The part of the frequency spectrum that is subjected to specified attenuation of signal strength by a filter.

Storage battery A device capable of transforming energy from electric to chemical form and vice versa. The reactions are almost completely reversible. During discharge, chemical energy is converted to electric energy and is consumed in an external circuit or apparatus.

Storage life The length of time a cell or battery can be stored on open circuit without permanent deterioration of its performance.

Storage temperature The range of ambient temperatures through which an inoperative power supply can remain in storage without degrading its subsequent operation.

Storage temperature range The minimum and maximum temperatures that the device can be stored or exposed to when in a non-oscillation state. After exposing or storing the device at the minimum or maximum temperatures for a length of time, all of the operating specifications are guaranteed over the specified operating temperature range.

Storage time In a BJT switching circuit, it is the time required for collector current to drop from 100% to 90% of its maximum value.

Strain The ratio of the change in length to the initial unstressed reference length.

Strain gage A measuring element for converting force, pressure, tension, etc., into an electrical signal.

Strand One of the wires that made up a stranded conductor.

Stranded conductor Conductor composed of a group of strands of wire twisted together.

Strap A type of interposer.

Stratification Over time, electrolyte tends to separate. The electrolyte at the top of the battery becomes watery while it becomes more acidic at the bottom. This effect is corrosive to the plates. Equalization of flooded (or wet) batteries helps reduce stratification.

Stray capacitance Undesirable capacitance that exists between two conductors such as two leads or one lead and a metal chassis.

Streaming An Internet derived expression for the one-way transmission of video and audio content.

Streamlined PCB design Streamline – v. Cause to be quick and efficient. Streamlined design

= accuracy plus speed. Streamlined Design, or SLD, is a set of policies that guide my design of printed circuit boards. The policies have been derived with the aim of simplifying and systematically eliminating errors from PCB design.

Stress corrosion The gradual deterioration of the mechanical properties of a material, usually accompanied by crack propagation, and caused by the acceleration of applied stress. This phenomenon usually occurs under high humidity conditions.

Strictly One Ampere hour is the charge transferred by one amp flowing for one hour. 1Ah = 3600 Coulombs.

Strike In the audio/video business, "strike" is a term for taking down an installation, for example at a show.

String A sequence of characters.

Stringing The act of installing overhead electrical wire or conductor.

Stringing block A sheave used to support and allow movement of a cable that is being installed. These are normally used overhead but there are also specialized designs used at the entrance to a conduit system.

Strip force The force required to remove a small section of insulation material from the conductor it covers. Usually measured in pounds.

Strip resonator A small AT cut crystal whose width is much smaller than its length. The crystal is a finite plate design requiring exact mathematical analysis to achieve good performance. The length of the resonator can be parallel to either the X or Z axis of quartz.

Stripping The removal of photoresist through either dry or wet processing.

Strobe A pulse or signal that clocks one or more digital data into a latch or memory circuit.

Strouhal number A nondimensional parameter important in vortex meter design defined as: s = Fh/V where f = frequency, V = velocity, and h = a reference length

Structural return loss Structural return loss is the measure of power loss on a cable or system and is caused by discontinuities in the cable conductor or dielectric. If these discontinuities are regularly spaced along a cable, they can cause severe transmission losses for frequencies whose wavelengths are twice that of the distance between these discontinuities.

Structural return loss is an unfavorable characteristic of poorly made cable, although careless installation can cause it as well. Liberty offers cable from manufacturing lines that are constantly computer monitored to avoid irregularities in the manufacturing process that could cause these flaws.

Studs Threaded bolt connectors used on high power cells

Stuff Attach and solder components to (a printed wiring board).

Sub harmonic The result of the division of the fundamental frequency by the desired number harmonic. The sub harmonic of 1000 Hz is 500 Hz.

Subassembly Components contained in a unit for convenience in assembling or servicing equipment.

Subatomic Particles such as electrons, protons and neutrons that are smaller than atoms.

Subatomic erosion A process in which structures on an integrated circuit's substrate are eroded by the flow of electrons in much the same way as land is eroded by a river (also known as electromigration)

Subatomic particle Any of various units of matter below the size of an atom, including the elementary particles such as electrons, protons, and neutrons.

Subcarrier A separate analogue or digital signal carried on a main radio transmission, which may carry extra information such as voice or data. It can be an monochromatic unmodulated continuous-wave (CW) signal or an already-modulated signal, which is then modulated into another signal of higher frequency and bandwidth. This is an established method of multiplexing.

Subcode Data, other than music, which is stored on a CD, MD, or other digital format.

Used to indicate track number, index number, use of emphasis, and other information.

Subnet mask The method used for splitting IP networks into a series of subgroups or subnets. The mask is a binary pattern that is matched up with the IP address to turn part of the host ID address field into a field for subnets.

Sub-panel A group of printed circuits (called modules) arrayed in a panel and handled by both the board house and the assembly house as though it were a single printed wiring board. The sub-panel is usually prepared at the board house by routing most of the material separating individual modules, leaving small tabs. The tabs are strong enough so that the sub-panel can be assembled as a unit, and weak enough so that final separation of assembled modules is easily done.

Sub pixel resolution An imaging technique that can yield a measurement that has a spatial resolution of less than one pixel.

Subscriber identity module An essential component of a GSM mobile phone. It contains the identity of the subscriber and assures the authentication during the access into the network and provides data storage for other subscriber related information, such as a personal address books.

Substation A substation is a small building or fenced in yard containing switches, transformers, and other equipment and structures for the purpose of adjusting voltage, monitoring circuits and other service functions. As electricity gets closer to where it is to be used, it goes through a substation where the voltage is lowered so it can be used by homes, schools, and factories.

Substrate Generic name for the base layer of an integrated circuit, hybrid, multichip module, or circuit board. Substrates may be formed from a wide variety of materials, including semiconductors, ceramics, FR4 (fiberglass), glass, sapphire, or diamond depending on the application. Note that the term substrate has traditionally not been widely used in the circuit board world, at least not by the people who manufacture the boards. However, there is an increasing tendency to refer to a circuit board as a substrate by the people who populate the boards. The main reason for this is that circuit boards are often used as substrates in hybrids and multichip modules, and there is a trend towards a standard terminology across all forms of interconnection technology.

Substrate geometry Substrate dimensions, typically dimensions critical to implement a successful bonding process, including the following: board size, bonding pad layout and dimensions, solder thickness, adjacent components, planarity, fiducial shape and dimensions, and board thickness and construction.

Subtractive process A process in which a substrate is first covered with conducting material, then any unwanted material is subsequently removed, or subtracted.

Sub-transmission system A high voltage system that takes power from the highest voltage transmission system, reduces it to a lower voltage for more convenient transmission to nearby load centers, delivering power to distribution substations or the largest industrial plants. Typically operating at voltages from approximately 35 kV to 100 kV.

Subwoofer A speaker that is dedicated to reproducing the lowest octaves on a soundtrack. Powered subwoofers include their own amplification, while passive models must be driven by external amplification. Most powered subwoofers include crossover circuitry that blends the subwoofer's output with the main speakers.

Successive approximation A technique used in analogue-to-digital conversion wherein a reference DAC is fed with a series of progressively finer digital data until its output matches the analogue input, at which point the input data to the reference DAC becomes the output of the ADC.

Sulfation (battery) The formation of lead sulfate of such physical properties that it is extreemly difficult, if not impossible, to reconvert it to active material.

Sulfation The formation of lead-sulfate crystals on the plates of a lead-acid battery. Commonly used to indicate the large crystals which form in partially discharged cells as the result of temperature cycling. These large crystals are more difficult to reduce by the charging current than are the smaller crystals that result from normal and self-discharge reactions. Sulfating can be caused by leaving the battery in a discharged state for long periods of time.

Sulfur-hexafluoride (SF6) A very dense, inert, non-conducting gas used inside high voltage equipment to insulate conducting components from surfaces at ground petential. It also is used as an interrupting medium in high voltage circuit breakers.

Sulphation Growth of lead sulphate crystals in Lead-Acid batteries which inhibits current flow. Sulphation is caused by storage at low state of charge.

Summing amplifier A device which combines the left and right channel audio signals into a single mono channel. It is useful in multi-speaker mono paging systems, or in large stadium and church environments where mono speakers are placed close to the presenter so that the sound image matches the presenter. Directly bridging the left and right channel outputs of a stereo audio device without using a summing amplifier may cause severe damage to the device and distortion to the signals.

Summing point The point at which two or more inputs of an operational amplifier are algebraically added.

Super cooling The cooling of a liquid below its freezing temperature without the formation of the solid phase.

Super heating 1. The heating of a liquid above its boiling temperature without the formation of the gaseous phase.

2. The heating of the gaseous phase considerably above the boiling-point temperature to improve the thermodynamic efficiency of a system.

Super heterodyne receiver Radio receiver that converts all radio frequencies to a fixed intermediate frequency to maximize gain and bandwidth before demodulation.

Super high frequency (SHF) Frequency band between 3 GHz and 30 GHz. So designated by federal communications commission (FCC).

Super invar shadow mask Achieves a breakthrough level of picture quality on a flat screen. By applying the optimal amount of vertical tension to the shadow mask inside the panel, this feature overcomes the colour shift problem common with conventional flat CRTs. It also raises the mask aperture by 20 percent, compared to previous models, to deliver extremely bright pictures.

Supercapacitor A capacitor that can store a large amount of energy. Also called Ultracapacitor or Booster capacitor.

Supercardioid microphone A unidirectional micro-phone with tighter front pickup angle (115 degrees) than a cardioid, but with some rear pickup. Angle of best rejection is 126 degrees from the front of the micro-phone, that is, 54 degrees from the rear.

Superconductivity A phenomenon occurring below a very low, characteristic critical temperature in certain materials (superconductors), characterised by the complete absence of electrical resistance and the damping of the interior magnetic field (the Meissner effect). Superconductors can carry currents that will not decay.

Superconductor Metal such as lead or niobium that, when cooled to within a few degrees of absolute zero, can conduct current with no resistance.

Superheterodyne receiver The most widely used type of radio receiver, in which all incoming modulated radio-frequency signals are converted to a common intermediate-frequency or IF carrier value for additional amplification and selectivity prior to demodulation. The advantage to this method is that most of the radio's signal path has to be sensitive to only a narrow range of frequencies.

Superimposition Placing one image over another so that both may be seen at the same time. The effect can be achieved in many ways: by more than one exposure on a single piece of film, by multiple printing, or by registered projection. Abbreviated "super."

Super-insulation Super-insulation is a means of constructing buildings to ensure minimal heat load. R-40 walls and R-60 ceilings, air-tight construction, and shuttered windows that face south are all examples of this.

Superposition theorem Theorem designed to simplify networks containing two or more sources. It states that in a network containing more than one source, the current at any one point is equal to the algebraic sum of the currents produced by each source acting separately.

Super-regenerative Receiver A receiver employing a detection principle based on injection of a supersonic signal for increased sensitivity.

Supertweeter A high frequency driver designed to reproduce very high frequencies, typically over 10 kHz.

Supply voltage The DC input voltage necessary for oscillator operation, specified in volts.

Suppressor grid A grid (usually between anode and screen grid) placed at ground potential for the purpose of eliminating secondary emission by redeflecting electrons to the anode.

Surface acoustic wave A technology used for automatic identification in which low power microwave radio frequency signals are converted to ultrasonic acoustic signals by a piezoelectric crystalline material in the transponder. Variations in the reflected signal can be used to provide a unique identity.

Surface insulation resistance The electrical resistance of an insulation material between a pair of contacts, conductors or grounding devices in various combinations, determined under specified environmental and electrical conditions.

Surface leakage current Diode reverse current that passes along the surface of the semiconductor materials.

Surface mount (SM) A component, either active or passive, having no separate leads but which is part of the component body to permit direct mounting on a printed circuit board.

Surface mount device (SMD) A component whose packaging is designed for use with surface mount technology.

Surface mount Surface mount technology. The technology of creating printed wiring wherein components are soldered to the board without using holes. The result is higher component density, allowing smaller PWB 's. Abbreviated SMT.

Surface mount technology (SMT) A technique for populating hybrids, multichip modules, and circuit boards, in which packaged components are mounted directly onto the surface of the substrate. A layer of solder paste is screen printed onto the pads and the components are attached by pushing their leads into the paste. When all of the components have been attached, the solder paste is melted using either reflow soldering or vapor-phase soldering.

Surface mounted packaging Semiconductor packages mounted on the surface of a printed circuit board or other substrate material.

Surface resistivity The resistance of a material between two opposite sides of a unit square of its surface. It is usually expressed on Ohms.

Surface roughness Disruption of the planarity of the semiconductor surface; measured as a difference between highest and deepest surface features; can be as low as 0.06 nm or high quality Si wafers with epitaxial layers..

Surface states Extra donors, acceptors, or traps, usually undesired, which may occur on a semiconductor surface because of crystal imperfections or contamination. These may vary with time.

Surface tension An effect of the forces of attraction between the molecules on the surface of a liquid. Surface tension is the reason water beads up better on the hood of your car when waxed versus unwaxed. The wax increases the surface tension of the water, and thus it beads up more readily.

Surface-barrier diode (Schottky diode)High speed diode that has very little junction capacitance. Also known as a "hot-carrier diode."

Surface-emitting laser diode A laser diode constricted on an integrated circuit's substrate such that, when power is applied, the resulting laser beam is emitted directly away from the surface of the substrate.

Surface mount technology or surface-mount toroidal (SMT) A surface mount device is a component, either active or passive, having no separate leads but which is part of the component body to permit direct mounting on a printed circuit board.

Surfactant A chemical added to any substance to lower its surface tension.

Surftape A punched carrier for tape and reel packaging of surface mount devices. It features a flexible, pressure-sensitive adhesive base, negating the need for cover tape.

Surge A sudden change (usually an increase) in the voltage of a power line; a surge is similar to a spike, but it lasts longer.

Surge capacity The amount of current an inverter can deliver for short periods of time. Most electric motors draw up to three times their rated current when starting. An inverter will "surge" to meet these motor-starting requirements.

Surge current A current of short duration that occurs when power is first applied to capacitive loads or temperature dependent resistive loads such as tungsten or molybdenum heaters-usually lasting no more than several cycles.

Surge withstand A measure of an electrical device's ability to withstand high-voltage or high-frequency transients of short duration without damage.

Surge withstand capability (SWC) test The SWC test wave is an oscillatory wave, frequency range of 1-1.5 MHz, voltage range of 2.5-3 kV crest value of first peak, envelope decaying to 50% of the crest value of the first peak in not less than 6 micro seconds from the start of the wave. The source impedance is from 150-200. The wave is to be applied to a test specimen at a repetition rate of not less than 50 tests per second for a period of not less than two seconds.

Surround output level Allows adjustment of the rear speaker level relative to the front speaker level according to viewing location and individual preference.

Surround sound (home theatre) Surround sound is the basis of a home theatre. It is created by a surround sound processor, multi-channel amplification and five or six speakers. Surround sound envelopes the viewer/listener with sound, includes special effects and simulates the ambience of the environments shown on the TV.

Surround sound An audio reproduction system using four or more speakers to simulate the three-dimensional effect of a live music performance or a cinematic environment. With surround sound, the phase of the original sound is delayed from a two-channel source, creating a surround effect and allowing you to simulate stereo sound even from a monaural source.

Surround sound modes A feature of A/V receivers that simulates the ambience of various sound environments with preset configurations for jazz, rock, concert and stadium.

Surround sound processor/amp Usually includes power amplification for the one centre speaker and two surround speakers in a home theatre system.

Surround speakers (surrounds) The two speakers in a home theatre system that are placed in the rear of the room or to the sides of the primary seating position. They are critical in producing surround sound.

Surveillance Random witnessing of testing or screening of devices to determine that the material meets all applicable requirements.

Susceptibility A measure of a device or system's ability to prevent undesirable responses when subjected to electromagnetic radiation.

Suspension brake A pneumatic clamp on the spherically compliant suspension which can be used to lock the head in a given orientation.

Suspension effect The source of error due to varied reference liquid junction potential depending upon whether the electrodes are immersed in the supernatant fluid or deeper in the sediment. Normally encountered with solutions containing resins or charged colloids.

Suspension, taut-band A mechanical arrangement of two ribbons under tension, one at each end of the moving element. The ribbons support the moving element, allow it to rotate freely, provide the restoring torque, and conduct current to the moving element of moving coil instrument.

SVGA super video graphics array A term used to denote resolutions higher than VGA (640 x 480). SVGA computer graphics cards have a resolution of 800 x 600 (480,000 pixels) but may be able to output resolutions of up to 1280 x 1024 and 16 million colours.

S video A composite video signal separated into the luma ("Y" is for luma, or black and white information; brightness) and the chroma ("C" is an abbreviation for chroma, or colour information).

S video input jack The 4-pin S-video input jack allows you to connect a DBS (Dolby Digital) system or DVD player directly to the luminance and chrominance circuits of the TV set. By keeping colour and luminance information separate, cross-colour interference is virtually eliminated from the picture.

SWAN (Structured wireless-aware network) A technology that incorporates a WLAN into a wired wide-area network (WAN). SWAN technology can enable an existing wired network to serve hundreds of users, organizations, corporations, or agencies over a large geographic area. A SWAN is said to be scalable, secure, and reliable.

Sweeling (battery) The swelling or bulging of a battery case that results from cell vents not allowing enough internal pressure to be relieved.

Sweep In audio, a sequence of puretone frequencies used to generate a frequency response curve.

Sweep generator Test instrument designed to produce a voltage that continously varies in frequency over a band of frequencies. Used as a souce to display frequency response of a circuit on an oscilloscope.

Sweep test Pertaining to cable, checking frequency response by generation an rf voltage whose frequency is varied back and forth through a given frequency range at a rapid constant rate and observing the results of an oscilloscope.

Swelling Distortion of cells caused by expansion of the active chemicals due to temperature and pressure effects.

Swing cell A lithium ion cell

Swing The maximum output voltage that an operational amplifier can deliver without saturation or clipping for a given load and operating supply voltage.

Swinging choke A filter inductor employed for voltage-regulation purposes in power supplies. It has saturable-reactor characteristics.

Switch A device for connecting and disconnecting power to a circuit.

Switch mode Uses a switching transistor and inductor to control/regulate the charging voltage/current.

Switch mode charger Charger which uses a switch mode regulator. More efficient but more costly than a Linear charger.

Switch mode regulator A switching regulator is a voltage regulator which uses an output stage, switched repetitively on and off, together with energy storage components (capacitors and inductors) to generate a DC output voltage. Regulation is achieved through Pulse Width Modulation (PWM). Output voltages can be generated that are greater than or less than the input voltage, and multiple output voltages can be generated with a single regulator.

Switch, general use A switch intended for use in general distribution and branch circuits. It is rated in amperes and is capable of interrupting its rated voltage.

Switch, general-use snap A type of general-use switch so constructed that it can be installed in flush device boxes or on outlet covers, or otherwise used in conjunction with wiring systems recognized by the National Electric Code.

Switch, isolating A switch intended for isolating an electrical circuit from the source of power. It has no interrupting rating and is intended to be operated only after the circuit has been opened by some other means.

Switch, knife A switch in which the circuit is closed by a moving blade engaging contact clips.

Switch, motor circuit A switch, rated in horsepower, capable of interrupting the maximum operating overload current of a motor of the same horsepower rating as the switch at the rated voltage.

Switch, network A Switch connects Client systems and servers together to create a nework.

Switch, transfer A transfer switch is an automatic or nonautomatic device for transferring one or more load conductor connections from one power source to another.

Switchboard A large single panel, frame, or assembly of panels having switches, overcurrent, and other protective devices, buses, and usually instruments mounted on the face or back or both. Switchboards are generally accessible from the rear and from the front and are not intended to be installed in cabinets.

Switcher A switch mode regulator.

Switchgear A general term covering switching and interrupting devices and their combination with associated control, metering, protective and regulating devices. Also, the assemblies of these devices with associated interrconection, accessories, enclosures and supporting structure, used primarily in connection with the generation, transmission, distribution and conversion of electric power.

Switching The connection of two points of a network at controllable instants of time. An alternative term is clamping.

Switching frequency The rate at which the dc voltage is switched in a converter or power supply.

Switching regulator A voltage regulator that uses a switching element to transform the supply into an alternating current, which is then converted to a different voltage using capacitors, inductors, and other elements, then converted back to DC. The circuit includes regulation and filtering components to insure a steady output. Advantages include the ability to generate voltages beyond the input supply range and efficiency; disadvantages include complexity.

Switching station A switching station is a type of substation where connections are made between several distribution and transmission lines. Voltage is not changed.

Switching surges High voltage spikes that occur when current flowing in a highly inductive circuit, or a long transmission line, is suddenly interrupted. As the magnetic field about the inductive conductor collapses, a brief by very high voltage can be generated at the terminal point of the circuit.

Switching transistor transistor designed to change rapidly between saturation and cut-off.

Switch leg That part of a circuit run from a lighting outlet box where a luminaire or lampholder is installed down to an outlet box that contains the wall switch that turns the light or other load on or off: it is a control leg of the branch circuit.

Symbian A company created by Psion, Nokia, Ericsson and Motorola in 1998 with the aim of developing and standardising an operating system which enable mobile phones from different manufacturers to exchange information.

Symbol A simplified design representing a part in a schematic circuit diagram.

Symbolic logic A mathematical form in which propositions and their relationships may be represented symbolically using Boolean equations, truth tables, Karnaugh maps, or similar techniques.

Symmetry (in an offset-zero instrument) The difference in indications when the polarity of a measured quantity is reversed expressed as a percentage of fiducial value.

Symmetry error (of an offset zero instrument) The difference in indications when the polarity of a measured quantity is reversed, expressed as a percentage of fiducial value.

Symmetry Symmetry is defined as the ratio of amount of time the voltage is in the logic "1" state compared to the time in the logic "0" state. The measurements are taken at the 50% points of the voltage transition between the two logic states. The time period of one cycle of the waveform is calculated first as below.

1/Frequency in Hertz = Time Period in Seconds

Next, the time period of the logic "1" state is measured from the 50% point of the waveform's positive voltage transition to the 50% point of the waveform's negative voltage transition, then compared to the total waveform period. The calculation for symmetry is shown below:

Logic "1" Time in Symmetry/Period of One Cycle X 100 = % Symmetry.

For the % symmetry of the logic "0" state, subtract the logic "1" symmetry from 100%. For example, 40/60% means that the waveform is in its logic "1" state 40% and in the logic "0' state 60% of the total waveform time period.

Sync Synchronization. In video, sync is a means of controlling the timing of an event with respect to other events. This is accomplished with timing pulses to insure that each step in a process occurs at the correct time. For example, horizontal sync determines exactly when to begin each horizontal scan line. Vertical sync determines when the image is to be refreshed to start a new field or frame. There are many other types of sync in a video system. (Also known as "sync signal" or "sync pulse.")

Sync generator A circuit that produces sync impulses used to control the time when certain events happen electronically. Also known as a "synchronizing pulse generator."

Sync polarity 1. A circuit can be designed to operate on the positive-going or negative-going part of the sync pulse. Some equipment has a sync polarity option switch to allow selecting which edge (plus or minus) to trigger on.

2. This refers to the duty cycle of the sync signal. A positive polarity sync signal is low most of the time, and high for a short time. Negative polarity sync is high most of the time and low for a short time.

Sync pulse Pulse used as a reference for synchronization.

Synchrotron A device used to produce high-energy X-rays that can inscribe features on a chip potentially as small as 100 nanometers.

Syntax The rules governing the structure of a language.

Synthesis The CAD operation of automatically generating an integrated circuit layout from a high-level description.

Syringe A container for adhesive dispensing via its narrow opening.

System Combination of several pieces of equipment to perform in a particular manner.

System level A behavioural description of a system that is an abstraction of the hardware/software implementation level.

System level prototyping station A rapid prototyping station configured for system design activities.

System losses The difference between the system net energy or power input and output resulting.

System on a chip A chip that is a self-contained system, including processing, memory and input-output functions.

System switcher An A/V switching device that also communicates with other components in a system including room lights and motorized devices. For instance, in addition to controlling a projector, a system switcher can turn lights on and off and raise and lower a motorized projector screen.

Systeme international d'unites (SI) The International System of Units comprised of Base Units, Supplementary Units and Derived Units.

T

T pad An attenuator whose components are wired to appear as the letter T.

T section A network connection which has two elements in series in one line and a third element in shunt from the junction of the two series elements to the opposite line, thereby giving a circuit diagram which looks like a T.

T&D Transmission and distribution.

T/H Track/hold

T/R Transmit/receive

T3 A type of data connection capable of transmitting a digital signal at 44Mbps. T3 lines are often used to link large computer networks, such as those that comprise the Internet.

TAB (tape automated bonding) A process in which transparent flexible tape has tracks created on its surface. The pads on unpackaged integrated circuits are attached to corresponding pads on the tape which is then stored in a reel. Silver-loaded epoxy is screen printed on the substrate at the site where the device is to be located and onto the pads to which the device's leads are to be connected. The reel of TAB tape is fed through an automatic machine which pushes the device and the TAB leads into the epoxy. When the silver-loaded epoxy is cured using reflow soldering or vapor-phase soldering, it forms electrical connections between the TAB leads and the pads on the substrate.

TAB component An IC mounted on a TAB site.

TAB tape A polymer film with patterned metal traces or leads. It is only a temporary support that is almost completely removed during the transfer of the chip from the tape to the substrate. The TAB tape can be handled in continuous tape or slide carrier format.

TAB tape automatic bonding The process where silicon chips are joined to patterned metal traces (leads) on polymer tape to form inner leads bonds and subsequently the leads are attached to the next level of the assembly, typically a substrate or board, to form outer lead bonds. TAB is the technique of interconnecting silicon with beam bonding as opposed to wire bonding.

Tachometer A transducer used for measuring the rate of revolution of a shaft.

Tacitron A thyratron capable of current interruption by grid action, which also generates much less noise than a conventional thyratron. Its anode current can be turned on or off in about a microsecond without the necessity of anode voltage removal, due to a special grid design which limits ion generation to the grid-anode region.

Tackiness The ability of solder paste to hold surface mount components in place after placement but before reflow soldering.

TACS Total Access Communications System (an AMPS variant deployed in a number of countries principally the UK).

TAD Total accumulated discharge (mA-hr)

Tafel equation The relationship between the internal electrode potentials in a battery and the current which flows. This is an exponential relationship based on empirical results which quantifies the elecrochemical reactions. It is analogueous to the Arrhenius equation which quantifies the thermochemical process relating the temperature to the rate at which a chemical action progresses.

Tag excitation device A term coined by the RFID Alliance Lab to refer to a device that sends signals to the tag regardless of the make or manufacturer. TED is used to measure the response of tags scientficially.

Tag line A rope used to control the position of equipment being lifted. This is not to be confused with the rope used to actually lift the equipment.

Tag talks first A means by which a reader in a passive UHF system identifies tags in the field. When tags enter the reader's field, they immediately communicate their presence by reflecting back a signal. This is useful when you want to know everything that is passing a reader, such as when items are moving quickly on a conveyor. In other cases, the reader wants to simply find specific tags in a field, in which case it wants to broadcast a signal and have only certain tags respond.

Talk time The length of time a battery can power a mobile phone when making or receiving calls.

Tally light A red light on the front of the camcorder that lets your subject know when the camcorder is recording.

Tank circuit Usually refers to the anode resonant circuit (L and C in parallel); it also applies to any parallel resonant circuit storing RF energy by virtue of its fly-wheel effect.

Tantalum capacitor Electrolytic capacitor having a tantalum foil anode. Able to have a large capacity in a small package.

TAP (Transferred account procedure) The essential charging methodology for international GSM roaming. There have been four TAP standards, TAP1, TAP2, TAP2+ and TAP3. The latter offers variable record length and is sufficiently flexible to support all future requirements arising from the move to 3G.

Tap changer A tap changer is a device that adjusts the voltage-capability of a transformer or a voltage regulator.

Tape A recording media for data or computer programs. Tape can be in permanent form, such as perforated paper tape, or erasable, such as magnetic tape. Generally, tape is used as a mass storage medium, in magnetic form, and has a much higher storage capacity than disk storage, but it takes much longer to write or recover data from tape than from a disk.

Tape and reel Refers to the packaging method used to accommodate automated pick-and-place equipment.

Tape automated bonding (TAB) A process in which transparent flexible tape has tracks created on its surface. The pads on unpackaged integrated circuits are attached to corresponding pads on the tape which is then stored in a reel. Silver-loaded epoxy is screen printed on the substrate at the site where the device is to be located and onto the pads to which the device's leads are to be connected. The reel of TAB tape is fed through an automatic machine which pushes the device and the TAB leads into the epoxy. When the silver-loaded epoxy is cured using reflow soldering or vapor-phase soldering, it forms electrical connections between the TAB leads and the pads on the substrate.

Tape bonding Utilization of a metal or plastic tape material as a support to a carrier of a component in a gang bonding process.

Tape wrap A spirally applied tape over an insulated or uninsulated wire.

Tape and reel Component packaging for placement via housing parts in cavities in a continuous strip. The cavities are covered by a plastic tape so that they can be wound in a reel for presentation to a component placement machine.

Taper charge In quick chargers the charging current is is progressively reduced in a controlled way by controlling the supply voltage. In slow chargers the voltage is fixed and the charging current reduces in an uncontrolled way due to increase in the cell voltage as the charge builds up.

Taper Refers to the configuration of the end-to-end resistor. It is usually reported as linear (evenly distributed) or logrithmic.

Tapered Nonunifrom distribution of resistance per unit length throughout the element of a potentiometer.

Tariff The schedule of all rates and services offered by CVPS or CVEC filed with the state PSB or PUC.

Technician Expert in troubleshooting circuit and system malfunctions. Along with a thorough knowledge of test equipment and how to use it to diagnose problems, the technician is also familiar with how to repair or replace faulty components. Technicians basically translate theory into action.

Telecommunications Any communication process that allows the transmission of information from a sender to a receiver by means of an electromagnetic or lightwave medium.

Teleforce Tesla's device for projecting concentrated non-dispersive energy by means of a narrow stream of metallic particles.

Telegeodynamics Tesla's system for prospecting and the location of underground mineralized structures through the transmission of mechanical energy through the subsurface. Data from reflected and refracted signals can be analyzed to deduce the location and characteristics of underground formations. Additional non-mechanical responses to the initial acoustic impulses may also be detected.

Telegraphy Communication between two points by sending and receiving a series of current pulses either through wire or by radio.

Telematics A wireless communications system designed for the collection and dissemination of information, particularly refers to vehicle-based electronic systems, vehicle tracking and positioning, on-line vehicle navigation and information systems and emergency assistance.

Telemetry The use of telecommunication for automatically indicating or recording measurements at a distance from the measuring instrument.

Telemetry device Devices used to transmit or receive data in a digital form.

Telemetry tracking and control (TT&C) Functions that provide for the monitoring and control of satellites.

Telephone Apparatus designed to convert sound waves into electrical waves which are sent to and reproduced ata distant point.

Telephone line Wires existing between subscribers and central stations in a telephone system.

Telephony Telecommunications system involving the transmission of speech information, allowing two or more persons to communicate verbally.

Teletypewriter Electric typewriter that like a teleprinter can produce coded signals corresponding to the keys pressed or print characters corresponding to the coded signals received.

Television System that converts both audio and visual information into corresponding electrical signals which are then transmitted through wires or by radio waves to a receiver which reproduces the original information.

Telex Teletypewriter exchange service.

Temper The softmess of a metal; terms such as soft-drawn, dead soft, annealed, and semi-annealed are used to decribe tempers used for conductor metals.

Temperature The average kinetic energy of the atoms or molecules of a body or substance, perceived as warmth or coldness. Measured in degrees Fahrenheit, Celsius, or Kelvin.

Temperature coefficient (TC) The change in the characteristic of a component which occurs because of a change in temperature. TC can be specified either as the number of parts per million (ppm) change per °C change in temperature, or as a percent change in value per °C change in temperature.

Temperature coefficient of frequency Rate at which frequency changes with temperature.

Temperature comparator An integrated circuit with a digital output that indicates whether a measured temperature is above or below a predetermined threshold.

Temperature compensation Optimal battery charging voltage is dependent on the battery temperature. As the ambient temperature falls, the proper voltage for each charge stage needs to be increased. When the ambient temperature increases, the proper voltage for each charge stage needs to be decreased. On some products, the Battery Temperature Sensor (BTS) allows the charger or inverter/charger to automatically scale charge-voltage settings to compensate for ambient temperatures. On others, there is a setting for hot, cold, and warm settings.

Temperature cut-off A temperature sensing method which detects heat rise in a cell at overcharge and switches the charger off or to a lower rate of charge.

Temperature cycle A test whereby devices are stored for short periods (15 minutes) alternately at high and low temperatures in gas filled chambers, with a maximum transfer time between chambers of one minute. Normally 10 cycles are performed from -65° to + 1 WC. This stresses device assembly because of the different thermal coefficients of expansion of the various materials used.

Temperature derating The amount by which power source or component ratings are decreased to permit operation at elevated temperatures.

Temperature error The maximum change in output, at any measurand value within the specified range, when the transducer temperature is changed from room temperature to specified temperature extremes.

Temperature lag The number of degrees above the actuation point that the media will be when the sensor operates. Lag is expressed for a designated rise rate (degrees per second), flow (feet per second), and system pressure. The lag is determined by multiplying the rise rate by the response time. Example: If a system with a constant flow, pressure, and rise rate of 10°F per second incorporated a sensor with a response time of 3 seconds, the lag would be 30 degrees.

Temperature range The minimum and maximum temperature for which the deviation from the nominal frequency will not vary by a given amount.

Temperature range, compensated The range of ambient temperatures within which all tolerances specified for Thermal Zero Shift and Thermal Sensitivity Shift are applicable (temperature error).

Temperature range, operable The range of ambient temperatures, given by their extremes, within which the transducer may be operated. Exceeding compensated range may require recalibration.

Temperature rise The increase in temperature that results when electrical load is carried by electrical equipment.

Temperature sensor Temperature sensor that uses an external diode-connected transistor as the sensing element to measure temperatures external to the sensor (for example, on a circuit board or on the die of a CPU). Generally produces a digital output.

Temperature switch A temperature switch is a sensor that upon the increase or decrease of a temperature, opens or closes one or more electrical switching element at a predetermined set point.

Temporal Relating to the sequence of time or to a particular time.

Temporary service Refers to electric service installed on a temporary basis to allow for construction of a new structure. Once a structure is completed, the temporary service is removed and permanent service installed.

Tensile strength The longitudinal stress required to break a prescribed specimen divided by the original cross-sectional area at the point of rupture (usually expressed in lbs. per square inch or PSI).

Tension The force in pounds of kilg grams on a conductor installed overhead. Too much tension on an overhead line can contribute to mechanical failure.

Tented via A via with dry film solder mask completely covering both its pad and its plated-thru hole. This completely insulates the via from foreign objects, thus protecting against accidental shorts, but it also renders the via unusable as a test point. Sometimes vias are tented on the top side of the board and left uncovered on the bottom side to permit probing from that side only with a test fixture.

Tera(T) Unit qualifier (symbol = T) representing one million million, or 1012. For example, 3THz stands for 3 x 1012 Hertz

TeraHertz The terahertz, abbreviated THz, is a unit of electromagnetic (EM) wave frequency equal to one trillion hertz (1012 Hz). The terahertz is used as an indicator of the frequency of infrared (IR), visible, and ultraviolet (UV) radiation. The terahertz is not commonly used in computer and wireless technology.

Terminal 1) The point at which electrical connections are made. 2) The mechanical device at such a point in a circuit, as at the end of a wire or cable, by means of which an electrical connection may be made.

Terminal A point of connection for two or more conductors in an electrical circuit; one of the conductors is usually an electrical contact, lead or electrode of a component.

Terminal capacity A conducting sphere or toroid that is connected to the high voltage terminal of a Tesla coil oscillator, the opposing terminal being grounded. It is known as an elevated terminal or elevated capacity when the oscillator is configured for use as a wireless transmitter in which case the terminal is mounted on the top of a insulated supporting structure. A large terminal capacity is illustrated in the patent Apparatus for Transmitting Electrical Energy.

Termination 1. A load or impedance at the end of a cable or signal line used to match the impedance of the equipment that generated the signal. The impedance absorbs signal energy to prevent signal reflections from going back toward the source. For video signals, termination impedance is typically 75 ohms; for sync signals, it is usually 510 ohms.

2. A connector at the end of a cable.

Termination impedance The impedance that should be presented to the source and load side of the filter to ensure proper performance.

Termination voltage The maximum voltage which can be tolerated by a cell during charging without damaging the cell. The cell voltage at which the charging process should be terminated.

Terminator A device that provides termination for a signal line or several signal lines at the end of a cable. Usually a close-tolerance resistor for each signal, a terminator is often mounted in its own enclosed connector, making it easy to install.

Tertiary logic An experimental technology in which logic gates are based on three distinct voltage levels. The three voltages are used to represent the tertiary digits 0, 1, and 2, and their logical equivalents FALSE, TRUE, and MAYBE.

Tesla (T) Unit of magnetic flux density equal to 1 weber per meter square or 10,000 gauss.

Tesla antenna A form of wireless antenna or wave launching structure developed by Nikola Tesla in which the transmitted energy propagates or is transferred to the receiver by a combination of electrical current flowing in the ground, electrostatic induction and electrical conduction through plasma with an embedded magnetic field.

Tesla coil An electrical oscillator developed by Nikola Tesla which produces high voltage, radio frequency alternating electric current. In its original form also known as the disruptive discharge coil. In their first application these devices served as power supplies for experimental high frequency light bulbs and tubes. They were also used as a source of radio frequency currents in Tesla's pioneering wireless experiments, offering a superior alternative to his high frequency alternators. The classic Tesla coil consists of an air core transformer with loose coupling between the primary and secondary. The primary winding is excited by the discharge of a high voltage capacitor through a mechanical switching device known as a "break." Operation is characterized by a series of individual primary pulses, each being followed by a more rapid series of diminishing secondary oscillations. The diminishing secondary oscillation is called ring down. The term "partially-damped wave" is often used in relation to the disruptive discharge Tesla coil. Vacuum-tube and solid-state Tesla coil (VTTC and SSTC) circuits have been developed in which the break is replaced with an electronic switching device. This can be a vacuum tube, power transistor, bipolar transistor, power MOSFET, or IGBT. Typically, these oscillators do not incorporate a primary capacitor, the excitation current being obtained directly from the electrical power supply. The operation of these oscillators is usually characterized by a rapid series of primary pulses at a rate identical to the resonant frequency of secondary resonator. The secondary oscillation is refreshed once or twice every cycle, depending on the configuration of the switching circuit. There is no ring down. The terms "undamped wave" and "continuous wave" are often used in relation to vacuum-tube and solid-state Tesla coils. In a properly tuned Tesla coil, the high potential appearing at the secondary's high voltage terminal is developed more or less through a process known as "resonant rise" and exceeds that which would be expected from simple ratio of transformation.

Tesla turbine A rotary engine consisting of multiple ported disks which are mounted in parallel on a shaft and placed within a cylindrical casing. In operation high velocity gases enter tangentially through a wide nozzle at the periphery of the disks, flow between them in free spiral paths to exit through central exhaust ports. The slight viscosity of the gas along with its adhesion to the faces of the disks combined to efficiently transfer the fuel's energy to the disks and on to the shaft.

Test Sequence of operations intended to verify the correct operation or malfunctioning of a piece of equipment or system.

Test board A printed circuit board deemed to be suitable for determining the acceptability of a group of boards produced with the same fabrication process.

Test equipment Finished wafers are tested for their electrical functionality using automated test equipment that makes electrical contact with the wafers by using wafer probes.

Test fixture A device that interfaces between test equipment and the unit being tested.

Test pads Contact points on TAB tape, located outside the OLB window, that are used to facilitate testing and burn-in of the IC while it is in the TAB frame. Test pads remain with the discarded portion of the carrier after excising.

Test pattern A pattern used for inspecting or testing purposes.

Test point A specific point of access to an electrical circuit used for electrical testing purposes.

Test protocol A sequence of control operations required to perform a test.

Test set A device used to measure the frequency and resistance characteristics of a quartz crystal unit. Often called a crystal impedance meter, it is abbreviated as "C.I.M."

Test site A position on a test head where a device-under-test (DUT) is tested.

Test site number A number that identifies a test site location on the test head.

Test specifications A document defining the operational parameters of a device.

Test tone A feature that offers individual adjustment of all four channels (left, centre, right and rear). While in the usual listening position, adjust the relative volume of each channel by using a remote control.

Tester The equipment used for electrically testing semiconductor devices to screen out all units that are not shippable to customers.

Testing collaboratory The National Microelectronics and Photonics Testing Collaboratory.

TETRA Abbreviation for Trans European Trunked Radio or Terrestrial Trunked Radio is a set of standards developed by the European Telecommunications Standardization Institute (ETSI) that describes a common mobile radio communications infrastructure throughout Europe.

Tetrapol A competitive digital PMR technology to TETRA developed by French vendors.

Tetrode A four-electrode tube or transistor.

TFE A Heat-resistant insulation compound made with tetrafluoroethylene (Teflon).

TFT (Thin film transistor)+A7485 LCD panel. A type of LCD flat panel display screen in which each pixel is controlled by one to four transistors. The TFT technology provides the best resolution of all the flat panel techniques, but it is also the most expensive. TFT screens are sometimes called active matrix LCDs.

TFTS Terrestrial Flight Telephone System.

THB Temperature/humidity bias.

THD total harmonic distortion (THD) A measure of signal distortion which assesses the energy that occurs on harmonics of the original signal. It is specified as a percentage of the signal amplitude.

As an example, if a 12kHz signal is applied to the input, THD would look at energy on the output occurring at 24kHz, 36kHz, 48kHz, etc. and compare it to the energy occurring at 12kHz.

THD+N total harmonic distortion plus noise THD+n is the most common audio measurement, a specification that includes both harmonic distortion of the sine wave and non-harmonic noise. THD+N tells the user what amount of hum, noise, and interference has been added to the audio signal by the equipment through which it is passing.

Theatre surround mode One of several different modes in audio receivers and systems that shape sound to more closely resemble a live performance. Theatre mode recreates the same intimate feeling you get in a theatre when watching plays, musicals and solo performances.

Theory of operation The entire earth possesses a naturally existing negative charge or DC electrostatic potential with respect to the conducting region of the atmosphere beginning at an elevation of about 50 kilometers. The potential difference between the earth and this region is on the order of 400,000 volts. Near the earth's surface there is a ubiquitous downward directed E-field of about 100 V/m. In operation, a grounded Tesla coil transmitter creates a local disturbance in this charge. The disturbance manifests itself as an annular distortion of the background electric field around the transmitters ground terminal. At a point in time when a measurement of the e-field component of the EM field at the ground terminal shows zero volts above the background potential, other measurements will show it rising in intensity until a point 1/4 wavelength (1/4 ë) away from the ground terminal is reached (axial projection). From there the e-field diminishes in intensity until, at 1/2 wavelength from the terminal, it again shows zero. At a measurement point approximately one wavelength away from the ground terminal, an induced e-field once again begins to emerge above the average background field level, again increasing in intensity until a second maxim is reached at 1 1/4 wavelength away from the oscillator. With a sufficiently powerful transmitter this phenomenon repeats itself over and over until the antipode is reached, at which point reflection takes place and the transmitted energy begins to travel back to its point of origin in the reverse direction. The elevated terminal of a type-one Tesla-coil transmitter functions as a capacitor plate. Opposite to this plate is every other electrically conducting body to which it is connected, including the earth and the receiving transformer's elevated terminal. The transmitter's elevated terminal serves two purposes: first, it acts as a charge reservoir in relation to the earth's surface in the immediate vicinity of the transmitter; second, it is one of two electrodes, the other electrode being the receiving facility's elevated terminal. The transfer of energy between a type-one transmitter and the requisite receiving transformer is by electrical conduction between the two respective ground terminals, and also between the two elevated terminals. In a low power system, the propagation of energy between the respective elevated terminals is the result of electrostatic induction or displacement currents, much like the transfer of electrical energy between the plates of a capacitor in an AC circuit. For a high-power system the coupling between elevated terminals is by electrical conduction through plasma. The two electrodes act as high voltage discharge terminals for the formation of capacitively coupled discharge plasma with interconnection taking place through the upper level atmosphere. In the operation of a type-two system, the ground between the driven oscillator and the nearby helical resonator is incorporated as a part of the transmitter. As a result, a powerful current flows through the earth between the two ground terminals. Coupling between the transmitter's two elevated terminals is by electrical conduction through plasma. There is also some magnetic inductive coupling between the two helical resonators. As a result, an electrical disturbance is impressed upon the earth. This can be detected "at great distance, or even all over the surface of the globe"

Thermal capacity The amount of energy required to raise the temperature of an object by one degree Celsius. Expressed in Joules/ Kg.

Thermal coefficient of resistance The change in resistance of a semiconductor per unit change in temperature over a specific range of temperature.

Thermal conductivity The property of a material that describes the rate at which heat will be conducted through a unit area of the material for a given period of time.

Thermal control circuit Circuit to monitor and control the temperature of something. For example the integrated temperature controller in Intel's processors.

Thermal cycling A method used to induce stresses on electrical components by means of sequential heating and cooling in an oven. It is used in accelerated reliability testing.

Thermal diffusion Process by which dopant atoms diffuse into the wafer surface by heating the wafer.

Thermal envelope A thermal envelope is a term generally used when describing the walls, windows, doors, ceilings, and floors around heated areas of a structure. By improving the thermal envelope, one can increase one's comfort and reduce energy requirements needed for heating and cooling.

Thermal expansion An increase in size due to an increase in temperature expressed in units of an increase in length or increase in size per degree, i.e. inches/inch/degree C.

Thermal fuse A safety device which interrupts a circuit when it detects excessive temperature.

Thermal gradient The distribution of a differential temperature through a body or across a surface.

Thermal imaging A photographic technique which displays the range of temperatures of a warm body in the form of a colour spectrum. Used as a design verification tool for detecting hot spots in battery and other equipment designs.

Thermal management The use of various temperature monitoring devices and cooling methods, such as forced air flow, within a processor or FPGA-based system, to control overall temperature of ICs and internal cabinet temperatures.

Thermal monitor The integrated thermal control system used in Intel's processor devices.

Thermal oxidation A process step used in the manufacturing of integrated circuits in which a layer of silicon dioxide is grown on the surface of a silicon wafer in a heated reaction chamber containing oxygen gas or water vapor.

Thermal profile A time versus temperature graph that displays the temperatures an assembly is subjected to over time in an oven during processes such as reflow soldering or the curing of adhesives, encapsulants, and conformal coatings.

Thermal protection A protective feature that shuts down a power supply if its internal temperature exceeds a predetermined limit.

Thermal relay Relay activated by a heating element.

Thermal relief pad A special pattern etched around a via or a plated through-hole to connect it into a power or ground plane. A thermal relief pad is necessary to prevent too much heat being absorbed into the power or ground plane when the board is being soldered.

Thermal resistance Normally stated in terms of °C/W, it is the indicator of the package's ability to dissipate the heat generated by the chip during operation. It is the indicator of the chip's ability to pass the heat generated by the semiconductor junctions to the package, that is the thermal resistance between the semiconductor junction and the case.

Thermal runaway Problem that can develop in an amplifier when an increase in temperature causes an increase in collector current. The increase in collector current causes a further increase in temperature and so on. Unless the circuit is designed to prevent this condition, the device can be driven into saturation.

Thermal secondary breakdown Burnout of a semiconductor junction area as the result of a reverse-bias voltage or current induced thermal runaway condition.

Thermal sensitivity shift The sensitivity shift due to changes of the ambient temperature from room temperature to the specified limits of the compensated temperature range.

Thermal shock Thermal shock is the effect of heat or cold applied at such a rate that non-uniform thermal expansion or contraction occur within a given material or combination of materials. In connectors, the effect can cause inserts and other insulation materials to pull away from metal parts.

Thermal shutdown Deactivating a circuit when a measured temperature is beyond a predetermined value.

Thermal stability The ability of a circuit to maintain stable characteristics in spite of increased temperature.

Thermal tracking Typically used to refer to the problems associated with optical interconnection systems whose alignment may be disturbed by changes in temperature.

Thermal zero shift An error due to changes in ambient temperature in which the zero pressure output shifts. Thus, the entire calibration curve moves in a parallel displacement.

THERMDC Thermal diode cathode pin on AMD and Intel processors.

Thermionic emission The production of electron emission by thermal means.

Thermistor A temperature-sensing element composed of sintered semiconductor material which exhibits a large change in resistance proportional to a small change in temperature. Thermistors usually have negative temperature coefficients.

Thermocompression bonding The joining of two materials without an intermediate material by the application of pressure and heat in the absence of electrical current.

Thermocouple A temperature sensor formed by the junction of two dissimilar metals. A thermocouple produces a voltage proportional to the difference in temperature between the hot junction and the lead wire (cold) junction.

Thermocouple meter A combination of the thermocouple and sensitive millivoltmeter which measures the electromotive force developed by the thermocouple as the result of its heating by the current passed through a resistance element in thermal contact with it. Thermocouple meters read rootmean-square values of alternating currents at frequencies extending as high as radio frequencies.

Thermode A set of blades or bars that are pulse-heated. A thermode is used to hold component leads in place and reflow solder them to bonding pads on a substrate during hot bar reflow soldering.

Thermometry Relating to the measuring of temperature.

Thermopile An arrangement of thermocouples in series such that alternate junctions are at the measuring temperature and the reference temperature. This arrangement amplifies the thermoelectric voltage. Thermopiles are usually used as infrared detectors in radiation pyrometry.

Thermoplastic A plastic compound that will soften and melt with sufficient heat. Thermoplastic insulation compounds are used to manufacture certain types of electrical cables.

Thermoset A plastic compound that will not remelt. Thermoset insulation compounds are used to manufacture certain types of cables.

Thermosetting plastic Polymer materials that cure (irreversably polymerize) at specific temperature and time conditions.

Thermostat An apparatus arranged so that it may be adjusted to maintain and keep constant any practicable temperature. A system enclosed in a thermostat is thus kept at a definite constant temperature. This is usually accomplished by actuating controls which Turn the heating mechanism on when an upper temperature limit is reached and which turn it off when a lower temperature limit is reached.

Thermowell A closed-end tube designed to protect temperature sensors from harsh environments, high pressure, and flows. They can be installed into a system by pipe thread or welded flange and are usually made of corrosion-resistant metal or ceramic material depending upon the application.

Thermtrip l Pin name of the thermal trip output pin of AMD processors. The pin is asserted at a nominal die temperature of 125'C.

Thevenin's theorem Theorem that replaces any complex network with a single voltage source in series with a single resistance.

THHN A thermoplastic-insulated, nylon-jacketed conductor designed for use in dry locations and an operating temperature of up to 90 degrees Celsius

Thick film resistor Fixed value resistor consisting of thick-film resistive element made from metal particles and glass powder.

Thick-film capacitor Capacitor consisting of two thick-film layers of conductive film separated by a deposited thick-layer dielectric film.

Thick-film process A process used in the manufacture of hybrids and, to a lesser extent, multichip modules in which signal and dielectric (insulating) layers are screen-printed onto the substrate.

Thin film A film of conductive or insulating material, usually deposited by sputtering or evaporation, that may be made in a pattern to form electronic components and conductors on a substrate or used as insulation between successive layers of components.

Thin film capacitor Capacitor in which both the electrodes and the dielectric are deposited in layers on a substrate.

Thin film detector (TFD) A temperature detector containing a thin layer of platinum and used for precise temperature readings.

Thin-film electroluminescence (TFEL) An advanced approach to the production of large-panel fluorescent displays that allows them to compete with other display technologies in data-display and graphics-display applications.

Thin-film head Currently, the most advanced type of magnetic recording head used to read and write information on disk drives. The two types of TFHs are inductive and the more advanced magneto-resistive.

Thin-film process A process used in the manufacture of hybrids and multichip modules in which signal layers and dielectric (insulating) layers are created using opto-lithographic techniques.

Thixotropic ratio An indication of thixotropy as a ratio of viscosities at two different shear rates.

Thombstoning A soldering defect in which a chip component is pulled into a vertical position leaving one side unsoldered.

Thomson effect When current flows through a conductor within a thermal gradient, a reversible absorption or evolution of heat will occur in the conductor at the gradient boundaries.

Three layer TAB tape Tape constructed with a conductive metal layer, typically 35 micron thick copper; a flexible heat resistant support layer, typically 125 micron thick polyimide; and an adhesive layer between the conductive layer and the flexible support layer. The polyimide thickness is not limited in the three layer tape manufacturing process because the openings in the tape are punched out rather than etched. The copper is usually tin or gold plated.

Three phase Multiple phase power supply or load that uses at least three wires where a different voltage phase from a common generator is carried between each pair of wires. The voltage level may be identical but the voltages will vary in phase relationship to each other by 120 degrees.

Three phase line A three phase line is capable of carrying heavy loads of electricity, usually to larger commercial customers.

Three phase supply AC supply that consists of three AC voltages 120° out of phase with each other.

Three terminal regulator A power integrated circuit in a 3-terminal standard transistor package. It can be either a series or shunt regulator IC.

Threshold Minimum point at which an effect is produced or detected.

Threshold limit value A guideline for the exposure of humans to solvents; it is expressed as a time weighted average (TWA) of the parts per million of vapor in air.

Threshold of pain (in dB SPL) The minimum value of sound pressure of a given frequency that will cause pain to a listener 50% of the time. Discomfort begins at 118 dB SPL Actual pain starts around 140 dB SPL within the frequency range between 200 Hz and 10,000 Hz.

Threshold voltage The voltage level at which a device will recognize a falling or rising voltage as a change in logic state.

Through hole A method for mounting components on a printed circuit board (PCB) in which pins on the component are inserted into holes in the board and soldered in place.

Through hole via A via that passes all the way through the substrate.

Through hole-mounting A mounting technique for semiconductor packages in which the device leads are passed through holes in the mounting surface. Attachment of leads may be accomplished with solder or with other mechanical means.

Throughput The measure of an equipment's capacity to process, usually expressed in wafers per hour.

Throughput rate The maximum repetitive rate at which a data converter can operate within a specified accuracy.

Throw The distance from projector lens to screen.

Thru hole A conductor used to make electrical and mechanical connection between conductive patterns on opposite sides of a printed circuit board.

Thumper A high voltage device used to locate an underground cable fault. The device applies a high voltage to the faulted cable with a resulting discharge to ground at the location of the fault. When the discharge occurs, there is an audible "Thump" which is used to locate the fault.

Thyratron A gas-tube rectifier with a grid electrode that has the ability to initiate current flow when the anode is positive. Once conduction occurs, the grid loses control until the plate current is interrupted.

Thyristor Also called a Silicon-Controlled Rectifier or SCR, it is a solid-state high current semiconductor switching device similar to a diode, with an extra terminal which is used to turn it on. Once turned on, the thyristor will remain on (conducting) as long as there is a significant current flowing through it. If the current falls to zero, the device switches off.

Terrestrial interference registration and identification system (TIRIS) Uses radio frequency identification to electronically control, detect and track objects by manipulating radio signals. Applications include traffic management, logistics systems, antitheft devices and security systems.

TIA (Transimpedance amplifier) A device used to convert input currents to output voltages.

Tilt angle A fixed angle measured from the horizontal to which a solar array is tilted. The tilt angle is chosen to maximize the array output. Depending upon latitude, season and time of day this angle will vary.

Timbre The "voice" or basic tonal quality of a speaker. To avoid distractions and achieve realism in a home theatre system, each of the system's speakers should have an identical, or at least similar, timbre.

Time base corrector (TBC) Digitally removes jitter from fluctuating video signals to deliver stable picture even with old tapes. Also eliminates colour distortion caused by jitter in the chroma signal.

Time base error Slight errors in the line-to-line position of video information that occur between recording and playback. At the time of playback, these appear as serrations, tending to make the edges of the image waver.

Time base generator A sync generator which puts a clock signal on the videotape to refer to for precise horizontal lock-up of an image.

Time code A digital or binary code used to label each frame of a video signal. This is very useful for editing the video since the time code is in the form of hours, minutes, seconds, and frames.

Time constant Time required for a capacitor in an RC circuit to charge to 63% of the remaining potential across the circuit. Also time required for current to reach 63% of maximum value in an RL circuit. Time constant of an RC circuit is the product of R and C. Time constant of an RL circuit is equal to inductance divided by resistance.

Time delay The introduction of an intentional delay to the opening function of a protective device.

Time division multiple access (TDMA) A communications technique that uses a common channel (multipoint or broadcast) for communications among multiple users by allocating unique time slots to different users. TDMA is used extensively in satellite systems, local area networks, physical security systems, and combat-net radio systems.

Time division multiplexing (TDM) Transmission of two or more signals on the same path, but at different times.

Time lapse Records a single frame of video footage at preset time intervals. This feature compresses the time of slow moving objects. When video is played back, events that take minutes or hours to unfold in normal time are compressed to seconds.

Time out of bag Time a component is exposed to a humid atmosphere.

Time scan/speedwatch Allows you to skip commercials and previews during playback.

Time search Locates the precise position of recordings by elapsed time on a VCR.

Time domain analysis A method of representing a waveform by plotting amplitude over time.

Time of day rate Also know as time-of-use (TOU) rate; charges customers according to when they use electricity. Customers pay more for electricity used during peak demand periods and pay less during the off-peak periods. These rates are structures to be reflective of the variation in the cost of service within the billing period.

Time of flight The time taken for a signal to propagate from one logic gate or opto-electronic component to another.

Time out error An error that occurs after the maximum time interval for an expected event to take place has lapsed.

Time varying fields Electric and magnetic fields that change in intensity or strength with time. Examples include 60 Hz, modulated, and transient fields.

Tin dip A term commonly misapplied to the solder dipping of device leads (see Solder dip).

Tin lead plating An electroless plating process in which exposed areas of copper on a circuit board are coated with a layer of tin-lead alloy. The alloy is used to prevent the copper from oxidizing and provides protection against contamination.

Tinning An abbreviation of tin-lead plating, which is an electroless plating process in which exposed areas of copper on a circuit board are coated with a layer of tin-lead alloy. The alloy is used to prevent the copper from oxidizing and provides protection against contamination.

Tip to tip dimension With respect to component geometry, the distance between the ends of the leads on opposite sides of a component after excising and forming.

Titler Allows you to add titles to your videos. Camcorders or VCRs can have either preprogrammed titles to choose from or an alphanumeric keypad (character generator) that allows you to create your own titles.

To package Cylindrical, metal can type of package of some semiconductor components.

Toggle switch A switch having two positions or two states. When an activating force is applied, the state changes.

Toggle To switch between alternate states. For example, between on and off, or caps and lower case.

Tolerance and stability Three main components of frequency control product specifications are:

1. Calibration tolerance at room temperature (25°C)

2. Stability over the temperature range

3. Aging

Calibration at room temperature is a measurement of the accuracy of the frequency at +25°C. Crystal frequencies are adjusted within the stated tolerance by changing the mass of the electrode. Lower frequencies are less sensitive to mass change and are therefore easier to hold tighter tolerances. Tolerance and stability are measured in parts per million (ppm).

Tolerance The allowable percentage variation of any component from that stated on its body. For instance: red=2%, gold=5% and silver=10%.

Tollsaver With the ring delay switch set to toll saver, an incoming call is answered on the fourth ring, if no previous messages have been received, and on the second ring otherwise. This feature tells the user if there are no recorded messages, and the user simply hangs up before the fourth ring.

Tombstoning The lifting of one end of a passive surface mount component during solder reflow caused by surface tension and unbalanced forces of solder wetting.

Toroidal coil An inductor or transformer whose core consists of a concentrically wound ribbon of magnetic material. Also used to filter noise on a DC supply line.

Torsional strength The torque required to separate adhesive bonded (and cured) materials and/or components.

Total clearing time The time elapsing form the initiation of overload current to final current interruption.

Total dose radiation Accumulated radiation due to prolonged exposure to a radiation field (as encountered, for example, by satellites in the earth's Van Allen belt).

Total dynamic head (TDH) When a pump is lifting or pumping water, the vertical distance from the elevation of the suction side of the pump to the elevation of the discharge side of the pump. The total dynamic head is the static head plus pipe friction losses. Static head is the vertical distance between water surfaces when the water is not moving.

Total effect The change in a stabilized output produced by concurrent worst case changes in all influence quantities within their rated range.

Total harmonic distortion (THD) 1.A measure of a signal's distortion content, represented by the harmonics of that signal expressed as a percentage of the signal amplitude;

2. the ratio of the rms sum of the first few (2nd to 6th) harmonic components of a device's output to its fundamental value, usually expressed in Db.

Totem pole A standard CMOS output structure where a P-channel MOSFET is connected in series with an N-Channel MOSFET and the connection point between the two is the output. The P-FET sits on top of the N-FET like a "totem pole." Both gates are driven by the same signal. When the signal is low, the P-FET is on; when the signal is high, the N-FET is on. This creates a push-pull output using just two transistors.

Touch panel A control panel with a flat surface (usually with graphic divisions or buttons) that functions as a switch or control. Also called a "touch screen."

Touch-up The identification and elimination of defects in a product.

Tower A tower is a steel structure found along transmission lines which is used to support conductors.

Trace A conducting connection between electronic components. May also be called a track or a signal. In the case of integrated circuits, such interconnections are often referred to collectively as metalization.

Traceability Traceability is the "chain" of instruments used to calibrate other items, provided through certification. Calibration certificates from NAMAS facilities are said to have direct traceability to National Standards.

Track and trace The process of retrieving information about the movement and location of goods.

Tracking A characteristic of a multiple-output power supply that describes the changes in the voltage of one output with respect to changes in the voltage or load of another.

Tracking array An array that is mounted on a movable structure that attempts to follow the path of the sun. Some tracking arrays are single axis while others are dual.

Tracking error The error indication at a scale mark, expressed in percentage of fiducial value, when the instrument is energized by the proportional value of the actuated end-scale excitation. On offset-zero indicators, the higher end-scale value should be used as the reference value.

Tracking regulator A plus or minus two-output supply in which one output tracks the other.

Tracking servo The control circuit used to keep the pick-up over the desired track.

Tracks A conducting connection between electronic components. May also be called a trace or a signal. In the case of integrated circuits, such interconnections are often referred to collectively as metalization.

Traction battery A high power deep cycle secondary battery designed to power electric vehicles or heavy mobile equipment.

Transceiver A combination of a transmitter and a receiver having a common frequency control and usually enclosed in a single package. Extensively used in two-way radio communications at all frequencies.

Transco A for profit power transmission company.

Transconductance Also called mutual conductance. Ratio of a change in output current to the change in input voltage that caused it.

Transconductance amplifier An amplifier that converts a voltage to a current.

Transconductance As most commonly used, the interelectrode transconductance between the control grid and the plate. At low frequencies, transconductance is the slope of the control-grid-to-plate transfer characteristic. More generally, the transconductance may be defined as the ratio of the change in the current flowing from one pair of terminals, these terminals being shortcircuited, to the change in the potential difference across another pair of terminals. All changes are assumed small.

Transconductance, conversion (of a heterodyne conversion transducer) The quotient of the magnitude of the desired output-frequency component of current by the magnitude of the input-frequency component of voltage when the impedance of the output external termination is negligible for all the frequencies which may affect the result. Unless otherwise stated, the term refers to the cases in which the input-frequency voltage is of infinitesimal magnitude. All direct electrode voltages and the magnitude of the local-oscillator voltage must be specified, fixed values.

Transcriber Equipment associated with a computing machine for the purpose of transferring input (or output) data from a record of information in a given language to the medium and the language used by a digital computing machine; or from a computing machine; or from a computing machine to a record of information.

Transducer Device that accepts an input of energy in one form and produces an output of energy in some other form, with a known, fixed relationship between the input and output. One widely used class of transducers consists of devices that produce an electric output signal, e.g., microphones, phonograph cartridges, and photoelectric cells

Transducer electronic data sheet A Transducer electronic data sheet, or TEDS, is a method for plug-and-play sensor and transducer hook-up in which the sensor's calibration information is stored within the device and downloaded to the master controller when requested.

Transducer factor The product of the current transformer ratio (CTR) and the voltage transformer ratio (VTR). Also called the power ratio.

Transducer vibration Generally, any device which converts movement, either shock or steady state vibration, into an electrical signal proportional to the movement; a sensor.

Transducer, self-contained A device internal to an instrument that changes an electrical input into an electrical output, of either the same or a different kind, in such a manner that the desired characteristics of the input are measured by a permanent –magnet moving-coil mechanism.

Transfer switch A switch designed to transfer electricity being supplied to loads (appliances, for example) from one source of power to another. A transfer switch may be used to designate whether power to a distribution panel will come from a generator or inverter.

Transformer An electromagnetic device for changing the voltage of alternating current electricity. Every transformer has a primary coil and one or more secondary coils. The primary coil receives electrical energy from a power source and couples this energy to the secondary coils by means of a changing magnetic field. The energy appears as an electromagnetic force across the coil, and if a load is connected to the secondary the energy is transferred to the load.

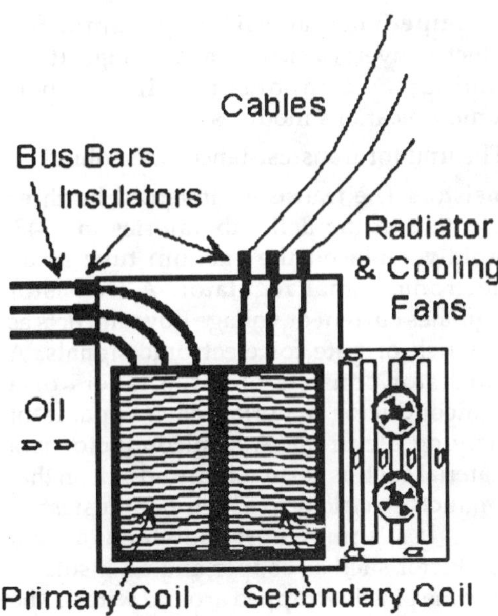

The voltage at which electric power is used in motors and lamps is less than that required for distribution. It is necessary to raise the voltage at the generating station to the value required for transmission, which is called "step up" the voltage. Then to lower it at the point of consumption to the values required by the motors and lamps, which is called "step down". The transformer is what makes these changes in voltage.

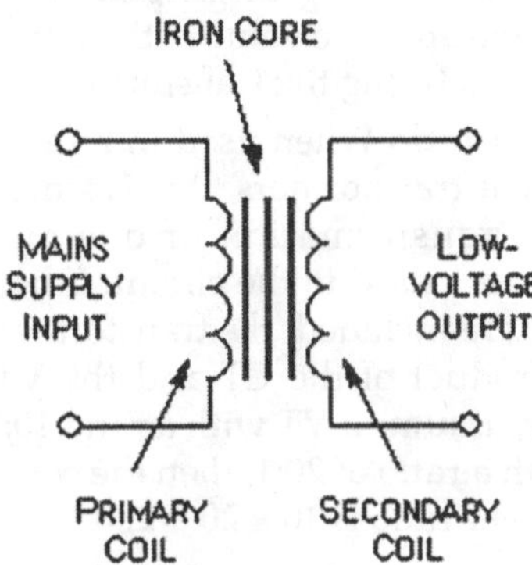

Fig. Transformer

Transformer coupling Also called inductive coupling. Coupling of two circuits by means of mutual inductance provided by a transformer.

Transformer insulation This is the material that is used to provide electrical insulation between transformer windings at different voltage levels and also between the energized parts and the metal tank of the transformer. Generally, for large transformers used in power applications, this is a combination of Kraft paper and mineral oil; however, in indoor building applications, transformers also may be insulated with plastic materials.

Transformer loss The ratio of the power that would be delivered to a specified load impedance if an ideal transformer were substituted for the actual transformer, to the power delivered to the specified load impedance by the transformer, under the condition that the impedance ratio of the ideal transformer is equal to that specified for the transformer. If the input and/or output power consist of more than one component, such as multifrequency signal or noise, then the particular components used and their weighting must be specified. This loss is usually expressed in decibels.

An older definition of transformer loss, which is now deprecated, is the loss which would be eliminated by the insertion, at any point in a transmission system, of an ideal transformer having an impedance ratio equal to the absolute value of the ratio of the impedances facing the transformer.

Transformer ratio When used in reference to instrument transformers, this is simply the ratio of transformation of one or more transformers used in the circuit. If both Cts and VTs are included, the transformer ratio is the product of the CT and the VT. For example, assume a VT with a ratio 10:1 and a CT with a ratio of 20:1, then the combined transformer ratio is 10 x 20 = 200:1.

Transformer taps A transformer is a simple device that "transforms" electricity from one voltage to another. Transformers are added to conventional, 8 ohm loudspeakers to adapt them for use in constant voltage, multi-speaker distributed audio systems. The transformers are "tapped", that is, designed with several different output points, to allow for different output wattages from the 70 V or 100 V constant voltage input. Taps are typically spaced at 2x increments, for example, 2 watts, 4 watts, 8 watts, 16 watts, etc.

Transformer voltage regulators Mechanisms that use multiple voltage taps on a transformer-like device to adjust voltage on a power line. As the voltage increases or decreases on the circuit, sensors in the voltage regulator call for the input or output of the regulator to connect to different voltage taps on the transformer winding. This adjusts the output voltage on the line to an acceptable level.

Transient A high-voltage surge or spike in an electrical transmission system caused by lightning strikes to nearby transformers, overhead lines, or the ground, and which may persist for a relatively short time after the phenomenon (sometimes called ringing). They may also result from the switching of motors, short circuits, or system switching.

Transient effect The result of a step change in an influence quantity on the steady state values of a circuit.

Transient recovery time The time required for the output voltage of a power supply to settle within specified output accuracy limits following a transient.

Transient response The ability of an electrical or other device to respond faithfully to sudden changes to the input conditions.

Transient response time The interval between the time a transient is introduced and the time it returns and remains within a specified amplitude range.

Transient vibration A temporary vibration or movement of a mechanical system.

Transimpedance The transfer function of a TIA; the output voltage divided by the input current.

Transimpedance amplifier An amplifier which converts a current to a voltage. It is a familiar component in fiber-communications modules.

The unit for transresistance is the ohm.

Transistors The transistor, invented by three scientists at the Bell Laboratories in 1947, rapidly replaced the vacuum tube as an electronic signal regulator. A transistor regulates current or voltage flow and acts as a switch or gate for electronic signals. A transistor consists of three layers of a semiconductor material, each capable of carrying a current. A semiconductor is a material such as germanium and silicon that conducts electricity in a "semi-enthusiastic" way. It's somewhere between a real conductor such as copper and an insulator (like the plastic wrapped around wires). The semiconductor material is given special properties by a chemical process called doping. The doping results in a material that either adds extra electrons to the material (which is then called N-type for the extra negative charge carriers) or creates "holes" in the material's crystal structure (which is then called P-type because it results in more positive charge carriers). The transistor's three-layer structure contains an N-type semiconductor layer sandwiched between P-type layers (a PNP configuration) or a P-type layer between N-type layers (an NPN configuration). As the current or voltage is changed in one of the outer semiconductor layers, it affects a larger current or voltage in the inner layer resulting in the opening or closing of an electronic gate. Today's computers use circuitry made with complementary metal oxide semiconductor (CMOS) technology. CMOS uses two complementary transistors per gate (one with N-type material; the other with P-type material). When one transistor is maintaining a logic state, it requires almost no power. Transistors are the basic elements in integrated circuits or ICs, which consist of very large numbers of transistors interconnected with circuitry and baked into a single silicon microchip or "chip."

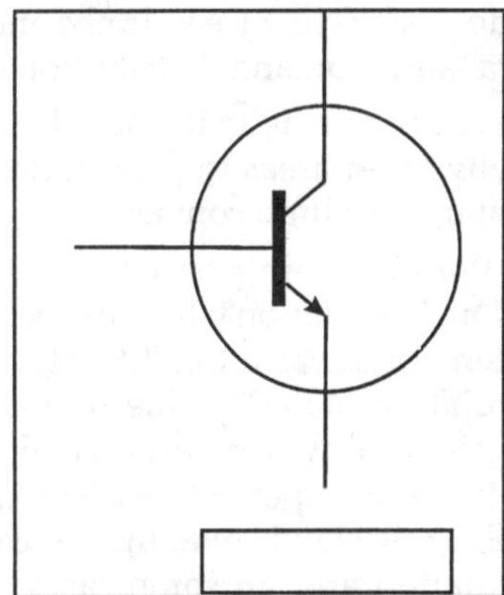

Fig. NPN Transistors

Transistor-transistor logic (TTL) Logic gates implemented using particular configurations of bipolar junction transistors.

Transition region The part of the spectrum between the passband and the stopband.

Transitional flow Flow between laminar and turbulent flow, usually between a pipe Reynolds number of 2000 and 4000.

Transmission The sending and receiving of telecommunications messages through appropriate channels.

Transmission cable Two or more transmission lines. If the structure is flat, it is sometimes called Flat Transmission Cable to differentiate it from a round structure such as a jacketed group of coaxial cables.

Transmission control protocol (TCP) In the Internet Protocol suite, a standard, connection-oriented, full-duplex, host-to-host protocol used over packet-switched computer communications,

Transmission lines Transmission lines are heavy wires that carry large amounts of electricity over long distances from a generating station to places where electricity is needed.. Transmission lines are held high above the ground on tall towers called transmission towers.

Transmission system Normally, the highest voltage network of an electric system. This is the portion of the system that carries high power over the longest distances. Typically operating at voltages in excess of 100 kV, and most usually at 200 kV and above.

Transmit to Is to deliver information between two or more locations. Transmitted information occurs through media such as Local Area Networks/Wide Area Networks and cable, coax, fiber optics and twisted pair wires, in addition to satellites and radio or optical wireless devices.

Transmitter Equipment which generates RF power to be fed to an antenna or other output interface for transmission. In modern radio equipment it consists of active components such as a modulator, driver and PA and passive components such as a TX filter. Taken together, these components impress an information carrying signal onto an RF carrier of the correct frequency by instantaneously adjusting its phase, frequency, or amplitude and provide enough gain to the signal to project it through the ether to its intended recipient.

Transponder A radio transmitter-receiver that is activated when it receives a predetermined signal. RFID transponders come in many forms, including smart labels, simple tags, smart cards and keychain fobs. RFID tags are sometimes referred to as transponders.

Transport, in plasmas The problem of understanding the motions of particles in a plasma (and the related flows of energy, momentum, and other physical quantities) is extremely important in many if not all areas of plasma research. The theory of transport in plasmas is highly complex, but an understanding of transport is vital to controlled fusion research (where insufficient energy confinement is a major obstacle to producing fusion energy), plasma astrophysics (where radiation transport through plasmas often plays a dominant role), and many other areas including high energy-density plasmas, plasma processing of materials, space plasmas, and more. Since plasmas are many-body systems, it is not possible to follow all 6 degrees of freedom of each particle in the plasma, and consequently statistical methods and fluid theories must be employed, though even these often prove barely tractable for realistic situations. The wide variety of possible plasma conditions (spanning over 30 orders of magnitude in density and over 6 orders of magnitude in temperature) leads to a wide range of phenomena, including flows, turbulence, waves and nonlinear wave-particle interactions, and shocks. Specific approximations are generally needed to treat specific classes of plasma conditions over specific time and distance scales. Some key topics in plasma transport research include the determination of transport coefficients such as viscosity and diffusivity, and related parameters such as electrical conductivity and particle and energy confinement times.

Transportability Those characteristics of hardware or software which allow its transferal from one system to another system and to interface compatibly with the hardware and software of the new system.

Transrectification factor The quotient of the change in average current of an electrode by the change in the amplitude of the alternating sinusoidal voltage applied to another electrode, the direct voltages of this and other electrodes being maintained constant. Unless otherwise stated, the term refers to cases in which the alternating sinusoidal voltage is of infinitesimal magnitude.

TRAU (Transcoder rate adapter unit) The transport unit for a 16kbit/s traffic channel on the A-bis interface.

Tray A cable tray system is a unit or assembly of units or sections, and associated fittings, made or metal or other noncombustible materials forming a rigid structural system used to support cables. Cable tray systems (previously termed continuous rigid cable supports) including ladders, troughs, channels, solid bottom trays, and similar structures.

Treble The upper range of audio frequencies, above approximately 4,000Hz.

Tree crews Tree crews are teams of employees or vendors who clear trees, limbs and brush from transmission and distribution lines.

Tree wire A tree wire is an insulated wire used in heavily treed areas to protect lines from momentary tree limb contact.

Treeing Water treeing is a form of cable insulation degradation where micochannels, that often appear as a tree-like structure in the insulation, develop due to a complex interaction of water, electrical stress, impurities and imperfections. The tree-like channels grow slowly over time, weakening the insulation and, in some cases, lead to cable failure.

Trench A semiconductor device feature that resembles a trench. A trench is typically formed as the repository of a differing material (than the trench walls), such as copper that will become an interconnect or an oxide in shallow trench isolation that will isolate gate structures from each other.

TRF tuned radio frequency An AM receiver with one or more stages of radio frequency before the detector.

TRIAC (triode AC switch) A three-terminal silicon device that functions as two SCRs configured in an inverse, parallel arrangement, providing a means of providing load current during both halves of the AC supply voltage. A TRIAC is generally used for motor speed control. Since load current (armature current) flows during both halves of the applied AC voltage, the motor rotates smoothly at all rotational speeds.

Triac Bidirectional gate controlled thyristor similar to an SCR, but capable of conducting in both directions. Provides full wave control of AC power.

Triangular wave A repeating wave that has equal positive going and negative going ramps. The ramps have linear rates of change with time.

Triaxial cable A cable with three conductors: one conductor surrounded by an inner shield and an isolated outer shield. Generally, the inner shield is connected to a guard potential and the outer shield to signal LOW or ground.

Tri-band Refers to a mobile phone able to operate on the three internationally designated GSM frequencies- 900, 1800 and 1900MHz.

Triboelectric effects The creation of electrostatic charge that occurs when two surfaces contact and then separate, leaving one positively charged and the other negatively charged.

Triboelectric noise The generation of electrical charges caused by layers of cable insulation. This is especially troublesome in high impedance accelerometers.

Tribo-electrification When two dissimilar substances are rubbed together, they become oppositely electrified; and if either is an insulator, it retains a charge. For example, if glass is rubbed with silk, the glass becomes positive and the silk negative. Careful experiments make it appear probable that this is a type of contact potential difference, and that the friction serves only to bring about surface contact over a larger area.

Trickle charge A continuous charge at low rate, balancing losses through local action and/ or periodic discharge, to maintain a cell or battery in a fully charged condition. Normally at a C/20 to C/30 rate.

Trigger An external stimulus that initiates one or more instrument functions. Trigger stimuli include; a front panel button (TAKE), an external input voltage pulse.

Trigger circuit Many of the recent developments in the application of electron tubes have utilized various types of triggering action. By this is meant a circuit which suddenly changes its electrical condition, just as if a trigger had tripped it. While there are many varieties of such circuits, they represent some form of unstable electrical equilibrium so the triggering disturbance, whatever it may be, causes it to break over into a new condition. These circuits range from very simple thyratron devices to elaborate vacuum-tube circuits which give some special type of output. Many of the modern developments would be impossible without these devices.

Triggering Initiation of an action in a circuit which then functions for a predetermined time. Example: The duration of one sweep in a cathode ray tube.

Tri-level sync A sync level scheme developed for HDTV in which the sync line first goes low, then transitions high while going through the reference voltage level, and then drops back down to the reference voltage. The transition of the positive-going sync signal through the reference voltage is the sync trigger.

Trillium A company that makes DUT or ATE systems.

Trim and form The step in the assembly process for devices assembled on a lead frame (such as flatpacks and dual-in-line packages) where the lead frame is trimmed off and the leads bent or formed into their specified positions.

Trim effect In a crystal oscillator, the degradation of frequency-vs.-temperature stability, and marked frequency offset, resulting from frequency adjustment which produces a rotation or distortion, or both, of the inherent frequency-vs.-temperature characteristic.

Trim sensitivity A measure of the incremental fractional frequency change for an incremental change in the value of load capacitance. Trim sensitivity (S) is expressed in terms of PPM/pF and is calculated with the following equation: where (Ct) is the sum of the shunt capacitance (CO) and the load capacitance (CL).

Trimmability The ability to adjust the frequency of an oscillator.

Trimmer Small value variable capacitor, resistor or inductor used to fine tune a larger value.

Trimmer capacitor A small variable capacitor used in parallel across a larger capacitor to adjust the total capacitance a small amount.

Triple point (water) The thermodynamic state where all three phases, solid, liquid, and gas may all be present in equilibrium. The triple point of water is .01°C.

Triple point The temperature and pressure at which solid, liquid, and gas phases of a given substance are all present simultaneously in varying amounts.

Trip out A trip-out is a disconnection of an electric circuit. When a line "trips out", the circuit breaker has opened and the line is out of service. The action of breaking a circuit usually refers to an automatic rather than a manual action.

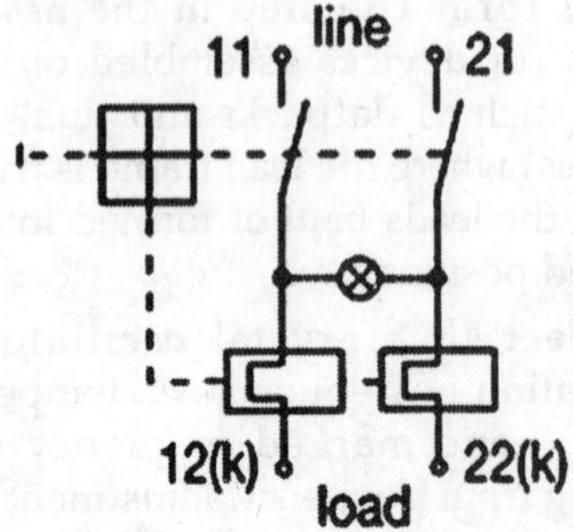

Fig. Trip free circuit breaker

Tri stage matrix A switching architecture that utilizes input stages, mid-stages and output stages in an efficient multi-stage matrix. Larger switching arrays are better served by this type of design since multiple signal paths are available for redundancy, the number of actual hardware signal crosspoints is reduced, and the physical size and cost of the unit is dramatically reduced. This design is not cost effective in switching arrays smaller than 32x32.

Tri state The tri-state function allows the oscillator to be isolated from the circuit upon application of a command signal. When this feature is activated, the output of the oscillator is in tri-state mode. The tri-state mode allows the customer to remove the oscillator from their circuit without physically removing it. Useful for tuning, testing or trouble shooting their board.

Tri state function A function whose output can adopt three states: 0, 1, and Z (high-impedance) The function does not drive any value in the Z state and, in many respects, may be considered to be disconnected from the rest of the circuit.

Tri state output An oscillator with this feature allows the output to be placed into a high impedance state. This feature is activated by the application of a logic control voltage to pin 1 of the oscillator.

Trivalent element One having three valence electrons. Used as an impurity in semiconductor material to produce p-type material. Most commonly used trivalent elements are: Aluminum, Gallium and Boron.

TRL Technology readiness level.

Troubleshooting Systematic approach to locating the cause of a fault in an electronic circuit or system.

True power In AC, the product EI cos  (the apparent power multiplied by the power factor).

True RMS Amps 1) The effective value of an ac signal. For an amp signal, true RMS is a precise method of stating the amp value regardless of waveform distortion. 2) An AC measurement which is equal in power transfer capability to a corresponding dc current.

True RMS The true root-mean-square value of an AC or AC-plus-DC signal, often used to determine power of a signal. For a perfect sine wave, the RMS value is 1.11072 times the rectified average value, which is utilized for low-cost metering. For significantly non-sinusoidal signals, a true RMS converter is required.

True RMS volts 1. The effective value of an ac voltage value regardless of the waveform distortion.

2. An ac measurement which is equal power transfer capability to a corresponding dc voltage.

TrueSync A technology which enables the optimal synchronisation of calendars, address books, action lists and memoranda. It enables multi-point, one-step synchronisation of wireless and wireline devices, desktop computers and server-based applications and services.

Tubular plate (battery) A positive plate which is composed of asembly of porous tubes of perforated metal or tissure with or without a central current collector spine. The active material is placed within the tube.

Tundershoot A transient change in output voltage in excess of specified output regulation limits.

Tuned circuit Circuit that can have its component values adjusted so that it responds to one selected frequency and rejects all others.

Tuner A component (or section of one) that receives radio signals and selects one broadcast from many.

Tungsten halogen lamp A gas-filled tungsten halogen lamp containing a certain proportion of halogens.

Tunnel diode Heavily doped junction diode that has negative resistance in the forward direction of its operating range.

Turbine A mechanical structure with rotatable blades mounted onto its assembly and mechanically coupled to an electrical generator. When a turbine is placed in the path of flowing water, steam, or moving air, the movement of the water, steam, or air across the blades causes them to turn. The generator's armature rotates within a magnetic field which produces electrical energy at the terminals of the generator.

Turbulent flow When forces due to inertia are more significant than forces due to viscosity. This typically occurs with a Reynolds number in excess of 4000.

Turning point An AT cut crystal has a temperature vs. frequency characteristic that can be represented by a third order polynomial. The turning points are points of zero slope where the slope reverses sign. Usually there is a lower turning point around -4° C and an upper turning point around +55° C. Often ovenized oscillators hold the temperature of the crystal at the upper turning point to obtain maximum temperature stability.

Turn on time Sum of delay time and rise time.

Turnover temperature The temperature at which the frequency is at the top of the parabolic curve.

Turns ratio Ratio of the number of turns in the secondary winding of a transformer to the number of turns in the primary winding.

TVM Test vector monitor

TVRO Television receive-only

TVS transient voltage suppressor Semiconductor device designed to protect a circuit from voltage and current transients. Typically implemented as a large silicon diode operating in avalanche mode to absorb large currents quickly.

TW 1. A thermosplastic insulated, moisture resistant conductor designed for use in wet or dry locations and an operatating temperature of up to 60 degrees celsius.

2. Trapezoidal Wire. Built as ACSR-TW or ACSS-TW, Trapezoidal Wire uses

trapezoidal formed strands in its construction to reduce overall diameter of the finished cable.

Tweak To adjust or fine-tune.

Tweeker A small screwdriver for making sensitive adjustments to audio/visual and other electronic equipment.

Twin dome sound system Provides high-quality audio with space-saving design. The dual-magnet speakers are located on either side of the picture tube to provide optimum stereo separation, while facilitating placement of the TV in a wall unit.

Twinaxial cable A cable with three conductors: one twisted pair of conductors surrounded by an outer shield.

Twinning A condition existing within a quartz stone wherein the optic and/or the electric axis suddenly reverses its natural order of polarity. A single piece of quartz material, which contains both left and right handed regions, is said to be "twinned".

Two layer TAB tape This tape is constructed with a conductive metal layer, typically 35 micron thick copper, and a flexible heat resistant support layer, typically 50 micron thick polyimide. The support layer thickness is limited to 50 microns or less due to chemical etching limitations of the two layer TAB tape manufacturing process. The copper is usually tin or gold plated.

Two phase Two repeating waveforms having a phase difference of 90°

TX Transmitter General abbreviation for items such as a digital communication modulator ICs, microwave point-to-point transmit modules, satellite downlink equipment and optical transmit components

Type i assembly A surface mount assembly with surface mount components on one or both sides of the substrate.

Type ii assembly A surface mount assembly with surface mount components on one or both sides of the substrate and through hole devices on the primary side.

Type iii assembly A surface mount assembly with surface mount components on the secondary side of a PCB and through hole devices on the primary side.

Type/angle of quartz cut The type and angle of a quartz cut affects the crystal device operating parameters, the most significant being frequency stability over temperature. The frequency stability is dependent upon the plane or the angle of the crystal element in relation to the crystalline axes of the crystal. The plane or angle is referred to as the crystal cut. A common type of thickness shear crystal fabricated from Y bar quartz is the AT cut. The frequency stability and operating temperature range required by the customer determine the angle of cut utilized.

Type-a transistor What Bell called the first point-contact transistors they built. These were installed into the Bell Phone system, and given to the military and scientists for other purposes as well -- but the Type-A transistor was never widely used.

Typical Error is within plus or minus one standard deviation (±1%) of the nominal specified value, as computed from the total population.

U

Ultimate trip current The minimum value of current that will cause tripping of a protective device.

Ultra fine pitch Center-to-center lead distances and conductor spacings of 0.010" or less.

Ultra high frequency (UHF) The part of the radio spectrum from 300 to 3000 MHz that includes TV channels 14-83, as well as many land mobile and satellite services.

Ultra high voltage (UFV) Electric systems in which the Root-Mean-Square ac voltage exceeds 800,000 volts.

Ultra-large-scale integration (ULSI) Refers to the number of logic gates in a device. By one convention, ultra-large-scale integration represents a device containing a million or more gates.

Ultra low frequency (ULF) The portion of the electromagnetic spectrum below 3 Hz.

Ultra wideband (UWB) A wireless technology for transmitting large amounts of digital data over a wide spectrum of frequency bands with very low power for a short distance. Ultra wideband broadcasts very precisely timed digital pulses on a carrier signal across a very wide spectrum (number of frequency channels) at the same time. UWB can carry a huge amount of data over a distance up to 230 feet at very low power (less than 0.5 milliwatts), and has the ability to carry signals through doors and other obstacles that tend to reflect signals at more limited bandwidths and a higher power.

Ultrasonic A frequency above the limits of the human ear - above 20kHz.

Ultrasonic bonding A bonding technique which utilizes a bonding head to apply ultrasonic vibration to the bonding wire and to convert the vibration to heat through pressure and through the friction between the wire and the pad created by the scrubbing action of the head. Since this technique requires no external heating, it results in minimal oxidation of aluminum bond wire and bonding pads. Used normally with aluminum wire, this is the most commonly employed bonding technique.

Ultrasonic cleaning The use of ultrasonic energy along with a chemical solvent to clean a component or a PCB assembly immersed in solvent. Mechanical oscillation is introduced by the ultrasonic energy to facilitate cleaning.

Ultrasonic welding Ultrasonic welding involves the use of high frequency sound energy to soften or melt the thermoplastic at the joint. Parts to be joined are held together under pressure and are then subjected to ultrasonic vibrations usually at a frequency of 20, 30 or 40kHz.

Ultraviolet That portion of the electromagnetic spectrum below blue light (380 nanometers).

Unbalance That condition which exists in a rotor when vibratory force or motion is imparted to its bearings as a result of centrifugal forces.

Unbalance tolerance The unbalance tolerance with respect to a radial plane (measuring plane or correction plane) is that amount of unbalance which is specified as the maximum below which the state of unbalance is considered acceptable.

Unbalanced audio An audio output where one of the two output terminals is at ground potential.

Unbalanced loads Refers to an unequal loading of the phases in a polyphase system.

Uncharged Material having atoms with the same number of electrons in orbit as the number of protons in the nucleus. Having no electrical charge.

Uncommitted logic array (ULA) One of the original names used to refer to gate array devices. This term has largely fallen into disuse.

Undercutting Inward sloping of die metallization or silicon such that its upper surface is wider than its base.

Underfill In flip chip applications, the material injected under the die after testing to ensure reliability. This material is particularly important for flip chips mounted on substrates with different CTEs than silicon, such as FR-4 and some ceramics.

Underground (UG) Underground is an electrical facility installed below the surface of the earth.

Underground residential distribution Refers to the system of electric equipment that is installed below grade.

Underscan A decreasing of the raster size (H and V) so that all four edges of the picture are visible on the screen. Underscanning allows viewing of skew and tracking that would not be visible in normal (overscanned) mode. It is also helpful when aligning test charts to be certain they touch all four corners of the raster. Likewise, when checking the alignment of multiplexer images from a film chain, underscan allows proper framing of the projected image going into the video camera.

Undershoot The difference in temperature between the temperature a process goes to, below the set point, after the cooling cycle is turned off and the set point temperature.

Undervoltage protection A circuit that inhibits the power supply when output voltage falls below a specified minimum.

Underwriters laboratories incorporated (UL) American association chartered to test and evaluate products, including power sources. The group has four locations so an applicant can interact with the office closest in the country to his/her own location.

Ungrounded junction A form of construction of a thermocouple probe where the hot or measuring junction is fully enclosed by and insulated from the sheath material.

Uni-directional antenna Antenna that radiates most of its power in one direction.

Unidirectional microphone A microphone that is most sensitive to sound coming from a single direction-in front of the microphone. Cardioid, supercardioid, and hypercardioid microphones are examples of unidirec-tional microphones.

Unidirectional unit Allows inputs to be measured in one direction only. The stated output range indicates the minimum and maximum input levels.

Unified remote control Enables the user to control other products (i.e., TV, VCR, Laser disc) from the same manufacturer with a single remote control.

Unijunction transistor Three terminal device that acts as a diode with its own internal voltage divider biasing circuit.

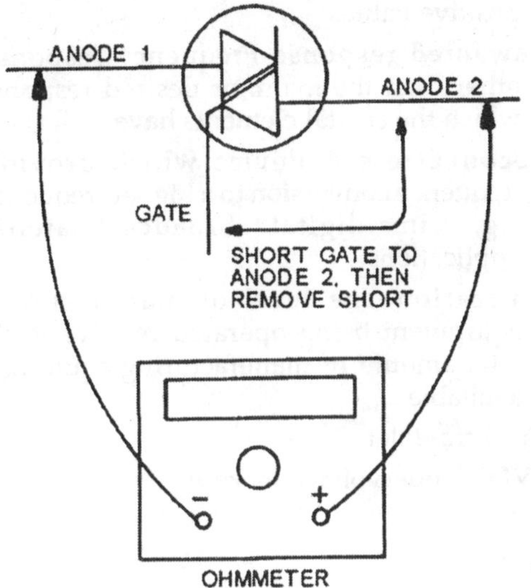

Fig. Unijunction transisto

Uninterruptible power supply (UPS) A type of power supply designed to support the load for specified periods when the line varies outside specified limits.

Union A form of pipe fitting where two extension pipes are joined at a separable coupling.

Unipolar Devices or processes in which current carrying areas and substrates are of similar polarities, usually MOS devices.

Unipolar mode The mode of operation of a converter wherein the zero to full-scale analogue value is of single polarity only

Unipolar offset The deviation of the ADC's actual first transition from its ideal level, i.e., 1/2 LSB above the analogue common

Unique identification A numbering scheme used by the U.S. Department of Defense to track high-value items and items, such as chemical suits, that have an expiration date.

Unique identifier A unique serial number that identifies the transponder.

Uniqueware A unique identification technique.

Uniqueware serialized A factory-programming service for 1-Wire EPROM chips with customer-specified data. Service provides one serialization file for customers to create identifiers in silicon.

Unity The factor of one.

Unity gain frequency Frequency of operation for a device where the gain of the component drops to unity.

Universal asynchronous receiver transmitter (UART) The microchip with programming that controls a computer's interface to its attached serial devices. Specifically, it provides the computer with the RS-232C Data Terminal Equipment (DTE) interface so that it can "talk" to and exchange data with modems and other serial devices.

Universal bushing well This 200 amp rated component is used as part of a system to terminate medium voltage cables to transformers, switchgear and other electrical equipment.

Universal mobile telecommunications service (UMTS) Universal mobile telecommunications service is a so-called "third-generation (3G)," broadband, packet-based transmission of text, digitized voice, video, and multimedia at data rates up to 2 megabits per second (Mbps) that will offer a consistent set of services to mobile computer and phone users no matter where they are located in the world. Based on the GSM (Global System for Mobile communication), UMTS, endorsed by major standards bodies and manufacturers, is the planned standard for mobile users around the world by 2002.

Universal multibrand remote Can be programmed to operate any TV, as well as a VCR and a cable converter.

Universal output transformer Usually refers to an audio output transformer having a number of taps on the secondary winding which provide a variety of impedances for matching the speaker impedance to that of the recommended load resistance.

Universal ports Universal Ports can accept a standard phone, an analogue proprietary phone, or a digital proprietary phone.

Universal product code A generic term that refers to the 12 digit data structure encoded in a UCC bar codes.

Universal serial bus An external peripheral interface standard for communication between a computer and external peripherals over an inexpensive cable. Many newer RFID readers can connect to computers via a USB port.

Unmodulated An RF carrier signal without modulation.

Unsaturated logic A form of logic containing transistors operated outside the region of saturation, which makes for very fast switching. An example is emitter-coupled logic (ECL).

Unsigned binary number A binary number in which all the bits are used to represent positive quantities. Thus, an unsigned binary number can only be used to represent positive values.

Unwanted response Frequency responses other than the main, or desired response, which the crystal elements have.

Upconverters A device which provides frequency conversion to a higher frequency, e.g., in digital broadcast-satellite applications.

Utilization The level of manufacturing equipment being operated relative to the total amount of manufacturing equipment available.

UV Ultraviolet

UVLO Undervoltage lockout

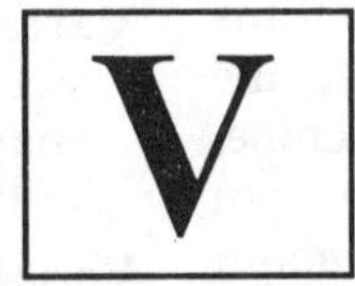

VA Electrical capacity or electrical load, expressed as Volts x Amps.

Vacuum Any pressure less than atmospheric pressure.

Vacuum circuit breakers Circuit breakers, normally applied at medium voltages, that use vacuum interrupters to extinguish the electrical arc and shut-off flowing current.

Vacuum interrupter A sealed "bottle" containing contacts of a switch inside a very high vacuum. When the contacts are parted in the vacuum, as there is no gas in the bottle to ionize, the current flow is quickly extinguished.

Vacuum tube Electron tube evacuated to such a degree that its electrical characteristics are essentially unaffected by the presence of residual gas or vapor. Have been essentially replaced by transistors for amplification and rectification. Cathode ray tubes are still used as display devices.

Vacuum-tube voltmeter A high input-impedance test instrument designed around a tube (or transistor) circuit.

Valence The combining capacity of an atom expressed as the number of single bonds the atom can form or the number of electrons an element gives up or accepts when reacting to form a compound.

Valence band The lower energy band in a semiconductor that is completely filled with electrons at 0 K; electrons cannot conduct in valence band.

Valence ring The outermost orbital ring of electrons in an atom.

Valence shell The outermost electron shell for a given atom. The number of electrons in this shell determines the conductivity of the atom.

Validation A "post-silicon" process that uses special purpose test hardware to prove that a product meets the design intent.

Valley of death The period during which a technology or product of research is too new to market; it shows commercial promise, but more research is needed to validate its potential.

Valuable final artwork A term used in "Streamlined_PCB_Design :" Artwork for electronic circuits which have been laid out and documented in forms perfectly suited to the photo-imaging and numeric-controlled tooling processes of printed circuit manufacture. It is termed "final" because it has been thoroughly checked for errors and any corrected as needed and is now ready for manufacture without further work by the PCB designer.

Valve regulated sealed cell (battery) A battery in which the cells are closed but have a valve which allows the excape of gas if the internal pressure exceeds a predetermined value (pressure).

Van de graaff generator An electrostatic generator used to produce high voltage DC, consisting of a large spherical terminal mounted on a hollow insulating support. An endless motor-driven dielectric belt runs through the support from the base to the pulley within the terminal. In the machine, charge is sprayed by point discharge from metal needles, held at a potential of about 10 kV, on to the bottom of the belt. A row of needles near the upper belt pulley removes the charge from the belt and passes it to the outer surface of the spherical terminal. Tesla designed a Van de Graaff generator in which the belt was replaced by desiccated air circulated by a disk-type blower.

Vapor barrier A vapor barrier is a material used to prevent the transfer of water vapor from one area to another.

Vapor deposition A processing step in the manufacture of optical glass fiber to reduce the inherent metallic impurities in the material, thereby minimizing light distortion.

Vapor phase reflow A solder reflow technique in which the solder joints are heated by the condensation of an inert vapor.

Vapor phase soldering A surface mount process in which a substrate carrying components attached by solder paste is lowered into the vapor-cloud of a tank containing boiling hydrocarbons. This melts the solder paste thereby forming good electrical connections. However, vapor-phase soldering is becoming increasingly less popular due to environmental concerns.

Vaporware Refers to either hardware or software that exist only in the minds of the people who are trying to sell them to you.

Varactor A semiconductor diode that acts as a variable capacitor whose value changes inversely to reverse bias voltage.

Varactor diode PN junction diode with a high junction capacitance when reverse biased. Most often used as a voltage controlled capacitor. The varactor is also called: varicap, tuning diode and epicap.

Variable capacitor Capacitor whose capacitance can be change by varying the effective area of the plates or the distance between the plates.

Variable frequency oscillator A signal-generating circuit where component values may be changed for altering the frequency of the output signal.

Variable level control This is a variable voltage level control similar to a contrast control on a data monitor. The level control increases or decreases the levels of red, green, and blue simultaneously, thus not affecting the adjusted grayscale of the monitor or projector.

Variable resistor Resistor whose resistance can be changed by turning a shaft.

Variables data Recorded parametric or delta values, traceable to individual devices, as opposed to lot data (or attributes data).

Varistor A metal (zinc) oxide over-voltage protective device.

Vault transformers A vault transformer is similar to a padmount transformer except that it is underground with a metal grill covering it.

V chip program lockout This feature allows parents to use television rating codes to block programs that they feel are inappropriate for their children to view. The V-Chip can be turned on and off through the on-screen menu, using a four digit, user-programmed code. Once you choose an appropriate ratings level, the V-Chip reads the ratings codes transmitted with each program, compares it with the level you set and automatically denies access to any programming that exceeds your rating. If a program is denied, a blocking message will appear on screen and the audio will be muted. Activating the V-Chip program lockout also temporarily disables the picture-in-picture function.

Vector Quantity having both magnitude and direction. Normally represented by a line. Length of the line indicates magnitude and orientation indicates direction.

Vector diagram Arrangement of vectors showing phase relationships between two or more AC quantities of the same frequency.

Vector photoplotter (also "vector plotter", or "gerber photoplotter" after gerber scientific co., which built the first vector photoplotters for commercial use) It plots images from a CAD database on photographic film in a darkroom by drawing each line with a continuous lamp shined through an annular-ring aperture, and creating each shape (or pad) by flashing the lamp through a specially sized and shaped aperture. The "apertures" are thin trapezoidal pieces of plastic which are mostly opaque, but with a transparent portion that controls the size and shape of the light pattern passing through it. The apertures are mounted on an " aperture wheel" which can hold up to 24 apertures (or 70 on certain models). The lamp and aperture wheel are fixed, and the table holding the film is moved in x and y dimensions (on small photoplotters), or vice versa (on very large photoplotters). A numeric datum sent to the control circuit of the photoplotter is either a D code or an X and/or Y dimension in inches, to the nearest thousandth. If it is a D code equal to D10 or above, the message tells the wheel to rotate the corresponding aperture location into position in front of the lamp. . Gerber photoplotters, if set up by an experienced craftsman, are well-suited for printed circuit artwork generation. Compare with laser photoplotter , which is faster, more accurate and has largely replaced the vector photoplotter.

There are still vector photoplotters in use. Some manufacturers take advantage of the large bed size of the largest Gerber photoplotters, roughly the size of a full-sized billiards table. This enables the production of very large photoplots. An example is Buckbee-Mears, which makes large antenna boards, and the USGS (United States Geological Survery) which has used them in map-making.

Vectorscope A special oscilloscope used in video systems to measure chroma.

Vehicle system The vehicle dissolves flux and imparts paste-like characteristics to solder paste.

Veiling luminance A luminance superimposed on the retinal image which reduces its contrast. It is this veiling effect produced by bright sources or areas in the visual field that results in reduced visual performance and visibility.

Velocity The time rate of change of displacement; dx/dt.

Velocity modulated scan An advanced circuitry located along the neck of the CRT that senses transitions from dark to light in the video signal. The black and white portions of the signal are sped up and slowed down respectively, resulting in sharp dark to light transitions and outstanding edge definition.

Velocity of propagation Nominal velocity of propagation is the speed of the signal in a given cable. In a vacuum, electromagnetic radiation (light, radio waves, etc.) travels at the speed of light. In a cable, it travels somewhat slower and in direct inverse proportion to the dielectric constant; the lower the dielectric constant, the closer to the speed of light the signal travels.

Vent cap (battery) The plug on top of a cell that can be removed to check and change the level of the electrolyte.

Vent valve (battery) A normally sealed mechanism which allows the controlled excape of gasses from within a cell.

Ventilation Ventilation is the circulation of air.

Venting The release of excessive internal pressure from a cell in a manner intended by design to preclude explosion.

Verification A "pre-silicon" test process done during development to gain confidence that a new design will produce the expected results.

Vertical blanking interval The blanking time at the beginning of each field. It contains equalizing pulses and vertical sync pulses.

Vertical centering control Adjusting the vertical centering control one way shifts the displayed image toward the bottom of the screen and the other way shifts the displayed image to the top of the screen. Also called "vertical shift," or "vertical position."

Vertical double images A video problem in which the display is split across the middle with two identical (squeezed) images on the top and bottom of the screen.

Vertical edge correction This circuit works in conjunction with the digital comb filter in a TV to sharpen the horizontal edges of objects in the picture.

Vertical filtering This is a feature in some scan converters that controls the number of lines to process, and the way they are processed. This affects the sharpness vs. flicker of the scan-converted picture.

Vertical interval The period of time between the end of one video field and the beginning of the next. During this time, the electron beam in a camera, monitor, or projector is turned off (blanked) while it returns from the bottom of the screen to the top. The portion of the video signal that represents this time period may also be called the vertical interval.

Vertical MOS Enhancement type MOSFET designed to handle much greater values of drain current than standard E-MOSFET.

Vertical position adjustment Allows you to raise or lower the image on the screen to suit you viewing preference.

Vertical resolution Also known as "vertical definition." The number of distinct horizontal lines, alternately black and white, that can be seen in a TV image. Vertical resolution is fixed by the number of horizontal lines used in scanning.

Vertical sync The portion of a video signal that indicates the end of a field of video information.

Vertical temporal A scaling process for video content that contains movement. This type of scaling process employs an averaging technique to merge the odd and even fields of video into a single frame. This type of processing treats the entire picture as if in motion and results in less motion artifacts. Disadvantages include blurring and loss of vertical resolution.

Very high frequency (VHF) Electromagnetic frequency band from 30 MHz to 300 MHz.

Very large scale integration (VLSI): Semiconductor fabrication technology that can create an approximate density of between 1000 and 1,000,000 devices on each individual die.

Very low frequency (VLF) Frequency band from 3 kHz to 30 kHz.

Very small aperture terminal (VSAT) Very small aperture terminal (satellite service) is satellite communications system that serves home and business users. A VSAT end user needs a box that interfaces between the user's computer and an outside antenna with a transceiver.

Vestigial side band (VSB) Modified AM transmission in which one sideband, the carrier, and only a portion of the other sideband are transmitted.

Via A vertical opening filled with conducting material that is used to connect wiring in one metallization layer to that in an adjacent layer.

Vibrating contacter A device which periodically makes and breaks contact in a cyclic fashion, used for converting d-c signals to a-c signals. (It is also called a vibrator or chopper.)

Vibration error The maximum change in output of a transducer when a specific amplitude and range of frequencies are applied to a specific axis at room temperature.

Vibration error band The error recorded in output of a transducer when subjected to a given set of amplitudes and frequencies.

Vibration, variable frequency A test which applies vibration to devices on a sample basis (since it is considered destructive) in order to subject them to the maximum possible construction stress. The device is vibrated with simple harmonic motion through a range of frequencies.

Vibrator A magnetically-operated, switch mechanism used to "chop" d-c into essentially square waves of a-c for the purposes of transformation. Two types are encountered in radio service; non-synchronous and synchronous. The non-synchronous vibrator merely performs the d-c to a-c conversion as mentioned above, while the synchronous vibrator is equipped with an additional set of contacts operating in synchronism with the first. The second switch set is used to convert the a-c, after transformation, back to d-c.

Video Pertaining to the visual component of a television signal. For computers, video refers to the rendering of text and graphics images on a display device.

Video amplifier A low-pass amplifier with a bandwidth of 2 to 10 MHz, used to strengthen the video signal for TV transmission and reception.

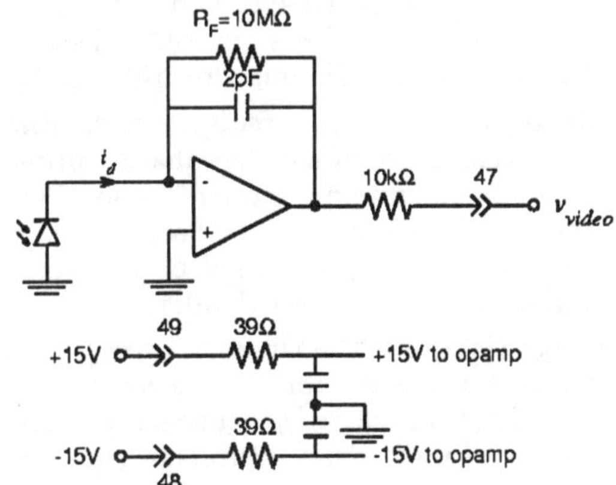

Fig. Video amplifier

Video card A circuit board that is usually mounted inside the computer that generates signals necessary to drive, or control a specific type of monitor.

Video connector The connector on the video card or computer's graphics output that is connected to the video input on the local monitor.

Video distribution amplifier An amplifier for strengthening the video signal so that it can be supplied to a number of video monitors at the same time.

Video encoder A device that converts RGB video into composite video.

Video gain The amplitude of a video signal.

Video input labelling Allows you to assign an on-screen label to each of the video inputs on the TV for easy identification of the video signal source. Video inputs can be labelled VCR, DVD or DBS instead of Video 1, Video 2, etc. Up to eight characters per label can be used.

Video noise reduction (VNR) In the event of weak signal conditions, video noise reduction reduces the noise, or "snow," in the luminance channel. It is important to note that VNR should remain off for maximum picture detail when receiving a strong signal or viewing high-quality video from sources such as DVDs or laser discs.

Video pair cable A transmission cable containing low-loss pairs with an impedance of 125 Ohms. Used for TV pick ups, closed circuit TV, telephone carrier circuits, etc.

Video projector A device that projects a video image onto a presentation surface.

Violet ray device An electrical appliance intended for the physical application of high potential, high frequency electrical current for electrotherapeutic purposes. It consists of two elements, a Tesla coil transformer assembly and a gas filled electrode. These instruments were very popular in the early part of the 20th century and were widely marketed. While the efficacy of the device has been disputed, it is said to work by stimulating blood circulation.

Virtual ground Point in a circuit that is always at approximately ground potential. Often a ground for voltage, but not for current.

Viscosity The inherent resistance of a substance to flow.

Voice activated recording The answering system automatically stops recording an incoming message if the caller is silent for eight seconds or more.

Voice coil Coil attatched to the diaphram of a moving coil loudspeaker. The coil is moved through an air gap between magnetic pole pieces.

Voice synthesizer Synthesizer that can simlate speech by stringing together phenomes.

Void A cavity inside the solder joint formed by gases released during reflow or by flux residues entrapped before solidification.

Volt A unit of electrical potential difference, abbreviation V or v. 1. The absolute volt is the steady potential difference which must exist across a conductor which carries a steady current of one absolute ampere and which dissipates thermal energy at the rate of one watt. The absolute volt has been the legal standard of potential difference since 1950.

2. The International volt, the legal standard prior to 1950, is the steady potential difference which must be maintained across a conductor which has a resistance of one International ohm and which carries a steady current of one International ampere.

1. Int. volt = 1.000330 Abs. volts.

Volt Unit of potential difference or electromotive force. One volt is the potential difference needed to produce one ampere of current through a resistance of one ohm.

Voltage Also called electromotive force. Voltage is a quantitative expression of the potential difference in charge between two points in an electrical field. The greater the voltage, the greater the flow of electrical current (that is, the quantity of charge carriers that pass a fixed point per unit of time) through a conducting or semiconducting medium for a given resistance to the flow. Voltage is symbolized by an uppercase italic letter V or E.

Voltage amplifier Amplifier designed to build up signal voltage. By design amplifiers can have a large voltage gain or a large current gain or a large power gain. Voltage amplifiers are designed to maximize voltage gain often at the expense of current gain or power gain.

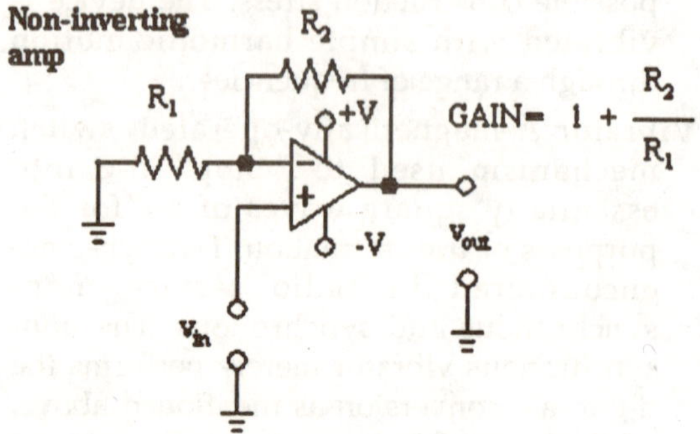

Fig. Voltage amplifier

Voltage arrestor A fast-acting, over-voltage protective device that can absorb or short a voltage to ground when the voltage is in excess of the device's rated value.

Voltage attenuation The ratio of the magnitude of the voltage across the input of a transducer to the magnitude of the voltage delivered to a specified load impedance connected to the transducer. If the input and/or the output voltage consist of more than one component, such as multifrequency signal or noise, then the particular components used and their weighting should be specified. By custom this attenuation is often expressed in decibels by multiplying its common logarithm by 20.

Voltage circuit (shunt circuit) A measuring circuit energized by a voltage that is a prime factor in determining the indication of the measured quantity. It may be the voltage directly involved supplied by a voltage transformer or a voltage divider.

Voltage clamping The circuitry necessary to protect relay or solid-state switching elements from excessive voltage. A possible source of this excessive voltage could be caused by switching current into inductive loads.

Voltage class The general strength of electrical insulation on a device, determining the maximum continuous voltage that can be applied between the conducting parts and ground potential, without damaging the insulation.

Voltage controlled oscillator Oscillator whose output frequency depends on an input control voltage.

Voltage cutoff A voltage sensing device which will end a charge or discharge at a preset voltage value.

Voltage divider Resistor sections placed across a voltage source for obtaining intermediate values of voltage.

Voltage divider biasing Biasing method used with amplifiers in which two series resistors connected across a source. The junction of the two biasing resistors provides correct bias voltage for the amplifier.

Voltage doubler A power-supply circuit so designed that the rectified voltage amplitude is doubled at the output.

Voltage drop The loss of voltage between the input to a device and the output from a device due to the internal impedance or resistance of the device. In all electrical systems, the conductors should be sized so that the voltage drop never exceeds 3% for power, heating, and lighting loads or combinations of these. Furthermore, the maximum total voltage drop for conductors for feeders and branch circuits combined should never exceed 5%.

Voltage feedback Feedback configuration where a portion of the output voltage is fed back to the input of an amplifier.

Voltage follower Operational amplifier circuit characterized by a high input impedance, low output impedance and unity voltage gain. Used as a buffer between a source and a low impedance load.

Voltage gain Also called voltage amplification. Ratio of amplifier output voltage to input voltage usually expressed in decibels.

Voltage limit A voltage value a battery is not permitted to rise above on charge and/or fall below on discharge

Voltage margining Setting the output voltage higher or lower than the nominal voltage so that the output voltage remains within the specification during all load conditions.

Voltage mode The functioning of a power supply so as to produce a stabilized output voltage.

Voltage monitor A circuit or device that determines whether or not an output voltage is within some specified limits.

Voltage multiplier A set of components designed to increase the voltage for a particular application.

Voltage of a circuit The electric pressure of a circuit in an electric system measured in volts. It is generally a nominal rating based on the maximum normal effective difference of potential between any two conductors of a circuit.

Voltage rating The highest voltage that may be continuously applied to a wire in conformance with standards or specifications.

Voltage regulation Automatic voltage regulators are relied upon for maintenance of constant generator voltage. Alternatorvoltage regulators have, for all practical purposes, become limited to four distinct types, the vibrating, the magnetic, the rheostatic, and electronic. D-c regulators are usually rheostatic.

Voltage regulator A circuit which provides a fixed or controlled voltage output from a variable voltage input. Used in power supplies and chargers. Switching regulators , Linear regulators and Shunt regulators are the most common types.

Voltage saturation The condition which exists in a thermionic electron tube when the space current (for a given cathode temperature) cannot be increased by further increase of electrode voltages (neglecting the increase due to the Schottky effect). The tube is said to be temperature limited when operating in the region of voltage saturation.

Voltage source A power source that tends to deliver constant voltage.

Voltage spread The difference between maximum and minimum voltages.

Voltage stabilization The use of a circuit or device to hold constant an output voltage within given limits.

Voltage standing wave ratio A standing wave may be formed when a wave is transmitted into one end of a transmission line and is reflected from the other end by an impedance mismatch. The Voltage standing wave ratio is the ratio of maximum to minimum voltage in a standing wave pattern. It may be stated as a ratio (VSWR) or in dB (return loss).

Voltage transformer ratio The ratio of primary volts divided by secondary volts

Voltage transformer Transformer used to accurately scale ac voltages up or down, or to provide isolation. Generally used to scale large primary or bus voltages to usable values for measuring purposes

Voltage trip A protective device that is factory calibrated to trip at a predetermined voltage value.

Voltage withstand test A field or factory test in which a conductor or electrical equipment is subjected to a higher than normal ac or dc voltage to test its insulation system.

Voltage, composite controlling The voltage of the anode of an equivalent diode, combining the effects of all individual electrode voltages in establishing the spacecharge-limited current.

Voltage, inverse peak The peak voltage which may be safely impressed across an electron tube in the inverse direction, i.e., with the anode negative and the cathode positive. If this voltage rating of the tube is exceeded there is danger of the tube conducting in the inverse direction and thus ceasing to act as a rectifier. In certain circuits this type of conduction will result in currents of destructive magnitude.

Voltage, maximum The specified voltage in an instrument that will not cause electrical breakdown or any observable physical degradation when applied continuously at the maximum operating temperature of the instrument and with any other circuit in the instrument energized at rated values.

Voltage, rated The specified voltage that an instrument (such as a wattmeter, power-factor meter, or frequency meter) is designed to carry continuously under usual service conditions. This is also the value of applied voltage used for test purposes.

Voltaic cell Primary cell having two unlike electrodes immersed in a solution that chemically interacts to produce a voltage.

Voltaic current Terms formerly used to designate the electricity furnished by the voltaic cell, before the identity of the electricity from different sources was recognized.

Voltaic efficiency The ratio (expressed as a percentage) between the voltage necessary to charge a secondary cell and the corresponding discharge voltage.

Volt ampere The product of effective voltage and effective current (see current, effective) in an alternating current load is its volt-ampere input.

Voltmeter The usual instruments of this class differ from ammeters used on the same type of service in only one essential respect: they are of very high resistance. Therefore, when connected across the terminals between which the voltage is to be measured, they take very little current and cause but a very slight drop in the potential difference. The current through the voltmeter is proportional to the voltage, and the scale may therefore be graduated to read directly in volts. Instruments are made which, with the proper change in connections, serve either as voltmeters or ammeters, the scale having two graduations. For high voltages, the voltmeter is placed in series with a large resistance, called a multiplier, so that the potential difference between its terminals is a known fraction of the voltage under test.

There are electrostatic voltmeters which may be used to measure electrostatic potentials of thousands of volts. A common form resembles a gold-leaf electrometer of large size, but with a brass pointer swinging on a scale in place of the gold-leaf.

Voltmeter, average A meter whose indication is proportional to the half-wave average value or to the full-wave average absolute value of an alternating voltage. The more generally used full-wave voltmeter employs some form of full-wave rectifier, as its name implies, and does not have turn-over error as does the half-wave circuit. The instruments are frequently calibrated in terms of the rms value of an assumed sinusoidal voltage.

Voltmeter, corona A meter which uses as an indicator the known relationship between the magnitude of voltage, electrode spacing, and the start of a corona discharge.

Voltmeter, electronic, peakreading Essentially some form of halfwave rectifier circuit which charges a capacitor to either the positive or negative peak of an electrical wave. Usually a d-c vacuumtube voltmeter is used to measure the average value of their capacitor voltage. The meter is subject to "turn-over" error in that it will give different readings, depending upon the actual connection employed, when measuring nonsinusoidal waves with different positive and negative peaks. The voltmeters are frequently calibrated in terms of the rms value of an assumed sinusoidal voltage.

Volt-second area That area under the curve of supply voltage vs. time which corresponds to the flux change in the core of a saturable reactor.

Volume flow rate Calculated using the area of the full closed conduit and the average fluid velocity in the form, Q = V x A, to arrive at the total volume quantity of flow. Q = volumetric flowrate, V = average fluid velocity, and A = cross sectional area of the pipe.

Volumetric energy density (Wh/L) The energy output per unit volume of a battery

Volumetric power density (W/L) -The power output per unit volume of a battery

VU (volume unit) A unit that is designed to measure perceived loudness changes in audio. 100 VU is 100 percent of the audio that is supposed to be present. VU is measured on a VU meter.

W Watt(s).

Wafer A thin slice, typically 10-30 mils thick, sawed from a cylindrical ingot (boule) of extremely pure, crystalline silicon, typically six to eight inches in diameter. Arrays of ICs or discrete devices are fabricated in the wafers during the manufacturing process.

Wafer bonding Process in which two semiconductor wafers are bonded to form a single substrate; commonly applied to form SOI substrates; bonding of wafers of different materials, e.g. GaAs on Si, or SiC on Si; is more dificult than bonding of similar materials. .

Wafer fabrication Process in which single crystal semiconductor ingot is fabricated and transformed by cutting, grinding, polishing, and cleaning into a circular wafer with desired diameter and physical properties. .

Wafer flat Flat area on the perimeter of the wafer; location and number of wafer flats contains information on crystal orientation of the wafer and the dopant type (n-type or p-type). .

Wafer lot A single lot of wafers processed through all processing steps including metallization together. A wafer run may consist of more than one device type, where the various device types differ only in the metallization pattern employed.

Wafer lot acceptance testing Testing of an integrated circuit wafer lot to determine its acceptability for the assembly of Class S devices.

Wafer map A plot of the viable dice on a wafer showing pass/fail information, parameter variations, or some other characteristics.

Wafer plane irradiance Exposure light at the substrate surface.

Wafer probing The process of testing individual integrated circuits while they still form part of a wafer. An automated tester places probes on the device's pads, applies power to the power pads, injects a series of signals into the input pads, and monitors the corresponding signals returned from the output pads.

Waffle pack A matrix tray for holding bare die. Typically, waffle packs are 2" x 2" or 4" x 4." The pockets for die in waffle packs are typically designed for specific die sizes; they are not standardized.

Wall thickness The thickness of the insulation or jacket.

Wardenclyffe Nikola Tesla's historic laboratory and wireless communications facility located about 65 miles east of New York City on the North Shore of Long Island. It was here that he worked out many of the final details for his "World System." Although the distinctive 187 foot tall tower was demolished in 1917, the sturdy 94 foot square building still stands in silent testimony to Tesla's unfulfilled dream.

Ward-Leonard controller A motor-generator system which uses a DC motor driving an AC generator to provide a variable power transmission. Used for high power load testing.

Warmup drift The change in output voltage of a power source from turn on until it reaches thermal equilibrium at specified operating conditions.

Warmup effect Magnitude of change of stabilized output quantities during warmup time.

Warmup time The time required after a power supply is initially turned on before it operates according to specified performance limits.

Warranty A contract between the purchaser of a product and the company that produced the product that details the components, duration and circumstances by which a defect in the product will be remedied.

Watchdog A feature of a microprocessor-supervisory circuit that monitors software execution in a microprocessor or microcontroller. It takes appropriate action (assert a reset or nonmaskable interrupt) if the processor gets stuck in an infinite execution loop.

Watertight So constructed that water/moisture will not enter the enclosure under specified test conditions.

Watt (amplifiers) A unit of measure for the output of an amplifier.

Watt (energy) A watt is a measure of how much electricity an appliance needs.

A watt is an electrical unit of power. This term is commonly used to rate appliances using relatively small amounts of electricity. Wattage is stamped on light bulbs and all appliances.

There is a mathematical relationship between watts, volts, and amps which is expressed as: Wattage = amps x voltage

For example, a 120 volt, 15 amp circuit will carry 1800 watts.

Wattage rating Maximum power a device can safely handle continously.

Watt hour (WH) A unit of energy equal to one watt of power being used for one hour.

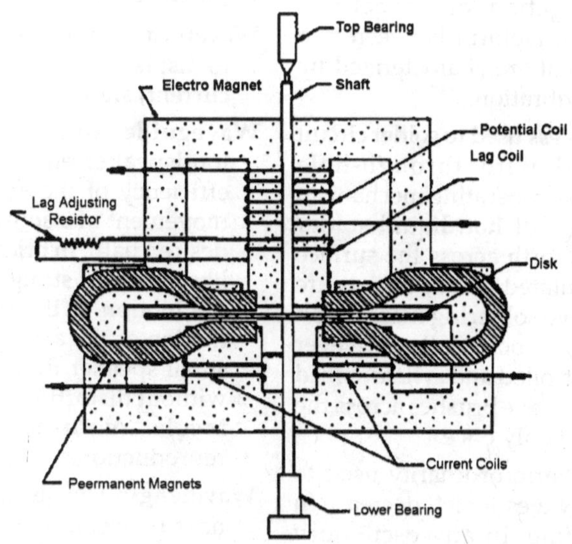

Fig.Watt hour

Wattmeter(S) A meter for measuring power, which usually has two coils, one fixed, the other capable of turning in the field of the first, both coils being without iron cores. The fixed coil is connected in series with the main circuit, so as to carry the whole current (or a known fraction of it, as determined by a shunt in d-c circuits or by a current transformer in a-c circuits). The movable coil, which is of high resistance, is connected across the terminals of the "load," that is, that portion of the circuit in which the power is to be measured, and the small current in the coil is therefore proportional to the voltage between these terminals. This coil turns against a hairspring, and since the torque, is proportional to the product of the instantaneous values of the currents in the two coils, it is proportional to the product of the main current by the terminal voltage, that is, to the required power. The scale may therefore be graduated directly in watts.

Wave Electric, electromagnetic, acoustic, mechanical or other form whose physical activity rises and falls or advances and retreats periodically as it travels through some medium.

Wave length The distance, measured in the direction of propagation, of a repetitive electrical pulse or waveform between two successive points that are characterized by the same phase of vibration.

Wave soldering A process used to solder circuit boards populated with through-hole components. A wave generating mechanism maintains a wave( ) of hot, liquid solder traveling back and forth across the surface of a tank. The populated circuit boards are passed over the wave soldering machine on a conveyer belt. The velocity of the conveyer belt is carefully controlled and synchronized such that the solder wave brushes across the bottom of the board only once.

Wave(S), damped A term ordinarily used to designate electric waves which decrease in amplitude with time. In any oscillatory circuit which contains resistance (and all practical ones will) the oscillations will be dissipated in resistance losses, and the amplitude of the oscillations will gradually decrease unless energy is continually added to the circuit. When energy is added to overcome this dissipation and maintain the amplitude constant, continuous waves result. A condenser discharging through an inductance will give rise to damped waves, and this was the basis of the old spark radio transmitters where the spark gap initiated the discharge and the oscillations continued until all the energy had been dissipated. Since these waves are not as effective for radio transmission as the continuous waves, and since they give rise to interference by virtue of their broad frequency band, they are no longer used for this purpose.

Wave, electromagnetic A wave characterized by variations of electric and magnetic fields. Electromagnetic waves are known as radio waves, heat waves, light waves, etc., depending on the frequency.

Waveform A display of a signal (on an oscilloscope) that shows the magnitude of current or voltage signals with respect to time. By displaying the waveform of a signal on an oscilloscope, the time between cycles can be measured and its frequency can be calculated.

Waveform monitor A special oscilloscope used to display and analyze electrical (voltage or current) signals.

Waveguide An acoustic device built into a loudspeaker enclosure that improves the efficiency of the speaker by confining the movement of a sound wave to travel over a desired path. In brief, a waveguide is a tube-like structure, straight or folded, that couples the motion of the loudspeaker cone to the motion of the air in the tube. This allows a small speaker driver to create clear sound, without distortion, even at the high volume levels required for low frequency reproduction.

Wavelength The distance from one peak to the next between identical points in adjacent waves of electromagnetic signals propagated in space or along a wire.

Wavelength is usually specified in meters, centimeters, or millimeters. In the case of infrared, visible light, ultraviolet, and gamma radiation, the wavelength is usually specified in nanometers (10-9 meter) or Angstroms (10-10 meter). Wavelength is inversely related to frequency. The higher the frequency of the signal, the shorter the wavelength.

Wave meter An instrument for measuring the frequency of a radio wave. The wave-meter is a mechanically tunable resonant circuit. Below 20 GHz, the wave-meter has been replaced by the frequency counter with much greater accuracy and ease of use.

Weatherhead Used to protect the entrance cable from the elements, i.e. water running down the cable into the meter socket.

Weatherproof So constructed or protected that exposure to the weather will not interfere with successful operation.

Weber The practical unit of magnetic flux. It is the amount of magnetic flux which, when linked at a uniform rate with a single-turn electric circuit during an interval of 1 second, will induce in this circuit an electromotive force of 1 volt.

Wedge bond A bond made by a bonding tool or capillary directly pressing against a round wire.

Weighting filter A special type of band-limiting filter used in measuring audio loudness levels that "weights", or gives more attention to, certain frequency bands. Common weighting filter designs include: A-weighting, a wide bandpass filter, centered at 2.5 kHz, that mimics the way we hear; and C-weighting, generally "flat" frequency response with -3 dB attenuation at 31.5 Hz and 8 kHz.

Weld seal Semiconductor device sealing accomplished by welding a metal lid to a metal package (typically employed for metal can devices).

Well to wheel efficiency The ratio between the mechanical energy ultimately delivered to the road wheels of a vehicle and the chemical energy content of the oil consumed in providing it. It is used to compare the fuel efficiencies of different methods of powering road vehicles and takes into account the refining process, the energy loss in the distribution process (in the case of hydrogen, the energy used to compress it) and the conversion efficiency of the vehicle's power unit.

Wet bench A system which includes several tanks for processing wafers in liquid chemicals.

Wet cell A cell with free flowing liquid electrolyte.

Wet cell battery A type of battery that uses liquid as an electrolyte. The wet cell battery requires periodic maintenance: cleaning the connections, checking the electrolyte level and performing an equalization cycle.

Wet clean Liquid process for removing contaminants from a wafer surface during integrated circuit fabrication.

Wet processing Etching using liquid acids, solvents, or water in process steps.

Wet solder mask Applied by means of distributing wet epoxy ink through a silk screen, a wet solder mask has a resolution suitable for single-track design, but is not accruate enough for fine-line design.

Wetted parts Materials in a sensor that are directly exposed to the media.

Wetting Term used in soldering to describe the condition that occurs when the metals being soldered are hot enough to melt the solder so it flowes over the surface.

Wetting balance An instrument used to measure wetting forces, and consequently, estimate solderability.

Wheatstone bridge A network of four resistances, an emf source, and a galvanometer connected such that when the four resistances are matched, the galvanometer will show a zero deflection or "null" reading.

Whip The vertical portion of the antenna assembly acting as the radiator of the radio frequency energy.

Whisker A metallic growth, needle-like in size, that appears on the surface of a PCB.

Whistler A plasma wave which propagates parallel to the magnetic field produced by currents outside the plasma, at a frequency less than that of the electron cyclotron frequency, and which is circularly polarized, rotating about the magnetic field in the same sense as the electron gyromotion. The whistler is also known as the electron cyclotron wave. The whistler was discovered accidentally during World War I by large ground-loop antennas intended for spying on enemy telephone signals. Ionospheric whistlers are produced by distant lightning, and get their name because of a characteristic descending audio-frequency tone, which is a result of the dispersion relation for the wave: lower frequencies travel somewhat slower, and therefore arrive at the detector later.

White level In television, the signal level that corresponds to the maximum picture brightness. The white level is set by the contrast control.

White noise Noise having a frequency spectrum that is continuous and uniform over a specified frequency band. Note: White noise has equal power per hertz over the specified frequency band. Synonym: additive white gaussian noise.

Wideband 1. An adjective describing the characteristics of a communications circuit or channel that can carry a large quantity of information at a high rate.

2. In video applications, a circuit or system with sufficient bandwidth to convey very high resolution information in an image (video) signal. For reconstructed video images from a computer, the required bandwidth is half the pixel clock rate.

Wideband amplifier Also called "broadband amplifier." Amplifier with a flat response over a wide range of frequencies.

Widescreen Widescreen is another term for a letterboxed program. Letterboxed programs generally fill the entire screen on a widescreen TV.

Width The size of the video display area in a horizontal direction.

Wiedemann-Franz law The ratio of the thermal to the electrical conductivity for all metals is proportional to the absolute temperature T, and has a value of 6.11 X 10-9T, calorie-ohms see-1 degree-1. For 0°C, this gives 0.00000147 calorie-ohm see-1 degree-1, which is very close to the observed value for platinum. The Wiedemann-Franz formula is theoretical and its coefficient involves both the Boltzmann con. stant and the electronic charge. (It is also called the Lorentz number.) For most metals the ratio as observed is a little higher than that given by the formula, doubtless because of thermal conduction due to other causes than electronic activity.

Wien-Bridge oscillator Oscillator that uses an RC low-pass filter and an RC high-pass filter to set the frequncy of oscillations.

Wi-Fi (wireless fidelity) A term for certain types of WLANs. Wi-Fi can apply to products that use any 802.11 standard. Wi-Fi has gained acceptance in many businesses, agencies, schools, and homes as an alternative to a wired LAN. Many airports, hotels, and fast-food facilities offer public access to Wi-Fi networks.

Wind energy conversion system Usually known as a windmill or a wind turbine. WECS convert energy from the wind into electricity. A complete set of components may include the following components: Wind Turbine, wiring, inverter, controller, batteries and other components depending upon the sophistication of the system.

Wind switch Suppresses the rumbling background noise that might be heard on recorded video when the wind is blowing.

Winding One or more turns of a conductor wound in the form of a coil.

Window comparator A device, usually consisting of a pair of voltage comparators, in which output indicates whether the measured signal is within the voltage range bounded by two different thresholds (an "upper" threshold and a "lower" threshold).

Wipe A visual transition between images during which the edge of one image moves across the screen revealing the next image.

Wiping action contacts Self-cleaning contacts that wipe or slide against each other when opening or closing a circuit.

Wire A strand or group of strands of electrically conductive material, normally copper or aluminum.

Wire bonding The process of connecting the pads on an unpackaged integrated circuit to corresponding pads on a substrate using wires that are finer than a human hair. Wire bonding may also be used to connect the pads on an unpackaged integrated circuit, hybrid, or multichip module to the leads of the component package.

Wire gauge American wire gauge (AWG) is a system of numerical designations of wire diameters.

Wire lubricant A chemical compound used to reduce pulling tension by lubricating a cable when pulled into a duct or conduit.

Wire wrap area A portion of a board riddled with plated-through holes on a 100-mil grid. Its purpose is for accepting circuits which may be found necessary after a PWB has been manufactured, stuffed, tested and debugged.

Wire wrapping Method of making a connection by wrapping wire around a rectangular pin.

Wired equivalent privacy (WEP) A security protocol specified in 802.11b, designed to provide a WLAN with a level of security and privacy comparable to what is usually expected of a wired LAN. Data encryption protects the vulnerable wireless link between clients and access points; once this measure has been taken, other typical LAN security mechanisms such as password protection, end-to-end encryption, virtual private networks (VPNs), and authentication can be put in place to ensure privacy.

Wireless Any system by which electrical energy is transmitted without wires from one location to another. The three principle methods are electrical induction, electromagnetic radiation, electrical conduction. The induction method involves the coupling of two widely separated loops of wire in much the same way as the primary and secondary windings of a transformer are linked. In the present system that we now know as "radio" the propagation of energy takes place by electromagnetic radiation. The third method – ground conduction – bears a great similarity to radio in that in both cases the transmitting element incorporates a radio frequency electrical oscillator. The difference exists in the fact that in the latter case the transmitter's output signal is sent to an underground connection rather that an elevated antenna.

Wireless communication Any broadcast or transmission that can be received through microwave or radio frequencies without the use of a cable connection for reception.

Wireless ISP (WISP) An Internet service provider (ISP) that allows subscribers to connect to a server at designated hot spots (access points) using a wireless connection such as Wi-Fi. This type of ISP offers broadband service and allows subscriber computers, called stations, to access the Internet and the Web from anywhere within the zone of coverage provided by the server antenna, usually a region with a radius of several kilometers.

Wireless local area network (WAN) A wireless data communications system that lies within a limited spatial area, has a specific user group, has a specific topology, and is not a public switched telecommunications network, but may be connected to one. LANs are usually restricted to relatively small areas, such as rooms, buildings, ships, and aircraft.

Wireless sensor network Wireless Sensor Network, or WSN, is a network of RF transceivers, sensors, machine controllers, microcontrollers, and user interface devices with at least two nodes communicating by means of wireless transmissions.

Wireless service provider A company that offers transmission services to users of wireless devices through radio frequency

(RF) signals rather than through end-to-end wire communication.

Wireless transport layer security (WTLS) The security level for Wireless Application Protocol (WAP) applications, developed to address the problematic issues surrounding mobile network devices such as limited processing power and memory capacity, and low bandwidth and to provide adequate authentication, data integrity, and privacy protection mechanisms.

Wirewound resistor Resistor in which the resistive element is a length of high resistance wire or ribbon usually nichrome wound onto an insulating form.

Wiring layer A layer carrying wires in a discrete wired board.

Withstand voltage The maximum voltage that can be applied between separate circuits without causing failure.

WLL (Wireless local loop) A technique for providing telephony and low speed data services to fixed customers using wireless. Regarded as having considerably potential for rapidly addressing the telecommunications gap in developing countries. A number of different WLL solutions have been marketed based on cellular and cordless technologies.

Woofer A loudspeaker designed to reproduce low frequencies.

Word A group of signals or logic functions performing a common task and carrying or storing similar data; for example, a value on the data bus could be referred to as a data word.

Work plane The plane at which work is usually done and on which the illuminance is specified and measured. Unless otherwise indicated, this is assumed to be a horizontal plane 30" above the floor.

Working standard A standard of unit measurement calibrated from either a primary or secondary standard which is used to calibrate other devices or make comparison measurements.

Working voltage The maximum voltage at which a device will operate continuously with safety.

Workstation A type of computer used in design or development work, such as engineering and CAD, requiring a moderate amount of computing power and high-resolution graphics.

Worst case condition A set of conditions where the combined influences on a system or device are most detrimental.

Wrap-around A video problem that occurs when the left picture information is displayed on the right side of the screen and the right picture information is displayed on the left side of the screen, separated by a vertical bar.

Write protect Any method that keeps data from being over-written. It may be a physical obstacle or a file attribute choice that prevents overwriting.

Wye A three phase, four-wire electrical configuration where each of the individual phases are connected to a common point, the "center" of the Y. This common point normally is connected to an electrical ground.

X Symbol for reactance.

XDP (extra device port) The XDP expands the number of telephones available in the system by allowing a digital proprietary phone (DPITS) and a single line telephone (SLT) to be connected to the same telephone jack but having different extension numbers.

XLR connector A type of audio connector featuring three leads: two for the signal and one for overall system grounding. A secure connector often found on high quality audio and video equipment. Also called a "cannon connector."

XML Extended Markup Language.

XO Crystal oscillator often used for computer timing. An oscillator in which the frequency is controlled by a piezoelectric crystal.

X-ray Radiographic analysis of the construction of a device. This is a less useful technique for devices with aluminum bond wires since only die attach and seal defects can be evaluated. For devices with gold wire it is a more valuable screen as it can detect damage done to wires during centrifuge and other such tests, as well as most assembly defects.

X-ray crystallography The use of the property of X-ray diffraction by crystals to determine their physical structure.

X-ray lithography Similar in principle to optical lithography, but capable of constructing much finer features due to the shorter wavelengths involved. However, X-ray lithography requires an intense source of X-rays, is more difficult to use, and is considerably more expensive than optical lithography.

X-ray lithography The lithographic process for transferring patterns to a silicon wafer. The electromagnetic radiation used is X-ray, rather than visible radiation.

X-ray tube A vacuum tube designed for producing X-rays by accelerating electrons to a high velocity by means of an electrostatic field and then suddenly stopping them by collision with a target.

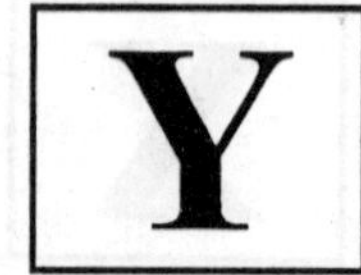

Y Symbol for admittance.

Y axis Vertical axis.

Y to C delay Relative delay or timing of the luminance channel compared to the chrominance channel in a video system.

Y/C The technical description of S-video. The luminance signal, Y, and the chrominance signal, C, are carried on separate signal/ ground pairs. Because the Y channel is carried separately, higher bandwidth is possible and colour subcarrier crosstalk is eliminated.

Yagi antenna An antenna system employing a basic antenna element as well as reflector and director rods. It is a highly directional antenna with excellent signal pick-up characteristics.

Y-connection Three-phase a-c equipment is wound with three wires whose currents differ 120' electrically in phase. The windings can be connected either in Y or delta. In balanced electrical condition, voltages and currents are the same in all coils. In the Y-connections one end of all three coils is connected in a common joint, and leads from each of the other ends constitute the threephase line.

Yield Yield refers to the percentage or absolute number of defect-free die on a silicon wafer or of packaged units that pass all device specifications. Because it costs the same to process a wafer with 10% good die and 90% good die, eliminating defects and improving yield become the critical variable in determining the cost per chip.

Yoke In television, an inductor arrangement around the neck of the picture tube that provides electromagnetic deflection of the beam both vertically and horizontally.

Young's modulus Young's Modulus (the Modulus of Elasticity) is equivalent to the ratio of normal stress to strain.

Z

Z A symbol for impedance.

Z axis Axis perpendicular to both X and Y axes.

Z cut A quartz wafer with the major surface of the wafer perpendicular to the Z crystallographic axis.

Zapping A desperation measure to revive a shorted cell suffering from dendrites. A very high current, low voltage pulse from a large capacitor used in an attempt to vaporise the dendrites.

Z-axis conductive epoxy Refer to anisotropic conductive adhesive.

Zebra battery A high temperature Sodium Nickel Chloride battery delivering high power.

Zener A semiconductor diode with the unique characteristic of providing a predictable value of voltage breakdown (called zener voltage) when in its reverse biased mode. At the breakdown mode, it exhibits a very sharp break from its nonconducting state into its breakdown state, maintaining a constant value of reverse voltage across it. The zener diode operates as a voltage regulator, voltage reference, and excess voltage circuit protection device.

Zener diode 1. A diode that makes use of the breakdown properties of a PN junction. If a reverse voltage across the diode is progressively increased, a point will be reached when the current will greatly increase beyond its normal cut-off value to maintain a relatively constant voltage. Either voltage point is called the Zener voltage.

2. The breakdown may be either the lower voltage Zener effect or the higher voltage avalanche effect.

Zener voltage 1. The field required to excite the Zener current, of the order of I volt per unit cell, or 107 volts/cm.

2. The voltage associated with that portion of the reverse volt-ampere characteristic of a semiconductor, wherein the voltage remains substantially constant over an appreciable range of current values.

Zenneck surface wave A low frequency transverse magnetic surface wave that travels along the interface between the ground and the air, in which the propagating energy does not radiate into space but is concentrated near the guiding surface. These waves do not contribute significantly to the field produced by a conventional dipole or quarter-wave radiator, however they can be strongly excited by a grounded quarter-wave helical resonator.

Zepto (z) A prefix meaning 10-21 or 0.000 000 000 000 000 000 001.

Zero adjustment The ability to adjust the display of a process or strain meter so that zero on the display corresponds to a non-zero signal, such as 4 mA, 10 mA, or 1 V dc.

The adjustment range is normally expressed in counts.

Zero angle The angle of cut that will result in a resonator that has zero temperature vs. frequency slope at the inflection point. For a fundamental, AT-cut crystal this angle is about 35 degrees and 15 minutes of arc.

Zero bit detection A circuit in a D/A converter that monitors the digital audio bit stream. upon encountering all bits low, or zero bits, the output of the D/A is disconnected from the preamp. This improves the signal-to-noise ratio specification.

Zero crossing The point at which a sinsoidal voltage or current waveform crosses the zero reference axis.

Zero offset The difference expressed in degrees between true zero and an indication given by a measuring instrument.

Zero output The aabsence of output signal or output power.

Zero point The electrical zero point where zero millivolts would be displayed. Used in conjunction with the slope control to provide a narrower range calibration.

Zero power resistance The resistance of a thermistor or RTD element with no power being dissipated.

Zero suppression The span of an indicator or chart recorder may be offset from zero (zero suppressed) such that neither limit of the span will be zero. For example, a temperature recorder which records a 100° span from 400° to 500° is said to have 400° zero suppression.

Zero voltage crossing (ZVC) The point in a plot of a voltage quantity against another parameter at which the voltage value equals zero.

Zero voltage switching The making or breaking of circuit timed such that the transition occurs when the voltage wave form crosses zero voltage; typically only found in solid state switching devices.

Zeroing Calibrating a meter so that it shows a value of zero when zero is being measured.

Zetta (Z) A prefix meaning 1021 or 1,000,000,000,000,000,000,000

ZIF Acronym for "zero insertion force" : a class of IC sockets that does not require device insertion, instead achieving contact with the IC pins through a clamping system that's controlled by a small lever on the side of the socket.

Zmax This parameter represents the speaker's impedance at resonance

Zobel A filter used to stabilize speaker impedance at a crossover frequency where impedance has risen to twice the nominal impedance.

Zone A specific area of a security system's coverage. A specific trigger input of an alarm.

Zone leveling (pertaining to semiconductor processing) The passage of one or more molten zones along a semiconductor body for the purpose of uniformly distributing impurities throughout the material.

Zoom lens A lens with a variable focal length.

ZS Zero scale.

ZVC Zero voltage crossing.

ZVS Zero voltage switching.